HUMAN DNA POLYMERASES

Biology, Medicine and Biotechnology

HUMAN DNA POLYMERASES

Biology, Medicine and Biotechnology

Giovanni Maga
Institute of Molecular Genetics IGM-CNR, Italy

Silvio Spadari
Institute of Molecular Genetics IGM-CNR, Italy

Giuseppe Villani
CNRS–Paul Sabatier University, France

Ulrich Hübscher
University of Zurich, Switzerland

World Scientific

NEW JERSEY • LONDON • SINGAPORE • BEIJING • SHANGHAI • HONG KONG • TAIPEI • CHENNAI

Published by

World Scientific Publishing Co. Pte. Ltd.
5 Toh Tuck Link, Singapore 596224
USA office: 27 Warren Street, Suite 401-402, Hackensack, NJ 07601
UK office: 57 Shelton Street, Covent Garden, London WC2H 9HE

Library of Congress Cataloging-in-Publication Data
Names: Maga, Giovanni, author. | Spadari, Silvio, author. | Villani, Giuseppe, 1949– author. | Hübscher, Ulrich, author.
Title: Human DNA polymerases : biology, medicine, and biotechnology / Giovanni Maga, Silvio Spadari, Giuseppe Villani, Ulrich Hübscher.
Description: New Jersey : World Scientific, 2017. | Includes bibliographical references.
Identifiers: LCCN 2017032766 | ISBN 9789813226401 (hardcover)
Subjects: | MESH: DNA-Directed DNA Polymerase--physiology | DNA-Directed DNA Polymerase--therapeutic use
Classification: LCC QH447 | NLM QU 141 | DDC 572.8/6--dc23
LC record available at https://lccn.loc.gov/2017032766

British Library Cataloguing-in-Publication Data
A catalogue record for this book is available from the British Library.

Cover image: The three-metal-ion mechanism of DNA synthesis

This image shows how DNA polymerase synthesizes DNA. DNA substrate and an incoming nucleotide (represented by yellow sticks) are converted to DNA product with one additional nucleotide (blue sticks). Binding and alignment of substrates in the polymerase active site (grey and red sticks) depend on two Mg^{2+} ions (A and B, shown as purple spheres). Capturing of the third C-site Mg^{2+} ion, which displaces a polymerase sidechain (yellow in the substrate state and blue in the product state), is essential for the new phosphodiester bond formation. Each Mg^{2+} ion has six ligands arranged in an octahedral geometry, some of which are water molecules (red spheres). This image is based on the X-ray structures reported by Gao and Yang in *Science* (volume 352, page 1334–1337) and prepared with PyMol (www.pymol.com).

Typeset by Stallion Press
Email: enquiries@stallionpress.com

Printed in Singapore

Preface

Maintenance of the information embedded in the genomic DNA sequence is crucial for the survival of any living species. DNA polymerases play pivotal roles in this complex process since they are involved in all DNA synthesis events occurring in nature. Besides their essential tasks *in vivo*, DNA polymerases are now the workhorses in numerous important molecular biological and medical core technologies such as the widely applied polymerase chain reaction (PCR), cDNA cloning, next-generation sequencing, nucleic acids–based diagnostics and in techniques to analyze ancient and otherwise damaged DNA (e.g. for forensic applications).

The history of DNA polymerases goes back 60 years to the mid-fifties in the last century. With the discovery of a DNA polymerase (now known as DNA polymerase I) in the bacterium *Escherichia coli* by Arthur Kornberg and his colleagues, it was for the first time possible to synthesize the genetic material, the DNA, in the test tube. A manuscript reporting the discovery submitted to the *Journal of Biological Chemistry* was rejected with comments such as "the researchers were incompetent" and DNA polymerase was "a poor name" for this enzyme. What a terrible judgment by an incompetent Editor! The observation by Kornberg and collaborators that a DNA polymerase synthesizes DNA according to the Watson–Crick base pair rule (A-T and G-C), needs activated bases (dNTPs), a template, a primer and $MgCl_2$ (or a divalent cation) is still true today for most DNA polymerases in all organisms tested.

Bruce Alberts in a feature article in *Nature* stated in 2003 (Alberts B. 2003. *Nature* 421: 431–435): "Knowledge of the structure of DNA enabled scientists to undertake the difficult task of deciphering the detailed molecular mechanisms of two dynamic processes that are central to life: the copying of the genetic information by DNA replication, and its reassortment and repair by DNA recombination. Despite dramatic advances towards this goal over the past five decades, many challenges remain for the next generation of molecular biologists."

A comprehensive book focusing on DNA polymerases appeared 30 years and covered exclusively animal DNA polymerases (Fry M., Loeb L.A. 1986 *Animal Cell DNA polymerases*, CRC Press, Inc. Baton Rota, Florida). At that time the animal DNA polymerase family was still small. Only three DNA polymerases, α, β and γ were known in detail and the scientific community just started to believe in a fourth DNA polymerase, called DNA polymerase δ. In the last 30 years, due to biochemical work, as well as the analysis of various genomes, including the one from humans, we have witnessed the discovery of an abundance of novel DNA polymerases in the three kingdoms of life (bacteria, archaea and eukaryotes) with specialized properties whose physiological functions are only beginning to be understood. For several decades the dogma was widely accepted that very accurate DNA polymerases guarantee the faithful duplication of DNA, while their limited capacity of making mistakes might be one of the drivers for evolution and a cause of disease. When by the end of last century the human DNA was sequenced nine newly discovered DNA polymerases appeared within three years in the literature (1999–2002). Many of these DNA polymerases appear to have distinct functions in translesion synthesis, in different DNA repair events or in immunoglobulin V(D)J recombination.

This book, which is entitled *Human DNA Polymerases: Biology, Medicine and Biotechnology,* is an updated version of our book *DNA Polymerases: Discovery, Characterization and Functions in Cellular DNA Transactions* by U. Hubscher, S. Spadari, G. Villani and G. Maga

(World Scientific Publishing, 2010, ISBN-13: 978-981-4299-16-9). In three sections we describe first, the biology of DNA polymerases (Chapters 1 to 4), second the medical applications of DNA polymerases (Chapters 5 and 6) and third, the biotechnical applications of DNA polymerases (Chapters 7–9).

Chapter 1 summarizes the history of DNA polymerases. General aspects of DNA polymerases are outlined in Chapter 2. Next the structural and functional aspects of the different DNA polymerase families are described (Chapter 3) and Chapter 4 discusses different DNA transactions of DNA polymerases. In the last few years DNA polymerases came also in the focus of human diseases (Chapter 5) and as targets for therapy by chemicals (Chapter 6).

With the discovery of the polymerase chain reaction (PCR) and the use of heat-stable DNA polymerases a door with an enormous potential opened to the scientific community. The history and the potential of evolved DNA polymerases are briefly introduced in Chapter 7. The field of Chemical Biology has developed techniques that allow the evolution of DNA polymerases in the test tube and thus a variety of novel applications were developed in the last ten years. This current status is summarized in Chapter 8. The final Chapter 9 is devoted to the billion USD market for evolved DNA polymerases in routine and medical applications.

In summary this book provides the arguments and evidence that characterized DNA polymerases as enzymes essential to life and also placed them among the most important tools in modern chemistry, biology, biotechnology, and medicine.

The chapters are written so that they can be read and understood individually. A certain redundancy of information was therefore done intentionally or references to other Chapters were made.

The authors would like to thank Ralph Imhof (DMMD Zürich), Wei Yang (NIH Bethesda), Andy Marx (University Konstanz) and Ramon Kranaster (MyPols Konstanz).

Pavia (Italy), Zurich (Switzerland), Toulouse (France)
January 2017

Giovanni Maga
Institute of Molecular Genetics IGM-CNR,
National Research Council, Pavia, Italy

Silvio Spadari
Institute of Molecular Genetics IGM-CNR,
National Research Council, Pavia, Italy

Giuseppe Villani
Institute de Pharmacologie et de Biologie
Structurale, CNRS-Université Paul Sabatier,
Toulouse, France

Ulrich Hübscher
Department of Molecular Mechanisms of
Diseases, University of Zurich, Switzerland

About the Authors

Giovanni Maga

Dr. Giovanni Maga is Senior Researcher and Head of the DNA Enzymology & Molecular Virology Unit at the Institute of Molecular Genetics IGM-CNR in Pavia, Italy, Contract Professor in Molecular Biology at the University of Pavia, Italy and member of the College of Supervisors of the PhD Programme in Genetics, Molecular and Cellular Biology at the University of Pavia, Italy.

He has been working since his graduation on the enzymology of genome duplication and repair. His research has focused on the characterization of different enzymatic systems responsible for nucleotide metabolism and the duplication of the genetic information in human cells and viruses as well as in protozoa parasites. He is also interested in the biochemical pathways involved in coupling the DNA replication and repair machineries to cell cycle regulation in normal vs. tumoral human cells. His interest is devoted both towards the elucidation of basic molecular processes and to the exploitation of novel enzymatic targets for antiviral and anticancer chemotherapy.

Presently his main lines of research are the following:

- Enzymology of genome replication and repair in eukaryotic cells and viruses;
- Novel targets and drugs for antiviral and anticancer chemotherapy.

Since 1990, Dr. Maga is author/co-author of 240 publications in international peer-reviewed journals (Web of Science), which obtained 9,212 citations. His H-index is 49 (Google Scholar) as at May 2017.

Silvio Spadari

Silvio Spadari was born in Italy in 1939 and obtained his PhD degree in Chemistry (1963), in Pharmacy (1968) and later a Medical Degree (1987) and a Specialization in Clinical Biology (1991) at the University of Pavia, Italy. In 1971 he was hired by the Institute of Molecular Genetics, National Research Council in Pavia where in 1983 he founded the section of "DNA Enzymology and Molecular Virology", which he has directed until 2006.

During his career he was Research Associate at the Department of Biochemistry, University of Iowa, USA (1963–1965), Visiting Scientist at the Department of Cell Biology, Roche Institute of Molecular Biology, Nutley, N.J., USA (1973–1975), Visiting Professor, for several periods of time from 1976 up to 2000, at the Department of Molecular Biology, Université Libre de Bruxelles, Belgium, at the Department of Molecular Biology, University of Hamburg, Germany, and at the Department of Pharmacology and Biochemistry, University of Zurich, Switzerland.

He was Editorial Board Member of *Nucleic Acids Research* (1986–1994), *In Vivo* (1987–2001), *Mutation Research: Genetic Instability and Aging* (1989–1993), Editorial Adviser of the *Biochemical Journal* (1981–2011), Official Reviewer of the Division of Physiology, Cellular and Molecular Biology, National Science Foundation, Washington, D.C., USA (1978–2000).

He is co-author of over 200 publications in international journals or books and his main research interests were as follows:

- Enzymology of DNA replication in eukaryotic cells and eukaryotic viruses (HSV-1, HSV-2, HIV and HCV). Among enzymes studied: human and herpesvirus DNA polymerases, HIV reverse transcriptase, HCV RNA polymerase and Helicase, DNA ligases, DNA topoisomerases, thymidine kinases, uracil-DNA glycosylases, dUTPases, thymidine phosphorylase, thymidylate synthase, etc.;

- Replicative strategies of human viruses (HSV, HIV, HCV);
- Development of novel anti-HSV, anti-HIV and anti-HCV compounds (nucleosides or non-nucleosides analogs);
- Control of herpes simplex virus latency and reactivation;
- Molecular basis for antiviral drug-resistance.

Giuseppe Villani

Giuseppe Villani was born in Italy in 1949 and obtained his Master's Degree in Biochemistry in 1973 at the University of Pavia, Italy under the supervision of Prof. Arturo Falaschi. From 1974 to 1978, he did his PhD at the Free University of Brussels, Belgium, working on the role of DNA polymerases in induced mutagenesis, under the supervision of Prof. Miroslav Radman. He then spent a postdoctoral period (1978–1980) at the Department of Biochemistry of the Stanford University, USA, in the group of Prof. Robert Lehman, where he studied the mechanisms of DNA replication in *Drosophila melanogaster*. Subsequently he moved to France, and became Assistant Professor of Toxicology at the University of Toulouse. In 1982, he was hired as Researcher at the French National Institute of Medical Research (INSERM), where he was promoted to the position of Research Director in 1995. He directed a group of researchers at the University of Toulouse until his retirement in 2015.

The main interest of Dr. Villani's entire scientific career was to investigate the molecular mechanisms underlying the action of DNA polymerases, with a particular emphasis on their role in replication of damaged DNA.

He is co-author of 95 articles in international journals and was Editor of the *Journal of Biological Chemistry* from 2002 to 2012.

Ulrich Hübscher

Ulrich Hübscher studied Veterinary Medicine in Fribourg, Bern, and Zürich in Switzerland and subsequently did postgraduate studies in Biochemistry and Molecular Biology in Zurich. After postdoc years in the laboratories of Arthur Kornberg (Stanford) and Robin Holliday (Mill Hill, London), he started his own work in Zurich since 1981, where he became Professor in 1989. He devoted his scientific research over the last 40 years on functions of DNA polymerases. In over 300 publications he contributed to biological functions of different DNA polymerases from *Escherichia Coli, Deinococcus radiodurans*, HIV and FIV viruses (RT) and eukaryotes (animals and human DNA polymerases α, β, γ, δ, ε, and λ) and their auxiliary proteins such as replication protein A, replication factor C, and proliferating cell nuclear antigen. He is a member of the Swiss Academy of Medical Sciences.

Contents

Chapter 1

History of DNA Polymerases

1.1 Imaging an enzyme that assembles the nucleotides into DNA

The beauty of the DNA double helix model[1] proposed by James Watson and Francis Crick (1916–2004) was that the structure immediately suggested its function. As they hinted in their *Nature* paper[2]: "*It has not escaped our notice that the specific pairing we have postulated suggests a possible copying mechanism for the genetic material.*" The chemical complexity of the molecule was thought to be sufficient to store the genetic information and definitely settled a decades-long controversy over whether DNA or protein was the "*life molecule*". The authors had from the beginning the idea that the structure of DNA should allow the molecule to copy itself during cell division, so that an exact copy could pass into each new cell. Indeed the DNA molecule was self-replicating following the unwinding of the two complementary and antiparallel strands. "*Each chain would act as a template for the formation on to itself of a new companion chain so that eventually we shall have two pairs of chains, where we only had one before.*" Moreover, the sequence of the pairs of bases will have been duplicated exactly because each base would attract its complementary one, by hydrogen bonding, so that two new double helices are assembled. Watson and Crick also proposed that one of the strand of each daughter molecule was newly synthesized and the other one was derived from the parental molecule. This *semiconservative mechanism of DNA replication* was soon (1958) demonstrated by Mathew Meselson and Franklin Stahl.[3]

The discovery of DNA structure started a new era in biology and the following years led to the complete elucidation of the genetic code, as reviewed in many articles and books such as those by Martynas Ycas (1917–2014)[4] and Carl Richard Woese (1928–2012),[5] and to the realisation that DNA also directs the synthesis of proteins. This was enunciated by Crick in the so called "central dogma of molecular biology" (DNA→RNA→Protein) that implies a one-way flow of genetic information between macromolecules through DNA→DNA *(DNA replication)*, DNA→mRNA (*transcription*) and mRNA→Protein (*translation*). Three kinds of RNA, synthesized by RNA polymerase, were found: that is messenger RNA (mRNA), transfer RNA (tRNA) and ribosomal RNA (rRNA). The mRNA transcript of DNA was indeed found to be the template for protein synthesis; tRNAs carried amino acids to the ribosomes and sequentially read on mRNA codons of three bases that specify an amino acid, and rRNA is the major component of ribosomes.

Robert W. Holley (1922–1993), Gobind Khorana (1922–1988) and Marshall Nirenberg (1927–2010) shared the Nobel Prize in Physiology or Medicine in 1968 for their interpretation of the genetic code and its function in protein synthesis. However, other Nobel laureates such as Crick and Sidney Brenner, who found that amino acids are coded by groups of three bases, Severo Ochoa (1905–1993), who isolated the polynucleotide phosphorylase, the enzyme that, under test-tube conditions, allowed the synthesis of high molecular weigth polyribo-nucleotides from nucleoside diphosphates, Francois Jacob (1920–2013) and Jacques Monod (1910–1976), with their proposal of the mRNA hypothesis, contributed significantly to this outstanding accomplishment.

Going back to the DNA structure, the simplicity of the model for DNA replication, thanks to the complementarity of A to T and of G to C, even induced Watson and Crick to think that nucleotides, paired to the DNA template, might be zippered together without any enzyme action. Perhaps it was sufficient for the nucleotides to align spontaneously, because of their reciprocal affinity, along each of the two separated complementary strands, and the DNA was duplicated! But to a biochemist such as Arthur Kornberg (1918–2007) this must have sounded like a heresy. And indeed only a scientist who knew how nucleotides were synthesized and activated in the cell could not only imagine an enzyme that would assemble nucleotides into DNA but also try to do it outside a living cell.

Kornberg, born in Brooklyn, N.Y., in 1918 and laureate in Medicine in 1941 from Rochester University School of Medicine, in his discovery of DNA polymerase was undoubtely guided by "*his unremitting fascination with enzymes, molecules which give the cell its life and personality and make things happen.*" He learned his love for enzymes from some of the best enzymologists of his time. He worked with Ochoa in 1946 at New York University Medical School "*learning the philosophy and practice of enzyme purification.*" He then spent six months in 1947 with Carl F. (1896–1984) and Gerty (1896–1957) Cori at Washington University Medical School in St. Louis, and was fascinated and inspired by their discovery of glycogen phosphorylase. They all contributed toward his scientific formation, made him fall in love with enzymes and convinced him that *enzymes were the key, the most effective way to understanding intracellular biochemical processes*.[6,7]

Thus, in the late 1940s–early 1950s, when it was becoming clear that pathways of degradation/energy production and biosynthesis were distinct, Kornberg returned to the NIH and started an "Enzymes and Metabolism Section" that included Leon Heppel (1912–2010) and Bernard Horecker (1914–2010). He started to purify and study enzymes that assembled various coenzymes such as NAD^+, NADP, FAD, thus greatly contributing to focusing the interest of biochemists in biosynthetic pathways. He first purified nucleotide pyrophosphatase from potatoes and found that the enzyme cleaved the pyrophosphate bond of NAD (nicotinamide adenin dinucleotide) releasing NMN (nicotinamide mononucleotide) and AMP. This led Kornberg to discover the enzyme NAD synthetase that utilized ATP to transform NMN into NAD with the release of pyrophosphate.

Then Kornberg discovered that FAD (flavine adenin dinucleotide) was synthesized by a similar mechanism as well as were the nucleotides. Thus, the mechanism involving nucleotidyl transfer from ATP (or any other nucleoside triphosphate such as UTP and CTP) and the release of pyrophosphate turned out to be a general mechanism for the synthesis of the precursors of each class of macromolecules in cells and tissues, proteins, lipids, carbohydrates, and, somehow, the nucleic acids. The building blocks, aminoacids, acetic acid, sugar, nucleotides first react with ATP (or UTP or CTP) to become activated precursors before being added to the corresponding molecule through this mechanism that involves the nucleotidyl transfer from a nucleoside triphosphate with the release of pyrophosphate. The pyrophosphate is then hydrolyzed to inorganic orthophosphate by an efficient inorganic pyrophosphatase present in all cells, thus driving the reaction in the direction of the synthesis.

The synthesis of DNA and RNA, in fact, also resembled that of the coenzymes NAD and FAD, each nucleotide being incorporated as the monophosphate with release of pyrophosphate. This research earned Kornberg the Annual Paul–Lewis award in enzymology (now the Pfizer Award) and started attracting applications from young and promising biochemists. But the mechanism of the synthesis of nucleic acids was still far away.

Then in the late 1940s–early 1950's, with John Buchanan (1918–2007) at the University of Pennsylvania and MIT, and Robert Greenberg (1918–2005) at the University of Michigan, who were investigating the biosynthesis of purine nucleotides A and G, Kornberg focused on pyrimidine biosynthesis. He studied orotic acid and showed first how a liver enzyme could produce an activated form of ribose, phosphoribosyl pyrophosphate (PRPP), from ribose phosphate and ATP, and then how two yeast enzymes could transfer the activated ribose to orotic acid and then decarboxylate orotic ribose-phosphate to uracil ribose phosphate (UMP). UMP and other nucleoside monophosphates were then converted to di- and triphosphates by nucleoside monophosphate kinases that utilize ATP as the phosphoryl donor. Thus, UMP could be converted to UDP and UTP, and CTP could be formed by amination of UTP. Rather than uracil, DNA contains thymine, the 5′-methyl derivative of uracil; thus uracil deoxyribose phosphate (deoxyuridylate, dUMP) is methylated to thymine deoxyribose phosphate (deoxythymidylate, dTMP). Contrary to the biosynthesis of pyrimidine nucleotide, where ribose phosphate from PRPP is added to the pyrimidine ring, for example orotic acid, in the byosinthesis of purine nucleotides such as AMP Buchanan *et al.* found that PRPP reacts with smaller molecules, such as aminoacids, in building the purine structure.

But soon Kornberg *et al.* found enzymes that could condense preformed adenine and guanine, such as those obtained from the breakdown of RNA and DNA by enzymes that degrade these molecules, with PRPP to make nucleotides directly. It thus became clear that there were two pathways for the biosynthesis of nucleotides: the *de novo pathway* in which the activated nucleotides are built from simpler molecules (aminoacids, carbon dioxide, sugar phosphate etc.) and a *salvage pathway* in which A, G, C, U and T obtained from the hydrolysis of RNA and DNA are phosphorylated and recycled to produce activated nucleotides.

1.1.1 *DNA polymerase activity in extracts of Escherichia coli*

In 1953, when he was only 35 years old, Kornberg resigned from NIH to become Professor and Chairman of the Department of Microbiology, Washington University, St. Louis. Having elucidated the biosynthesis of several coenzymes and learned how nucleotides are synthesized and activated in the cells, he was ready to try the synthesis of DNA in a broken-cell extract and to search for enzyme(s) that assemble the nucleotides into DNA. Of course Watson and Crick had just proposed in 1953 a mechanical model for DNA replication that did not exclude a *spontaneous* assembly of nucleotides in the synthesis of a DNA chain directed by base pairing with each strand of the parental duplex, and anyhow to scientists the chemical mechanism of DNA synthesis in the cell appeared too complicated to be accurately performed outside a living cell. On the contrary, Kornberg, educated by Carl and Gerty Cori and by Ochoa, among the most famous enzymologists of that time, was well aware that one needed enzymes to understand biological events, and he was convinced that a *biochemist devoted to enzymes could reconstitute any metabolic event in the test tube as well as, or even better, the cell does it.*

Kornberg thus hypothesized the need of an enzyme assembling the nucleotides into DNA in cell extracts, just like the synthesis of glycogen and fats were achieved outside living cells. And the search for such an enzyme, then named DNA polymerase (pol), was indeed soon successful in Kornberg's laboratory, thanks also to some young scientists such as Robert Lehman, Maurice Bessman, and the technician Ernie Simms, later joined by Julius Adler and Steve Zimmermann. They made several appropriate observations and choices at the beginning and during earlier experiments with a broken-cell extract of *E. coli*, a rapidly growing bacterium and, therefore, a good source of the enzyme. A brief summary of these observations is given below:

1. First, in December 1955 Kornberg's group obtained some [^{14}C]thymidine (a known constituent of DNA) from Morris Friedkin (1918–2002), a colleague in the Pharmacology Department at the Washington University School of Medicine. They showed that after one hour of incubation with the bacterial cell extract, ATP, Mg^{++} and buffer, a very tiny amount of radioactivity was trapped in an acid insoluble pellet, presumably DNA because of its sensitivity to deoxyribonuclease (DNase).
2. Second, they identified dTMP in the reaction mixture as well as additional products of phosphorylation such as dTDP and dTTP. This suggested that they should look in *E. coli* extracts and purify an activity converting thymidine into dTMP (thymidine kinase) useful for the synthesis of [^{32}P]-TMP. The incorporation of [^{32}P]-TMP, rather than [^{14}C]thymidine, led to a dramatic increase in terms of counts/min in the acid insoluble pellet, and also suggested that they synthesize [^{32}P]-dTTP as a possible direct precursor for DNA synthesis.

   Thymidine diphosphate (dTDP) was also a likely possibility because of the recent finding by Marianne Grunberg-Manago (1921–2013) and Ochoa that ribonucleoside diphosphates, rather than triphosphates, were the substrates *in vitro* for a ribonucleotide-polymerizing enzyme (polyribonucleotide phosphorylase) from the bacterium *Azotobacter vinelandii*.[8] But there was no evidence of a template-directed synthesis of RNA, and the reaction was indeed reversible only *in vitro*. Thus polynucleotide phosphorylase *in vivo* turned out to be the enzyme involved in messenger RNA degradation rather than in RNA synthesis. Experimental evidence supporting the conclusion, in analogy with the synthesis of DNA, that the four rNTPs, a DNA template and a DNA dependent RNA polymerase were required for RNA synthesis, came in the following years by the laboratories of Audrey Stevens (1932–2010), Sam Weiss, and Jerard Hurwitz.[9]

However, as we said previously, the fact that this enzyme, in the *in vitro* reverse reaction, was template independent and the synthesized RNAs had a composition depending on the ratios of the ribonucleoside diphosphates present, proved quite useful for the *in vitro* synthesis of different RNAs that helped to define the genetic code.

This was a time when more was known about degradation rather than biosynthetic pathways. The glycogen phosphorylase of Carl and Gerty Cori, the enzyme which inspired Arthur Kornberg, could only elongate glycogen with glucose-1-phosphate *in vitro*. Later, in 1957, the Argentinian doctor and biochemist Luis Leloir (1906–1987), a 1970 Nobel laureate in Chemistry, demonstrated that the synthesis of glycogen *in vivo* was not the reverse of the degradation and was catalyzed by glycogen synthase, an enzyme that elongates glycogen using UDP-glucose. Thus *in vivo* synthesis and degradation of glycogen, as well as of other important biological molecules, are coordinated and one is switched off when the other one is active and vice versa.[10]

3. Kornberg and his group correctly sensed that the DNA present in the extract was degraded to deoxyribonucleoside monophosphates (dNMPs) by endogenous endonucleases and that dNMPs were then converted to dNTPs by the corresponding kinases in the presence of ATP. This was in agreement with the pathways of purine and pyrimidine biosynthesis leading to nucleoside-5′-phosphates and with his previous finding that ATP was the activated intermediate in the synthesis of coenzymes such as NAD, FAD and CoA. Thus the Kornberg group purified the four [^{32}P]-dNMPs generated from [^{32}P]-labeled DNA (isolated from [^{32}P]-labeled *E. coli*) and converted them to the corresponding labeled dNTPs with ATP and the nucleotide kinases and nucleoside-diphosphate kinase they had partially purified from *E. coli*.
4. Kornberg originally thought that DNA was needed in the reaction as a primer that was elongated by DNA pol, by analogy with the work of Carl and Gerty Cori on the elongation of glycogen by glycogen

phosphorylase. He did not immediately think about DNA functioning as a template, as earlier suggested by Watson and Crick on the basis of its structure and of the opening of the double helix. (The initial lack of acknowledgement by Kornberg of the template function of DNA — and of the work of Watson and Crick — was probably one of the reasons for the caustic comments of one referee who rejected the *J. Biol. Chem.* papers in 1957! See below). When Kornberg found that all four dNTPs were required, he admitted that the added DNA was serving as a primer and template, and indeed he subsequently demonstrated that the DNA synthesized by DNA pol was a faithful copy of the added DNA template (see next section).

5. Finally, with the recognition that [α-^{32}P]-dNTPs were the true substrates for a DNA pol, the enzyme that converted them into an acid-insoluble product (i.e. DNA) was purified in quantities that allowed its further characterization.

By analogy with Friedrich Miescher's (1844–1895) difficulties in publishing his discovery of *nuclein* essentially because of the skepticism of his own mentor Hoppe-Seyler (1825–1895), who wanted to confirm himself the results before publication, the first two reports by Lehman *et al.* on the enzymatic preparation of the four dNTPs and the partial purification *E. coli* DNA pol, and by Bessman *et al.* on the general properties of the reaction, raised much criticism by the reviewers and were declined by the *J. Biol. Chem.* in the fall of 1957. "*It is very doubtful that authors are entitled to speak of the enzymatic synthesis of DNA*" and "*Polymerase is a poor name*" etc.... were some of the critical comments. The two papers[11,12] were accepted several months later by John Edsall (1902–2002), the new Editor-in-chief.

Thus Kornberg correctly imagined the DNA pol as an enzyme extending a DNA chain by successive additions of a properly activated nucleotide, but he added DNA to the reaction mixture *never imagining at the beginning to detect an enzyme that, unlike any others, took directions and instructions from its DNA substrate thus assembling nucleotides by Watson–Crick base pairing to create a complementary copy of the parental DNA chain.*

1.1.2 *Escherichia coli DNA polymerase can synthesize DNA with genetic activity: creating life in the test tube*

In 1959 Kornberg moved to Stanford University School of Medicine as Professor and Chairman of the Department of Biochemistry and he also shared the Nobel Prize in Medicine or Physiology with Ochoa. Among the referees who declined the JBC papers in 1957, one, "*whose caustic literary style unmistakably identified him*" to Kornberg, stated that "*the authors had to show genetic activity in the synthetic substance to have it qualify as DNA.*"

Why? Said Kornberg: "*This criterion is met by less than 2 percent of the papers on DNA appearing in biochemical journals.*"

As we have mentioned, he withdrew the papers and waited for the next Editor-in-chief. But certainly when he found that all four dNTPs were required, it became clear that the added DNA was serving not only as a primer but also as a template. Thus it was important to show that the DNA synthesized by the newly discovered DNA pol was a faithful copy of the added DNA template. This was proved by showing that 1) the newly synthesized DNA contained equal amount of A and T, and of G and C; 2) using DNAs of various species with A+T/G+C ratios ranging from 0.5 to 1.9 as primers, the ratio in the product DNA closely corresponded to that of the added DNA and was independent of the relative concentrations of the dNTPs.

In summary, the bacterial DNA pol could copy any DNA, just for its ability to recognize the bases rather than their sequence in the template chain, and the above results suggested, that "*the enzymatic synthesis of DNA by the DNA polymerase of E. coli represented the replication of a DNA template*". It should be noted however that for some years the question of the "primer" for DNA pol remained open until it was found that all DNA pols also require a primer to initiate a DNA chain (see below).

Still after each of the many seminars Kornberg gave in those years, where he described the adequate performance as template-primers of DNAs from *Pneumococcus, Hemophilus* and *B. subtilis* but with a net loss rather than an increase of their measurable transforming activity, he was invariably asked the same question: "*Why have you been unable to replicate the genetic activity of the DNA template?*".

The solution came from studying the replication of a small bacterial virus ΦX174 containing a single-stranded circular DNA of about 5000 nucleotides. Robert Sinsheimer had found that the ΦX174 DNA was single-stranded for a part of its life cycle and, after entering *E. coli*, the DNA is converted by host enzymes to a double-stranded circle (a *replicative form* that serves as template for the synthesis of DNA of the progeny virus), and within 20 minutes the bacterial cells released hundreds of viral particles. Finally he showed that, although with a much lower efficiency, naked viral DNA was still infectious.

Kornberg and Mickey Goulian obtained some ΦX174 single stranded DNA from Sinsheimer, and copied it with the purified *E. coli* DNA pol. By using appropriate radioactive and "heavy" precursors, (i.e. replacing deoxythymidine with 5-bromodeoxyuridine), and a novel enzyme DNA ligase (independently discovered in 1967, with different assays), by Martin Gellert at NIH[13], Charles Richardson at Harvard,[14] Lehman at Stanford,[15] Hurwitz at Albert Einstein Medical School,[16] and Kornberg and Nicholas Cozzarelli (1938–2006) at Stanford,[17] an enzyme able to join or ligate adjacent ends of a DNA chain, with a free OH group at the 3′ end and a phosphate group at the 5′ end to create a circle), they isolated fully synthetic circles of viral DNA and showed that they were infectious.

After a decade they had finally synthesized a DNA with genetic activity. The results published in the *Proc. Natl. Acad. Sci. USA* in 1967 as the 24th paper of the series "Enzymatic synthesis of DNA"[18] raised the interest of many newspaper and television reporters, not only in the United States but also in many other countries, essentially asking "*Is the DNA you have made a living molecule?*". It was quite difficult for Kornberg to explain that "*viral DNA has no life of its own, nor does the virus that bears it, in the sense of being able to grow and reproduce outside the cells of an organism. Yet the DNA has a potent life force. Once inside a bacteria, plant or animal cell, viral DNA can divert the host machinery to doing little else but making thousands of viruses identical to itself*".

1.1.3 *Bacteria contain many DNA polymerases*

The DNA pol first discovered by Kornberg and co-workers in *E. coli* is now called DNA pol I, because other DNA pols (II, III, IV and V) were found later. As a matter of fact the simple mechanism of DNA replication suggested by the DNA structure proposed by Watson and Crick requires not only several DNA pols but also many other different enzymes and specific proteins to assure the extremely high number of polymerization steps on a single natural DNA template molecule and a high fidelity of DNA replication, thus avoiding a too high level of mutations unacceptable for individuals and species. Although the problem of DNA replication was solved intuitively by Watson and Crick in 1953, determining the actual molecular basis of DNA replication has required considerable more time and efforts.

Indeed, some ten years after Kornberg's discovery of DNA pol I, the DNA replication process became more complicated with the discovery that *E. coli* contained other DNA pols and the first one discovered by Kornberg was not the most important for the replication of the DNA molecule (see Chapter 2). In 1969 Paula De Lucia and John Cairns, then Director of the Cold Spring Harbor Laboratory, by analyzing thousands colonies of mutagenized *E. coli* for DNA pol activity, found one with only 0.5 to 1 percent of the normal level of DNA pol. Impaired only by an increased sensitivity to ultraviolet light, the mutant, named polA1 mutant for its assonance with Paula, grew and multiplied at a normal rate.[19]

A year later, in 1970, it was just Kornberg's son, Thomas, who had joined Malcolm Gefter's laboratory in the Biology Department at Columbia University, who found in Cairns' mutant a DNA pol activity distinct from the one his father had discovered. He named the purified enzyme DNA pol II.[20,21] Similar results were reported the same year by Rolf Knippers[22] and by Robb Moses and Charles Richardson.[23]

In the course of chromatographic separation Tom Kornberg also noticed another band of DNA pol activity in the position where it would have been obscured by DNA pol I in wild type *E. coli*. He named it DNA pol III.[24] Gefter *et al.* located the gene for DNA pol III and were able to demonstrate that conditionally lethal mutations in this gene blocked DNA replication. Subsequent studies revealed that DNA pol III is the core of the DNA pol III holoenzyme.

More bacterial DNA pols, named DNA pols IV and V were revealed almost 30 years later, in 1999, by studying inaccurate DNA replication across UV lesions during the so called process of translesion DNA synthesis (see Chapter 4).

DNA pols I, II and III differed each other in several properties, such as optimal template-primer, sensitivity to sulfhydryl-blocking agents, ethanol and salt, K_m for dNTPs, and interaction with DNA binding proteins. They, however, shared among themselves (and with DNA pols subsequently discovered not only in bacterial, but also in animal and plant cells and in viruses) basic and fundamental features such as: 1) requirement for a DNA template, guiding the DNA pol to select any deoxynucleotide by base-pairing to the one on the DNA template (a mismatch nucleotide at the 3′ terminus is usually removed by an intrinsic 3′→5′ exonuclease proofreading activity present in several DNA pols); 2) need for 2′-deoxyribonucleoside 5′-triphosphates (dNTPs), which are incorporated as monophosphate to the 3′-hydroxyl group at the growing end of a DNA strand (the *primer*) with formation of a new phosphodiester bond following nucleophilic attack of the α-phosphate of the incoming dNTP by the 3′-OH group on the primer terminus, with release of pyrophosphate (*no DNA pol is able to initiate new chains*); 3) chain growth exclusively in the 5′→3′ direction and antiparallel with respect to the template strand; 4) a 3′→5′ exonuclease activity, also called proofreading activity, and 5) a 5′→3′ nuclease activity, also called editing activity, important, for example, in the excision of pyrimidine dimers, is also present in DNA pol I as well as in other DNA pols. These and other features, such as binding sites for dNTPs, the template and the growing chains, will be subsequently discussed for each DNA pol as well as the ways they probably function *in vivo*.

We shall just mention here that Kornberg also demonstrated, by equilibrium dialysis, that dNTPs bind to a single site on DNA pol I with the triphosphate moiety being of primary importance for the specificity of binding. Thus he also found that 2′,3′-dideoxyribonucleoside 5′-triphosphates (ddNTPs), the analogs of the natural 2′-deoxyribonucleotide substrates of DNA pol, but lacking the 3′-hydroxyl group functioning as a primer, bind as well to the dNTPs binding site. In fact, ddTTP binds even better than dTTP. In the presence of an excess of DNA pol, ddNTPs are also incorporated and then function as DNA chain terminators. These results inspired Frederick Sanger (1918–2013) to develop his chain-termination technique of determining the sequence of a DNA chain,[25,26] and for this in 1980, he received with Walter Gilbert and Paul Berg a second Nobel Prize in Chemistry, 22 years after earning his first Nobel Prize in Chemistry for determining the complete amino acid sequence of insulin.

1.1.4 *How is a new DNA chain started? Discontinuous DNA synthesis and the need for a RNA primer*

The replication of infectious circular single-stranded ΦX174 DNA, demonstrated in 1967, implied that DNA pol I could start a new chain. If this were the case the initiating nucleotide should retain the triphosphate group, but this was never found. A stimulatory effect of added boiled *E. coli* extract, removed by digestion with DNase, rather suggested that a fragment of *E. coli* DNA, later removed by the 3′→5′ proofreading and 5′→3′ editing activities associated to DNA pol I, initiated the reaction. But the problem of the *primer*, i.e. how a new DNA chain started *in vivo*, remained, because microscopic and genetic analysis indicated a synchronous replication of both daughter helices. How could one reconcile the synchronous replication of the two chains of the DNA helix, both serving as template, but oriented in opposite directions, with the limitation of DNA pols that only proceeded in the 5′→3′ direction, just opposite that of the template?

The dilemma was resolved by Reiji Okazaki (1930–1975), a former visiting scientist in Kornberg's laboratory. He found that much of the newly synthesized DNA momentarily existed as small fragments, about 1000 nucleotide long (Okazaki fragments) near the replicating fork. These became covalently joined into longer daughter chains by DNA ligase as the replication process proceeded, suggesting that DNA replication is a discontinuous process.[27] The direction of DNA synthesis is thus always in the 5′→3′ direction and might be continuous on the 3′→5′ leading strand but needed to be discontinuous in 5′→3′ lagging strand.

1.1.5 *RNA priming as a mechanism for initiation: DNA primase*

The problem of how DNA chains are started was finally solved when Kornberg had the idea to study the conversion of the circular single-stranded chromosomal DNA of phage M13 into the double-stranded DNA replicative form. Such conversion, mediated by bacterial enzymes (including RNA polymerase), occurs upon phage entering the bacterial cell, and Kornberg hypothesized that RNA polymerase, which is able to start chains *de novo*, could make a short fragment of RNA on single strand M13 DNA that could be utilized by DNA pol to start a DNA chain. The small RNA fragment would then be removed by the proofreading and editing properties of the DNA replication machinery. The process should be inhibited by rifampicin, an inhibitor of RNA polymerase. Based on these hypotheses the results were obtained in a few days,[28] and such *RNA priming* was subsequently found to be the most general mechanism for initiation of DNA chains.

There are, however, exceptions to RNA priming such as: 1) a 3′-hydroxyl primer can be created by a specific endonuclease, as in rolling circle replication of some phages and plasmids or by folding back of the 3′OH end of a linear, single-stranded DNA (such as that of parvovirus); 2) a virus-encoded protein may serve as primer by pairing a single deoxyribonucleotide (dCMP) with the end of a template strand as in the replication of the linear chromosomes of Adenovirus and *B. subtilis* phage Φ29.

The specialized RNA polymerases that synthesize minitranscripts serving as primers are called DNA primases. Their variety is fascinating: some read sequences of a few bases, others synthesize a primer of an exact length but are indifferent to their sequence, some even accept deoxyribonucleotides as well as ribonucleotides for elongation. Some are tightly bound to DNA pols such as in eukaryotic DNA pol α, or PrimPol, others to a helicase (phage T4) and some contain the primase and helicase activities within a single polypeptide (phage T7). *E. coli* primase associates with many other proteins to form a mobile multifunctional primosome. In no instance it is clear how a primase, by analogy with RNA polymerase, recognizes its "promoter" or terminates "transcription".

The primers are ultimately removed, in *E. coli* by DNA pol I, in eukaryotes by RNase H, Dna2 nuclease and flap endonuclease 1 when the DNA growing chain reaches the previous primer, just before another enzyme, DNA ligase, joins together the DNA fragments thus restoring the phosphodiester backbone.

1.2 Late 1960s to early 1970s DNA replication shows its complexity

In 1967 *in vitro* replication of a short circle of ΦX174 DNA appeared to be accomplished by just two enzymes, a DNA pol to start and extend the chain and a DNA ligase to join the ends. However, only a few years later the replication process definitely appeared much more complex, requiring the need of a complex network of interacting proteins and enzymes for the replication both of a simple bacteriophage DNA and of the much more complex host bacterial chromosome.

We have already mentioned that the problem to reconcile the synchronous growth of both daughter chains with the $5' \rightarrow 3'$ direction of synthesis by all DNA pols was solved by Okazaki, whereas the need for a primer was solved again by Kornberg.

1.2.1 *The DNA structure is much more complex, rich of conformational flexibility and thus full of functional potentialities than the one proposed by Watson and Crick*

The need for a complex network of many proteins and enzymes interacting among themselves and with the DNA helices during DNA replication is not at all surprising. The structure proposed by Watson and Crick in 1953 immediately suggested the function of DNA and appeared to everybody so naturally logic and unquestionable that there was no time for astonishment and for looking back at the times when the genetic secret of our life was unknown.

However the structure proposed by Watson and Crick in 1953 was a static, three dimensional structure, and as time went by it became clear that the structure of DNA is much more complex, rich in conformational flexibility and full of functional potentialities compared to what it appeared at first sight. DNA is indeed a dynamic molecule whose movement and conformational changes are essential features for its function. The DNA molecule not only contains a type of genetic information written in the base sequence (nucleotide triplet) that, being a genetic code, in the end specifies the aminoacids sequence, and, therefore, the proteins structure. It also contains a quite important conformational type of genetic information, a kind of phenotypic code, that allows a specific interaction with proteins.

For the double helix itself, different variants were soon hypothesized and demonstrated, the best characterized being the A, B and Z structures, depending on the environment such as ionic strength, ions, relative humidity, organic solvents and temperature. The capacity to assume alternative conformational forms is further increased by the presence of particular sequences in the DNA, such as the highly repetitive AT rich sequences of satellite DNA, the inverted repetitive "hairpin" sequences, the GC rich telomeric sequences synthesized by a special DNA pol (Telomerase), discovered in 1985 by Carol Greider and Elizabeth Blackburn,[29] that uses a short RNA, rather than a DNA, as a template, just to mention a few. Blackburn and Greider shared the 2009 Nobel Prize in Physiology or Medicine with Jack Szostak for the discovery of telomerase and of how chromosomes are protected by telomeres.

Furthermore DNA has a kind of tertiary structure defined by the interchangeable terms supercoiling, supertwisting, superhelicity, that define the spatial coiling or the twisting upon itself of a duplex DNA helix. This property was first observed and it is better understood in small circular molecules such as bacterial plasmids, small viral chromosomes, mitochondrial DNA. However, it is an integral feature of linear chromosomes as well as eukaryotic ones, when ends are united to a solid matrix to form circles or too remote to rotate rapidly and free rotation of the ends of DNA is restricted. The degree of supercoiling is crucial in replication, transcription, recombination and repair and it is regulated by DNA topoisomerases (see next section).

1.2.2 *DNA binding proteins, DNA helicases, DNA topoisomerases*

While many groups began studying and finding several DNA pols in eukaryotic cells as well, another problem remained: what opens up the double helix giving rise to the replication fork in front of the advancing DNA pol(s)? We have seen that the structure of the double helix resembles a zipper that cannot be open unless a slider is moved in the right direction. And, importantly, energy is needed for this.

The most obstinate hunter of the slider was Bruce Alberts who ultimately did not discover the slider but a new family of important proteins, the DNA binding or extending proteins, exemplified by T4 gp32 protein. These proteins promptly cover the bared single strands of DNA, protecting them against nuclease action, preventing them from reannealing and configuring them (by keeping them in an extended form) to serve as templates for DNA pols.[30–32]

The problem of the slider needed to open up the zipper was solved once again by Kornberg. In 1975 he found in *E. coli* a protein (the *rep* protein) that by utilizing ATP could move along the DNA opening up the double helix in front of the advancing DNA pol. It was a DNA-dependent ATPase and was named DNA helicase. The trick it used is the one utilized by the muscle motility proteins: ATP-dependent periodical changes in its shape enable the helicase to enter the double helix opening it to the advancing DNA pol(s).

If DNA binding proteins and DNA helicase solved one problem, another one immediately appeared: both the advancing of the replication complex and the fact that the daughter chains synthesized at the replication fork have to wind around each other to reform a double helix created topological problems downstream and upstream from the replication fork. For example, a great torsion would develop along the unreplicated DNA chains that should rotate continuously at the expense of a lot of energy. If we assume that a cell nucleus corresponds to a tennis ball, then its DNA would be a thread of 20 Km, and it is easy to imagine what would happen if one tries to rotate a thread of 20 Km in a tennis ball.

These problems were resolved with the discovery by James C. Wang of the first DNA topoisomerase, an enzyme that control the topological state of the DNA.[33] There are different types of DNA topoisomerases and with simple and economic mechanisms they can rotate DNA chains. As in Wangs's words:[34] "*DNA topoisomerases are magicians among magicians; they open and close gates in DNA without leaving a trace, and they enable two DNA strands or duplexes to pass each other as if the physical laws of spatial exclusion do not exist. Because the biological functions of the DNA topoisomerases are deeply rooted in the double helix structure of DNA, it should not be surprising that these enzymes participate in nearly all biological processes involving DNA*".

1.3 Concluding remarks, parts 1.1–1.2

After the discovery of the DNA structure by Watson and Crick in 1953, the subsequent work of Kornberg and coworkers first led to the discovery of the DNA pols, enzymes that synthesize long chains of DNA, fill up gaps in DNA molecules and repair raw ends of DNA. Other important enzymes and proteins were then discovered and found essential for DNA replication (Table 1.1). Some of them allowed the manipulation of DNA molecules such as *DNA ligases*, which join contiguous ends of

Table 1.1 DNA replication proteins in *E. coli* in 1975

Protein	Function
DNA polymerase I	Repair
DNA polymerase II	Repair
DNA polymerase III	Replicase
DNA primase	Synthesis of initiator RNA
DNA ligase	Replication — Repair (DNA joining)
DNA helicase	Unwinding of double-stranded DNA
SSB	Binding and protecting single-stranded DNA
DNA topoisomerases	Resolving topological issues of DNA

DNA chains, various *DNA exonucleases*, and special DNA endonucleases, called *restriction nucleases,* that can modify DNA ends or cut DNA molecules at specific sequences respectively. The latter studies earned the Nobel Prize to Werner Arber, Hamilton Smith and Daniel Nathans (1928–1999) in 1978 for their applications to problems of molecular genetics. Finally a unique DNA template-independent DNA pol was found in calf thymus, *terminal transferase*, that can add nucleotides to the end of a DNA chain.[35]

The combined use of such enzymes has allowed the artificial manipulation of DNA molecules that soon led to important technical advances such as sequencing of DNA, that made it possible to sequence the genome of many organisms, including the human genome; gene cloning and genetic engineering, that have led and will lead to a cascade of progress and practical benefits in medicine, agriculture and chemical industry.

What are regarded as the first experiments of genetic engineering are those of Stanley N. Cohen, a physician in the Department of Medicine at Stanford University, and Herbert Boyer of the University of California, San Francisco. They were studying bacterial resistance to antibiotics treatments and wanted to understand how the genes of plasmids could made bacteria resistant to antibiotics. Back in 1972–73, by using the restriction enzyme EcoR1 that cuts DNA molecules creating protruding complementary or cohesive ("sticky") ends which can be joined by DNA ligase, they were able to cut, combine and transplant genes in plasmids and, following the introduction of the plasmid into *E. coli*, to perpetuate an animal gene in the bacterium, opening the way for the introduction and propagation of any gene.[36] Furthermore, in 1976 Boyer, with the venture capitalist Robert Swanson, founded *Genentech* (a composite of *Genetic Engineering Technology*, *Inc*), the leading biotechnology corporation that was the first to express human genes in bacteria such as the hormone *somatostatin* in 1977 and *insulin* in 1978.

1.4 Multiple DNA polymerases in eukaryotic cells: DNA polymerases α, β and γ as the first ones

From what we have so far presented in a rather simple way, it is clear that, although the problem of DNA replication was solved intuitively by

Watson and Crick in 1953, determining the actual molecular basis of replication has required considerable more time and effort. No wonder that the complexity of DNA replication forced things to be worked out first in prokaryotes, thanks to simple *in vitro* replication systems relying on bacteriophages pioneered by Kornberg (ΦX174, M13), Alberts (T4), and Richardson (T7). Such *in vitro* systems provided the means to identify the products of several genes involved in DNA replication, because DNA synthesis was observed in extracts prepared from wild-type strains but not in extracts of mutants. Using *in vitro* complementation and reconstitution assays, biochemists were able to fractionate the complex extracts and to define the individual enzymes and proteins necessary for replication, some of which have been mentioned above such as the main actors, the DNA pols, but also several accessory proteins, RNA polymerase and DNA primases, single-stranded DNA binding proteins, DNA helicase, DNA topoisomerases, RNase H and DNA ligase.

The most important insight was that rather than carry out a set of sequential individual reactions with one class of proteins completing its function, dissociating from DNA and the next protein further processing an intermediate, the proteins of the replication apparatus form a stable complex that remains associated with the DNA for thousands of nucleotide during chain growth. Some proteins are important to recognize an origin of replication, to alter its structural conformation and to lead to its further opening by the action of a helicase. Others are needed to suppress potential origins elsewhere on the chromosome, and finally others might appear more important or more specific to replicate the DNA such as the DNA pols, DNA primase, DNA helicase, DNA topoisomerase, single-stranded DNA binding proteins and DNA ligase because they are needed to prime and elongate DNA chains and to propel the two forks in bidirectional replication.

So if in prokaryotes different DNA pols are present and the one most directly involved in DNA replication has a higher complexity (See *E. coli* pol III holoenzyme in Chapter 2) it was quite likely that many DNA pols were also present in eukaryotes, and a similar or higher degree of complexity could be expected in functionally operating replicative DNA pols in eukaryotes. The excellent book "*Animal cells DNA polymerases*" published in 1986 by Michael Fry and Laurence A. Loeb still contains the most complete list of references, including reviews, regarding DNA pols α, β, γ and δ until 1985. Here, we will briefly review the most important steps leading to the discovery of these first DNA polymerases.

1.4.1 *DNA polymerase α*

Following the first demonstration, in 1956, by Kornberg *et al.* of an activity (named *E. coli* DNA pol I) capable of incorporating radioactive thymidine in extracts of *E. coli*, Frederick Bollum and Van Rensselaer Potter (1911–2001) in 1958 were the first to present evidence for a DNA pol activity in the high speed supernatant of homogenates from regenerating rat liver.[37] A similar activity was also present in nuclei, in agreement with two earlier reports in 1956 by Morris Friedkin, Donald Tilson and DeWayne Roberts[38] and by Friedkin and Horatio Wood[39] on the incorporation of [^{14}C]-thymidine into DNA of chick embryos, bone marrow cells and isolated thymus nuclei, respectively.

But, in truth, back in 1951, Peter Reichard and Bengt Estborn[40] had already demonstrated incorporation of [^{15}N]-thymidine into DNA, and in 1954 Friedkin and Roberts[41] had observed uptake of [^{14}C]-thymidine in Yoshida ascite tumor cells incubated *in vitro* and Kiao-Hung Lu and Theodore Winnick (1912–2007) in embryonic tissue cultures.[42] Friedkin was the scientist who, in 1955, gave some [^{14}C]-thymidine to Kornberg for his first experiments in *E. coli* extracts, before it became commercially available in 1957. But Friedkin was no longer inclined to search for incorporation of thymidine into DNA with extracts from broken eukaryotic cells. Thus Bollum, certainly inspired by the preliminary experiments of Friedkin and then guided by the convincing and definite results of Kornberg in *E. coli*, can be considered as the pioneer of eukaryotic DNA pols. Upon the commercial availability of [^{3}H]-thymidine in 1958, Bollum reported the incorporation of thymidine into DNA in extracts of calf thymus, and, in 1960, just after the full characterization of *E. coli* DNA pol I by Kornberg's group in 1957, he also reported a partial description and purification in calf thymus of a DNA pol activity and its requirements for the four dNTPs, a primer DNA and Mg++.[43] An extensive purification of a high molecular weight DNA pol from calf thymus appeared in 1965 by Masahiko Yoneda and Bollum.[44]

In the following years a high molecular weight DNA pol activity was demonstrated to be ubiquitous in growing eukaryotic cells, such as human, murine, hamster, sea urchins, chick embryo, avian tissues and yeast, by various groups, including Bollum's and those of Arthur Weissbach, Lawrence Loeb, David Korn, Robert Gallo, David Baltimore, Irving Johnston, Samuel Wilson, Hamish M. Keir (1931–2013), Francois Chapeville, Erhard Wintersberger. A complete list of references can be found in the 1986 book of Michael Fry and Loeb "*Animal cell DNA polymerases*".[45]

This high molecular weight DNA pol activity showed a characteristic sensitivity to sulfhydryl group blockers such as N-ethylmaleimide and to NaCl concentration above 25 mM and has been referred to by Bollum as "high molecular weight cytoplasmic DNA polymerase" because most of the activity (>80%) was present in the cytoplasm of cells broken by an aqueous procedure. This was clearly an artifact due to the use of aqueous extraction, because in the following years different techniques, including mechanical dissection of nuclei and use of immunoperoxidase staining in fixed cells labelled with monoclonal antibodies against DNA pol α, confined this polymerase to the nuclei of all the examined cell types. Until a uniform nomenclature was adopted in 1975 (see later), it was called with different names such as "high MW cytoplasmic or maxi DNA pol" by Bollum, DNA pol II by Weissbach, cytoplasmic N2 by Korn, DNA pol I by Gallo, and DNA pol C by Baltimore.

1.4.2 *DNA polymerase β*

For many years, in analogy with *E. coli* DNA pol I, the high molecular weight cytoplasmic DNA pol (6–8 S) was assumed to be the only type of DNA pol in eukaryotic cells. However, a low molecular weight DNA pol, much less abundant than the high molecular weight enzyme in proliferating cells, was first identified in 1971 in HeLa cells by Weissbach *et al.*[46] This enzyme was then found in rat liver nuclei by Earl Baril *et al.*,[47] in calf thymus by Lucy Chang and Bollum,[48] and in rat liver by Haines *et al.*[49] and by Berger *et al.*[50] Other studies indicated that perhaps mitochondria contained a DNA pol that differed from these two high- and low-molecular weight DNA pols.

The new low molecular weight (3–4 S) DNA pol (or DNA pol I for Weissbach, "mini" for Bollum, DNA pol II for Gallo, cytoplasmic N1 for Korn and DNA pol N for Baltimore) was considered to be located exclusively in the nucleus and to be present in all eukaryotes examined. It showed resistance to sulfhydryl group blockers such as N-ethylmaleimide, to NaCl or KCl concentration up to 200 mM in 50 mM K-phosphate buffer, and to a number of chemical agents such as ethanol, acetone (20–25%), urea (5 M) and phosphonoacetic acid. An interesting study on the phylogeny of DNA pol β is that of Chang (1976) showing its wide distribution in multicellular animals and its absence in bacteria, plants and protozoa.[51]

Distinguishing among different DNA pols in eukaryotic cells was rather difficult because one had to completely rely on biochemical approaches, whereas we have previously seen that almost in the same years (1969–1972) conditional mutants of DNA replication allowed the unambiguous identification in *E. coli* of DNA pols II and III by T. Kornberg, Malcom Gefter, Knippers, Moses and Richardson, although they were much less abundant than DNA pol I.

1.4.3 *Lack of relationship between high- and low-molecular-weight DNA polymerases*

However, the low-MW DNA pol was not immediately accepted as a distinct DNA pol by the scientific community. Immediately following its discovery, and with the possible identification of another different DNA pol in mitochondria, Chang and Bollum in 1972 reported an antigenic relationship between the high- and low-molecular weight DNA pols from a variety of cells and suggested that they could share common peptides sequences or even subunits[52] and others, in particular Norman Hecht[53] and Lawrence Lazarus,[54] in sedimentation studies of 1973, reported a salt-mediated conversion of the two DNA pols.

On the contrary Silvio Spadari and Weissbach in 1974[55] and Graham Smith *et al.* in 1975[56] with antibodies prepared against human high molecular weight DNA pol and Gilbert Brun *et al.* in 1975[57] with antibodies against avian high- and low-molecular weight DNA pols, showed an absence of cross-reaction between the two DNA pols. No cross-reactivity was also found with any of the oncornavirus reverse transcriptases, just discovered by Howard Temin (1942–1994) and Baltimore, or with another DNA pol found in Weissbach's laboratory (1972, 1973) in human cells and named R-DNA pol for is ability to copy, much better than the low-molecular weight enzyme, a deoxyprimed homoribotemplate such as oligo(dT)/poly(A) (see section on DNA pol γ).

1.4.4 *1975: First nomenclature system for eukaryotic DNA polymerases*

Because of the proliferation of systems for naming the DNA pols in various laboratories, Weissbach, Baltimore, Bollum, Gallo and Korn devised the nomenclature reported in Table 1.2 at a meeting held at the Massachusetts Institute of Technology (MIT) on May 29, 1974.[58] On May 1975, the 48 scientists attending the international conference on Eukaryotic DNA pols organized by Korn of Stanford University, at the Asilomar Conference Center, Pacific Grove, California, decided to accept this uniform system of nomenclature.

Thus the size, cellular localization or template preference for naming DNA pols were abandoned in favor of a system of Greek letters, proposed by Weissbach, assigned to the DNA pols according to the order of discovery, a system acceptable to everybody because it was different from any of the systems previously used.

The high-molecular weight DNA pol first discovered by Bollum (1958, 1960 and 1965) became DNA pol α; the low molecular weight DNA pol first discovered in 1971 in human cells in Weissbach's laboratory became DNA pol β; the other DNA pol again described and

Table 1.2 Nomenclature of eukaryotic DNA polymerases in 1975

	Formerly used systems (identified by the proposer's name)				
New system	**Baltimore**	**Bollum**	**Gallo**	**Korn**	**Weissbach**
DNA pol α	C	6-8S (maxi)	I	cytoplasmic N2	II
DNA pol β	N	3-4S (mini)	II	cytoplasmic N1	I
DNA pol γ	A		III		R

characterized in Weissbach's laboratory in 1972–1974 and called R-DNA pol for its preference to copy a synthetic deoxyprimed polyribotemplate, such as oligo(dT)/poly(A), but not natural RNA, was called DNA pol γ.

All these pols, in particular DNA pols α and β, were active in copying "activated DNA" that is a DNA prepared by limited digestion of duplex DNA with DNase I, providing, in the DNA substrate, 3′OH primers and gaps to be filled up by DNA pols. Double-stranded and single-stranded DNA's, as expected, were inactive as template because of the lack of 3′-OH primers.

The retroviruses reverse transcriptase, discovered in 1970 by Howard Temin[59] at the University of Wisconsin-Madison and independently by David Baltimore[60] at MIT, for which they shared the 1975 Nobel Prize in Physiology or Medicine with Renato Dulbecco (1914–2012), remained separate from this nomenclature because they represented distinct entities. Other viral-induced DNA pols such as Herpes and Vaccinia that could be referred to by the name of the virus, also remained separate.

Finally terminal deoxynucleotidyltransferase (TdT), an unusual template-independent DNA pol, found in 1960 by Bollum in calf thymus, was also not included (see Chapter 3).

1.4.5 *DNA polymerase γ*

If, at the Asilomar meeting of 1975, the attending scientists were all convinced of the existence of two distinct DNA pols, α and β, in eukaryotic cells, a word of reservation was still present about DNA pol γ, the last of the DNA pols described in eukaryotic cells included in Table 1.2. This reservation was shared by many scientists, whether either attending or not attending the meeting, because it represented only about 1% of the total cellular DNA pol activity and, most intriguingly, the preferred template were deoxyprimed ribohomopolymers, in particular oligo(dT)/polyrA that in 1972 Chang and Bollum had found to be a substrate for DNA pol β as well.[61]

Rather than a novel cellular DNA pol, it was regarded more like a cellular "reverse transcriptase-like" activity with a possible function in normal cellular processes like cytodifferentiation or gene amplification. It was also suggested that it was an activity due to DNA pol β, because

Chang and Bollum reported in 1972 that, in the presence of Mn^{++} ions and salt, DNA pol β was able to copy oligo(dT)/poly(rA). After all, despite a marked preference for deoxyribohomopolymers, the *E. coli* DNA pol I was also able to use ribohomopolymers as template.

At the Asilomar meeting one of us (S. Spadari), visiting scientist in Weissbach's laboratory in 1973–75, reported his results on DNA pol γ just after the previous speaker had said that DNA pols α and β represented approximately 95% and 5%, respectively, of the DNA pol activity in dividing cells, the "rest" or 0% being the other DNA pols! By modifying the previous purification procedure, Spadari and Weissbach purified DNA pol γ from cytoplasm (over 2000-fold) and from nuclei (almost 3500-fold). The two forms of DNA pol had a similar MW (110–120,000 daltons) as measured by sedimentation studies, much higher than the approximately 70,000 daltons of reverse transcriptases; they could use, although with lower efficiency, activated salmon sperm DNA as substrate and this activity was not inhibited by an antibody against DNA pol α, thus demonstrating that it was not due to a contamination with this DNA pol; they copied, in contrast with the less purified preparation first reported by Weissbach in HeLa cells in 1972,[62] all the homoribopolymers -poly(C), poly(I), poly(U) but with a preference for poly (A)-; contrary to DNA pol β, they utilized oligo(dT)/poly(rA) as primer-template in the presence of phosphate that completely inhibited the similar activity of DNA pol β; they were both unable, in contrast to reverse transcriptases, to copy natural RNA; they were not inhibited by antisera prepared against several reverse transcriptases as those obtained from Woolly monkey virus, Rauscher murine leukemia, or the Mason-Pfizer monkey virus. Interestingly both forms showed K_m values for deoxyribonucleoside triphosphates (approximately 5×10^{-7} M) that were one order of magnitude or more lower than the K_m values of DNA pols α and β. Finally similar forms of DNA pols γ were purified from the normal human diploid fibroblast cell line WI-38, suggesting that DNA pol γ was not unique to human cells derived from tumor tissues.[63]

Although the data presented by Spadari in Asilomar claiming that DNA pol γ was a distinct third DNA pol present in eukaryotic cells were not immediately accepted, subsequent studies very soon revealed, as we will see later, that DNA pol γ is the mitochondrial enzyme and that it replicates mitochondrial DNA.

1.5 Early attempts to ascribe an *in vivo* function to DNA polymerases α, β and γ

The lack of eukaryotic conditional mutants in DNA replication made it difficult to assign a function to DNA pols α, β and γ. Even in prokaryotes, where mutants were available and the properties of enzymes and proteins involved in DNA transactions were much better known, assigning a functional role to a molecule was not without doubts. While it seemed unreasonable to assume that the DNA pols are geared for anything other that some form of DNA synthesis (replicative or repair synthesis), in prokaryotes there is no absolute division of labor. Thus, DNA pol III is responsible for chromosomal replicative synthesis and DNA pols I and II are mainly involved in repair-type synthesis. However, we know that DNA pol I is needed for the removal of the RNA primers from the Okazaki pieces and for the resynthesis of removed sequences, and DNA pol III is involved in some forms of repair synthesis including recombination. Nevertheless, separation of main functions still remains in bacteria. Thus, following the identification of the first three DNA pols in eukaryotes, considered for several years as the sole eukaryotic DNA pols, scientists began wondering which of the animal DNA pols was more involved in either processes (see also Chapter 4).

1.5.1 *Positive correlation of DNA polymerase α with cellular DNA replication and development*

The catalytic properties of the first three DNA pols present in eukaryotic cells were different enough to allow, in the early 1970's, their simultaneous assay without appreciable interference. The assay for DNA pol β was absolutely specific because of its insensitivity to N-ethylmaleimide (NEM) and its high pH optimum and preference for high ionic strength. If enzyme samples were preincubated with 5 mM NEM at 0°C for 30 min to inactivate DNA pols α and γ, the β-assay was strictly specific for the DNA pol β. The γ-assay, using oligo(dT)/poly(rA) in the presence of 50 mM K-phosphate (pH 8.5) that completely inhibits the activity of DNA pol β on this template-primer, was also specific. Under the assay conditions used for DNA pol α, DNA pol γ was inactive and DNA pol β responded with 30% efficiency, and, therefore the DNA pol α results could be corrected for its contribution.

Thus studies with cultures of mouse L cells in different growth conditions by Chang and Bollum in 1973,[64] in synchronized HeLa cells by Spadari and Weissbach in 1974[63] and by Rosanna Chiu and Earl Baril in 1975,[65] with regenerating rat liver by Baril *et al.* in 1973,[66] in phytoaemoagglutinin-stimulated lymphocytes by Mayer *et al.* in 1975[67] and Umberto Bertazzoni *et al.* in 1976,[68] and many others,[45] all showed a positive correlation between activity of DNA pol α and rate of DNA synthesis, with much lower changes of DNA pols β and γ. Because Okazaki fragments were also observed as intermediates during DNA synthesis in eukaryotes, Spadari and Weissbach in 1975[69] showed that only DNA pol α could synthesize DNA covalently bound to RNA. The RNA-DNA molecule could then be extended by either DNA pols α, β or γ indicating that the DNA pol α was capable of both RNA- and DNA-primed DNA synthesis, whereas DNA pols β and γ were capable of only DNA-primed DNA synthesis.

In the late 1970s when DNA pol δ was not yet widely recognized as a novel DNA pol (a destiny common to all DNA pols so far discovered after the identification of the major high MW DNA pol by Bollum) changes in DNA pols α, β and γ activities were studied in several systems to provide insights about their possible function. Among them fertilized eggs and developing embryos (from Xenopus, sea urchin, Drosophila) which showed very rapid DNA synthesis, mouse and rat testis, spleen, heart, brain and liver with important tissue changes during perinatal development and in senescence or, like in liver, a programmed proliferation of hepatocytes after partial hepatectomy. A positive correlation between DNA pol α activity and cell proliferation and development was always found.[45]

As an example we report the impressive patterns of the three DNA pols α, β and γ observed by Ulrich Hübscher and Spadari in 1977 in rat brain neurons [Figure 1.1(a)], spleen [Figure 1.1(b)] and ventricular cardiac muscle [Figure 1.1(c)] during perinatal development.[70] Comparison

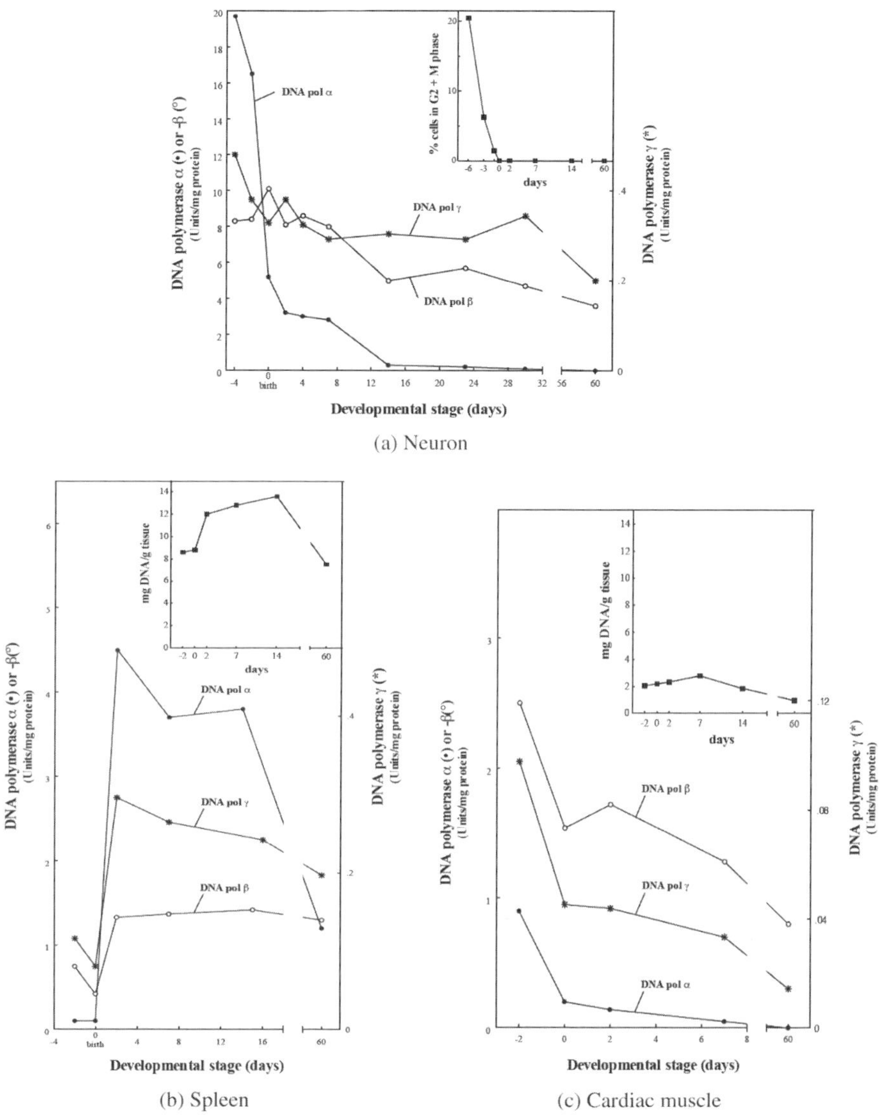

Figure 1.1 **Developmental profiles of DNA pols α, β and γ in (a) forebrain neurons, (b) in spleen and in (c) ventricular cardiac muscle**. The insert shows the percentage of cells in G_2 and M phase as determined by cell-cycle analysis (a) or the amount of DNA per g tissue chosen as a parameter of the rate of cell multiplication (b and c). Redrawn from Hübscher *et al.*[70]

of the relative amounts of the three DNA pols present in late gestation in forebrain neurons showed that approximately 65–70% of the total activity is due to DNA pol α, 25–30% to β, and 2% to γ. After birth, the relative amounts of DNA pols β and γ increase to the adult proportions of 92–95% and 5–7%, respectively. The loss in DNA pol α activity correlates temporally very well with the decline in mitotic activity [Figure 1.1(a), insert], which drops to zero at birth. Later studies by Spadari *et al.* in 1988[71] showed that the DNA pols δ and ε levels during the course of neuronal development are coincident with those of DNA pol α, suggesting a coordinated function of these DNA pols in DNA replication. In spleen DNA pols α, β, and γ activities changed from 1%, 85%, and 14% prenatally (day −2) to 74%, 21%, and 5%, respectively, postnatally (day 2). In adults, DNA pol α is present in proportions (45%) much higher than in the other tissues studied (approximately 1% in neurons and cardiac muscle). The developmental variation of DNA pol α activity in spleen closely mirrors the rate of cell multiplication in this organ [Figure 1.1(b), insert]. The high activity of DNA pol α in adults seems to be ascribable mainly to dividing and differentiating lymphocytes in the germinal centers.

Finally the developmental profiles of DNA pols α, β and γ in ventricular cardiac muscle [Figure 1.1(c)] were quantitatively similar to those in neurons and showed a good correlation with the virtually complete cessation of ventricular cellular proliferation before birth. All three DNA pols are already low by the second prenatal day where approximately 70% of the total activity is due to DNA pol β and only 26% to DNA pol α. DNA pol α declined more than β and γ during further development and can no longer be assayed by the seventh postnatal day, whereas the activities of DNA pols β and γ persist into adulthood. In summary these experiments convincingly demonstrated the strong correlation between cell proliferation and DNA pol α.

1.5.2 *DNA polymerase γ is the mitochondrial DNA polymerase and replicates mitochondrial DNA*

The mitochondrial DNA pol, which had never been highly purified before 1977, was considered a fourth cellular DNA pol, distinct from DNA pol γ.

Arthur Bolden *et al.*[72] in Weissbach's laboratory then purified a unique DNA pol from mycoplasma-free, HeLa cells mitochondria which was very similar to cytoplasmic and nuclear cellular DNA pol γ. This result suggested that the DNA pol γ and the mitochondrial DNA pol were identical.

In the same year, Hübscher and Spadari[73] aware in advance of this finding thanks to the personal communication of Weissbach, approached this problem by using an elegant system particularly suited to separating nuclear and mitochondrial compartments. Taking advantage of the fact that, in the brain, synaptic end knobs contain mitochondria anatomically away from and thus completely unassociated with nuclei, and that these structures can be isolated as distinct particles (synaptosomes) surrounded by a close membrane, they isolated mt-DNA pol from synaptosomes [Figure 1.2 and 1.3 part (a)] and the DNA pol γ from brain nuclei [Figure 1.3 part (b)].

Almost 98% of the activity was recovered from synaptosomes, and comparison of the DNA pol activities purified from nuclei and mitochondria showed that a DNA pol with the properties of DNA pol γ occurred in both locations. Because DNA pol γ was the only DNA pol found in mitochondria, they concluded that DNA pol γ and mt-polymerase were identical. Bertazzoni *et al.* (1977) also made this observation on analogous enzymes from chick embryo nuclei and mitochondria.[75]

In the following two years (1978–9) Hübscher and Spadari provided clear evidence that DNA pol γ replicated mt-DNA in purified synaptosomes, a *quasi in vivo* mitochondrial DNA synthesizing system.[74,76] They showed that synaptosomes permeabilized with 0.5% Brij 58 incorporated [^{3}H]dTTP into DNA and that the labeled DNA was mitochondrial DNA by several criteria, such as: 1) expected density following analytical ultracentrifugation in CsCl gradient, 2) expected 6 bands upon digestion with Hind III, 3) presence of both fully duplex circular DNA molecules and duplex circular DNA molecules containing replication structures in electron micrographs, 4) a rate of DNA synthesis (around 10% of the input DNA in 1 hr) several thousand fold higher than the rate of repair synthesis in control nuclei and finally 5) a significant reduction of the number of replication structures in the presence of dideoxyTTP, an inhibitor of DNA pol γ.

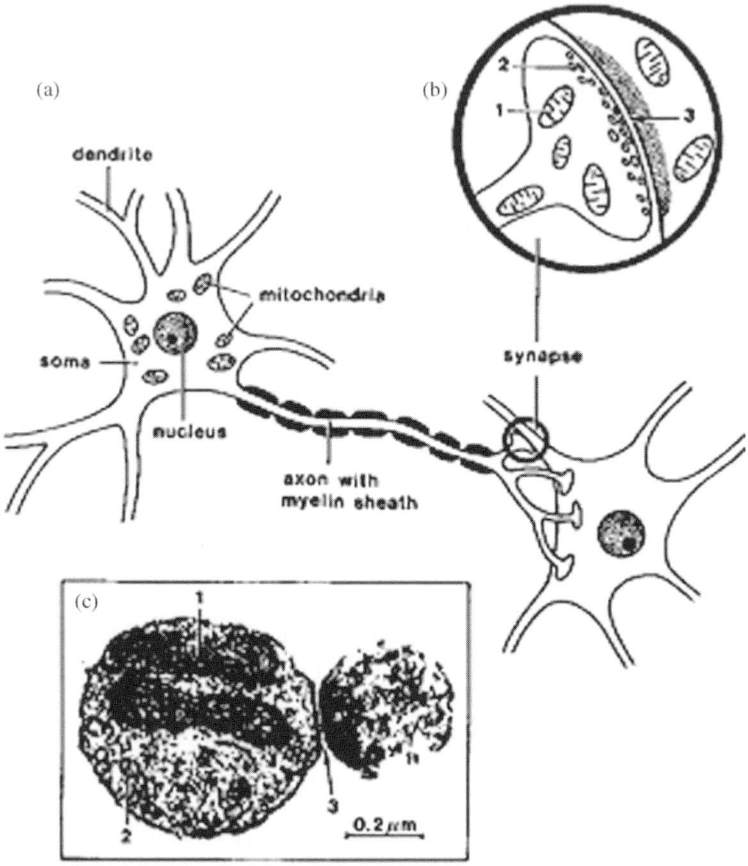

Figure 1.2 **Structure of neurons and their synapses simplified from electron micrographs**. (a) Schematic representation of a neuron and its connections. (b) Enlarge synaptic end knob with mitochondria (1), synaptic vesicles (2), and synaptic cleft (3). (c) Electron micrographs of synaptosomes; identification of structures as in (b). Reproduced from Hübscher *et al.*[74]

1.5.3 *Further evidence for a major involvement of DNA polymerase α in DNA replication and of DNA polymerase β in DNA repair*

In the years 1978–1980, other interesting evidence for a major involvement of DNA pol α in DNA replication and of DNA pol β in DNA repair

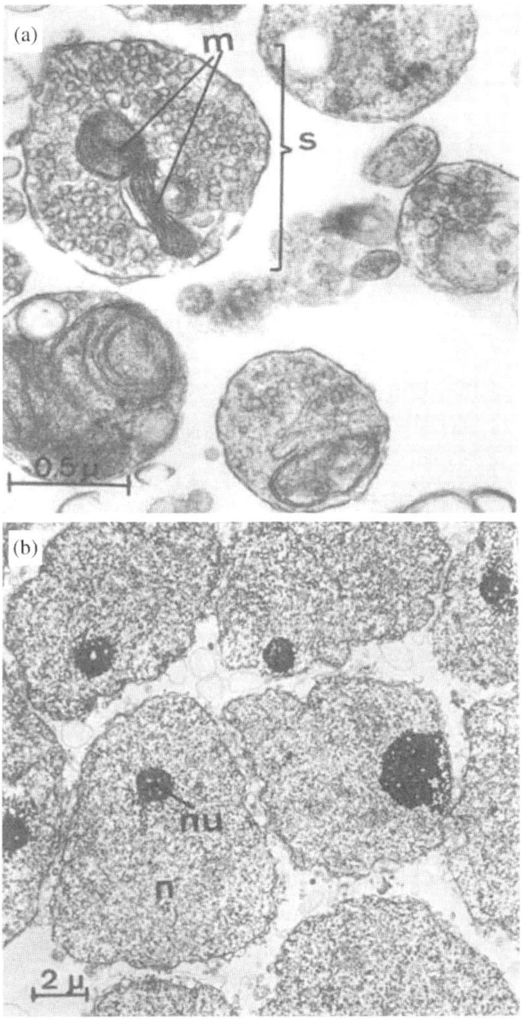

Figure 1.3 **Electron micrographs of synaptosomes (a) and nuclei (b) from rat brain cortex**. N = nucleus; nu = nucleolus; s = synaptosome; m = mitochondria. The bar represents 0.5 μm (a) or 2 μm (b). Reproduced from Hübscher *et al.*[73]

came from the laboratories of Spadari and Hübscher. Following a preliminary report in 1973 that aphidicolin, a tetracyclic diterpenoid obtained from *Cephalosporium aphidicola*, inhibited incorporation of thymidine into DNA of cultured human embryonic lung cells and the growth of Herpes simplex and Vaccinia viruses, with no effect on RNA and protein synthesis,[77] it was found in 1978 that aphidicolin prevented mitotic division of sea urchin embryos,[77] which required replicative DNA synthesis. In particular Guido Pedrali-Noy and Spadari in 1979 showed that aphidicolin selectively inhibited purified DNA pol α as well as DNA pols induced by Herpes simplex and Vaccinia viruses.[78] DNA pols β and γ were completely resistant to aphidicolin. The mechanism of inhibition of DNA pols by aphidicolin was also investigated by Pedrali-Noy and Spadari.[78]

With recognition that DNA pols δ and ε, later recognized as the 4th and 5th eukaryotic DNA pols with a critical role in DNA replication, were also found sensitive to aphidicolin, Spadari's group demonstrated that the selective inhibitory effect of aphidicolin on the replicative DNA pol α and the reversibility of its action *in vivo* allowed the synchronization of HeLa (and other eukaryotic cells) in culture.[79–81] They showed that the inhibition of DNA pol α by aphidicolin prevented G1 cells from entering the DNA synthetic period (S-phase), blocked cells in the S phase and allowed G2, M and G1 cells to continue the cell cycle and to accumulate at the G1/S border.

Aphidicolin thus appeared a more useful and physiological reagent than hydroxyurea and thymidine to synchronize cells because it did not affect cell viability or S phase duration. Furthermore the 24-hour treatment with aphidicolin did not interfere with the synthesis of dNTPs or of DNA pols. Cells exposed to the drug continued to synthesize all three DNA pols α , β, and γ as well as all dNTPs which, when the block was removed, were present at levels optimal for DNA initiation and replication.[79] This simple technique could be applied to cells growing in suspension or monolayers (including plant cells — *see plant cells DNA polymerases section*) and allowed to harvest large quantities of synchronized cells.

Plant cells α-like DNA pol was also found sensitive to aphidicolin by Spadari's group. Figure 1.4 depicts electron microscope autoradiographs of mammalian and plant cells incubated with aphidicolin, showing that aphidicolin stopped nuclear DNA synthesis in both types of cells while replication of mitochondrial or chloroplast DNA, respectively, was not affected.[80,82,83]

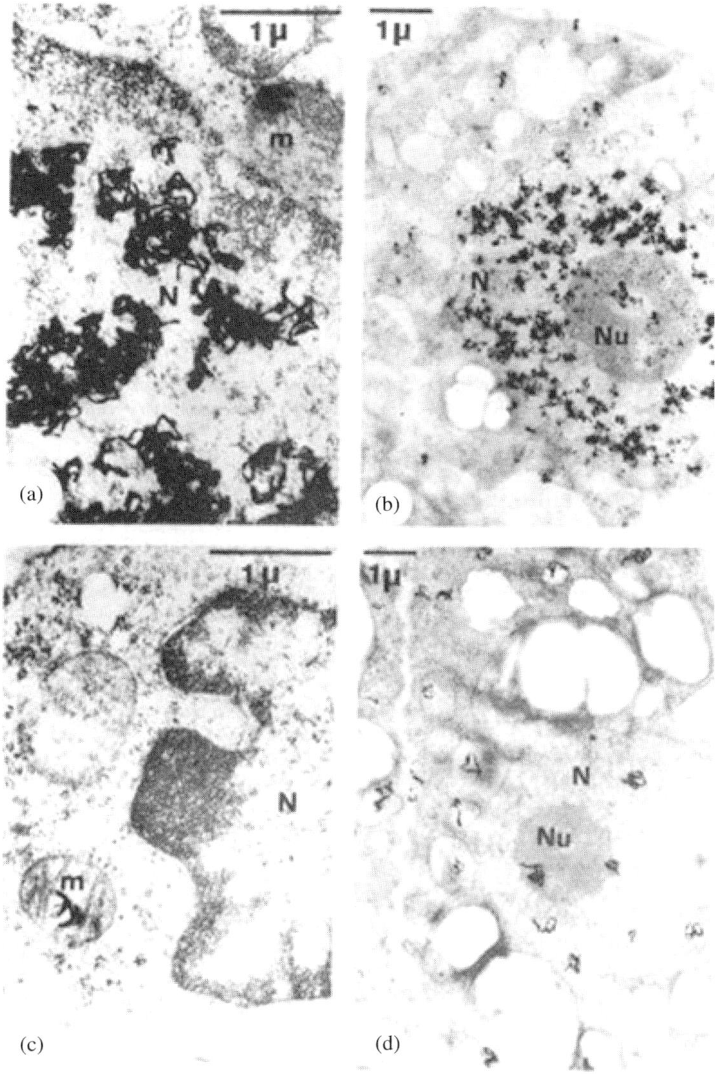

Figure 1.4 **Electron microscopic autoradiographs of eukaryotic (a and c) and plant cells (b and d) incubated with labeled thymidine without (a and b) or with aphidicolin (c and d).** Both types of cells, in the absence of aphidicolin, carry out nuclear and organellar (mitochondrial or chloroplast) DNA synthesis. The addition of aphidicolin completely inhibits the incorporation of thymidine into nuclear DNA but has no effect on organellar DNA synthesis. N = nucleus; Nu = nucleolus; m = mitochondria. Reproduced from Spadari *et al.*[82]

In 1978–1979 a major role of DNA pol β in DNA repair was assigned more by exclusion, following the demonstration of the role of DNA pol α and γ in the replication of nuclear and mitochondrial DNA, respectively, rather than by a direct proof. Hübscher and Spadari clearly demonstrated, however, that rat nuclei from adult nondividing neurons that they had shown to contain virtually no DNA pol other than β (β, 99.2%; γ, 0.8%) were able to repair UV-damaged DNA. Incorporation of [^{3}H]dTTP in such nuclei was stimulated approximately 7–10 fold by irradiation with UV light, and this stimulation was suppressed when DNA synthesis was allowed to proceed in the presence of dideoxyTTP, an inhibitor of DNA pol β. These data for the first time suggested a role of DNA pol β in DNA repair.[84] In 1981 a clear radioautographic demonstration of DNA repair synthesis in HeLa cells in the presence of aphidicolin was obtained by Spadari's group, again suggesting a major role of DNA pol β in DNA repair.[85] Of course the above experiments did not rule out that aphidicolin could suppress some modes of DNA repair due to DNA pol α and that some other types of DNA damage might require DNA pol α. The extent of damage, nature of damaging agent and many other factors might determine the relative contribution of DNA pols β and α to repair of DNA.

1.6 DNA polymerases δ and ε

The father of DNA pol δ was Antero So. In 1976, a year after the proposal to name eukaryotic DNA pols with Greek letters in order of discovery, John Byrnes *et al.* purified from erythroid hyperplastic bone marrow a new species of DNA pol.[86] In contrast to DNA pols α, β and γ, the novel DNA pol was associated with a very active 3′→5′ proofreading exonuclease activity, similar to the one associated with prokaryotic DNA pols and not separable from the DNA pol activity by several chromatographic steps and by sucrose gradient centrifugation. Both exonuclease and DNA pol activities had identical rate of heat inactivation suggesting the coexistence of the two activities in a single polypeptide chain. The novel DNA pol was named δ but for 4–5 years it was neglected by the scientific community, its uniqueness being only the inseparability from the 3′→5′ exonuclease. All other activities and properties (such as reaction requirements, sensitivity to NEN and aphidicolin, and resistance to dideoxy-NTPs) resembled those of DNA pol α which was the predominant DNA pol activity in the cells.

DNA pol δ was generally assayed with polydA/oligoT as template-primer because of a rather feeble activity observed with primed single-stranded DNA, and this property did not help to immediately recognize it as a distinct DNA pol. In the 1980s however all doubts about DNA pol δ as a polymerase structurally and immunological distinct from DNA pol α were dispelled, and its fundamental role as a complex with several auxiliary proteins in DNA replication in conjunction with DNA pol ε and DNA pol α/primase was then widely recognized (see Chapter 2).

A distinctive feature of DNA pol δ, revealed more than 10 years later in 1987 by Cheng-Keat Tan, again in So's laboratory,[87] was its dependence on proliferating cell nuclear antigen (PCNA), a protein that increased the DNA pol δ processivity by 40 fold (see also Chapter 2). Thus the high processivity of DNA pol δ and the much lower processivity of DNA pol α, which was found to possess a primase activity, suggested a role for DNA pol δ in continuous replication of the leading strand and for DNA pol α/primase for the discontinuous replication of the lagging strand.

Soon, however, Federico Focher *et al.* in Hübscher's laboratory in 1988 established a novel procedure to purify calf thymus DNA pol δ from cytoplasmic extracts that allowed the simultaneous isolation and separation of DNA pol δ, DNA pol α, a DNA dependent ATPase and PCNA.[88] The final preparation had typical properties of a DNA pol δ including a 3′→5′ exonuclease activity and efficiently replicated naturally occurring genomes such as primed single-stranded M13 DNA and single-stranded porcine circovirus DNA, this last one thanks to an associated or contaminating primase activity. Processivity of at least a thousand nucleotides was evident for this pol and even in the absence of PCNA, a property that was in disagreement with the dependence of the processivity of DNA pol δ on PCNA.

A similar PCNA-independent DNA pol was identified in a variety of cells, e.g. HeLa cells,[89] purified in several ways and recognized as a DNA pol distinct from δ and, therefore, named DNA pol ε. Recent experiments indicate that DNA pol δ is responsible for the replication of the lagging strand of DNA while DNA pol ε replicates the leading strand (see Chapter 4).

1.7 Yeast DNA polymerases

Yeasts are eukaryotic microorganisms with about 1,500 species. Commercial baker's yeast *S. cerevisiae* is a very important model organism in cell biology research, and for a long time has been used by researchers to obtain information about the biology of eukaryotic cells. Thus, following the discovery of a DNA pol activity in bacteria by Kornberg and in eukaryotic cells by Bollum, the abundance of yeast and the power of yeast genetics and molecular analysis immediately led researchers to look for a similar DNA synthesizing enzyme in yeast.

Erhard Wintersberger, at the Institut fur Krebsforschung der Universitat of Wien started to look for a DNA pol activity in yeast mitochondria in 1966,[90] particles that in eukaryotes have their own replicating machinery. Later, in 1970 he wondered whether the enzyme he had found in mitochondria was different from the DNA pol active in the yeast nucleus.[91] Following the report by Hans Eckstein *et al.* in 1967 from Hamburg that extracts from synchronized cells of *S. cerevisiae* exhibited a DNA pol activity oscillating during the cell cycle with a maximum just before the onset of DNA synthesis,[92] Ulrike and Erhard Wintersberger partially purified and characterized from a mitochondria-free cell extract of *S. cerevisiae* two high MW DNA pols: a major one, called **A**, and a minor one, called **B**, both apparently distinct from the high MW mt-DNA pol they had further purified at the same time both from wild type and from a "*petite*" mutant.[91] A low MW DNA pol activity similar to DNA pol β found in those years in higher eukaryotic cells was never detected in *S. cerevisiae*.

A better purification and characterization of the two DNA pols present in extracts of commercial bakers' yeast and wild type *S. cerevisiae* grown aerobically to late log phase was done by Chang in 1977 in order to compare them with the properties of the well-characterized bacterial and mammalian DNA pols. Thus, she purified two immunologically distinct DNA pols and called them I and II, corresponding to the previously described **A** and **B** enzymes.[93] DNA pol I carried out pyrophosphate exchange and pyrophosphorolysis reaction and had no associated nuclease activity, in general resembling DNA pol α. However DNA pol II had an associated 3′→5′ exonuclease activity and could excise mismatched 3′

nucleotides from suitable template systems, resembling more prokaryotic DNA pols II and III.

Like other DNA pols so far discovered, neither of the two yeast DNA pols was capable of initiating *de novo* DNA synthesis but Chang had shown that only DNA pol I could use oligoribonucleotides as initiators. Paolo Plevani and Gianfranco Badaracco, following the finding by Ronald Conaway and Lehman that DNA pol α was associated with a DNA primase in Drosophila melanogaster embryos[94] which was later extended to several other eukaryotes, isolated a complex of yeast DNA pol I and DNA primase in 1984.[95] In addition, Lianna Johnson *et al.* in Juidth Campbell's laboratory clearly demonstrated in 1985 that the gene coding for this enzyme was essential, because disruption of the gene in germinated spores gave a terminal phenotype consistent with a defect in the S phase.[96]

The discovery of another yeast nuclear DNA pol, biochemically and immunologically distinct from DNA pols I and II, came in 1988, from the laboratory of Peter Burgers at the Washington University School of Medicine, St. Louis.[97] By using yeast cells from a protease-deficient strain, he purified a complex containing three subunits, one of them with a DNA pol and a $3' \rightarrow 5'$ exonuclease activity that could excise single nucleotide mismatches providing a base-paired primer-template that could be elongated by the DNA pol, consistent with a proofreading function during *in vivo* DNA replication. Thus this novel DNA pol, that he called DNA pol III, appeared functionally related to mammalian DNA pol δ. With appropriate mutants Campbell's group in 1989 showed that not only DNA pols II and III were genetically distinct, but also that DNA pol II, the one present in lower amounts, was clearly distinct from pol I. Highly purified DNA pol II contained a $3' \rightarrow 5'$ exonuclease but not DNA primase activity.[98,99]

Thus, four DNA pols as well as their corresponding genes had been identified in yeast cells: DNA pol I was analogous to DNA pol α; DNA pols II and III resembled DNA pols ε and δ, and a mitochondrial DNA pol was akin to DNA pol γ. Still another DNA pol was inferred by Alan Morrison *et al.* in 1989 from the sequence of gene REV3, a non essential gene required for the mutagenic repair of UV damage[100] (see also Chapter 3).

Table 1.3 Nomenclature of eukaryotic DNA polymerases in 1990

DNA polymerase	Yeast enzyme	Yeast gene	Mammalian enzyme
α	I	*POL1*[a]	α
β	—	—	β
γ	m	*MIP1*	γ
δ	III	*POL3*[b]	δ
ε	II	—	ε

[a]Former designation, *CDC17*; [b]Former designation *CDC2*; m = mitochondrial.

1.7.1 *Revised nomenclature for eukaryotic DNA polymerases*

Burgers *et al.* in 1990 proposed a revised nomenclature for eukaryotic DNA pols (Table 1.3) and the use of a separate letter of the Greek alphabet to designate each genetically distinct yeast DNA pol.[101] Later, the availability of yeast genome sequence has allowed the identification of others DNA pols for a total of at least eight (see Table 1.4 and Chapter 2).

1.8 Plant cell DNA polymerases

Following the discovery of multiple DNA pols in prokaryotes (DNA pols I, II and III) and eukaryotes, (DNA pols α, β and γ), several groups started to study DNA pol activities in plant cells in the mid-1970s. Simon Litvak's group at the Université de Bordeaux II was one of the most active, studying the DNA pol activities in ungerminated or germinating wheat embryos. In 1975 they purified and partially characterized three cytoplasmic DNA pols (A, B and C) from ungerminated wheat embryos and another DNA pol from purified mitochondria that behaves like DNA pol B.[102] Considering the possibility that cytoplasmic DNA pol B was a release product of broken mitochondria, on the basis of chromatographic separation, template-specificity and enzyme properties, the conclusions emerging from their and others' studies[103–107] were that: 1) plants might contain two high MW DNA pols; 2) such enzymes possessed associated exonuclease activities like those of bacterial DNA pols, and 3) that no low MW DNA pol analogous to animal DNA pol β was found in plant cells.

In an attempt to gather experimental data on the DNA pols of plant cells, which could allow a better and meaningful comparison with mammalian cells, Spadari's group in Pavia in 1978 used plant cells in culture where the relationship between DNA synthesis and DNA pols can be more easily examined under strictly controlled experimental conditions (proliferating or resting cells, normal or tumorous cells, mutant cell lines). The major DNA pol activity they purified from cultured *Oryza sativa L.* cells shared most of the properties of animal DNA pol α, and, at the stage of purification employed, possessed an associated 3′→5′ exonuclease activity.[108] Because in *Oryza sativa L.* cells as well as in other plant cells lines like *Parthenocissus tricuspidata*, *Acer pseudoplatanus L.* and *Medicago sativa L*, the purified DNA pol was the most abundant in proliferating cells and responded to changes in the rate of cell multiplication, they postulated that this activity corresponded functionally to the animal DNA pol α and propose to call it α-*like* DNA pol.

In 1980 the same group[109] showed that spinach leaves contained at least two DNA pols, one of them localized in the chloroplasts. The one present and predominant in crude leaf extracts and practically absent from purified chloroplasts was the previously described α-*like* DNA pol. They purified a distinct DNA pol from purified chloroplasts whose properties significantly resembled those of the animal mitochondrial DNA pol γ and thus proposed to name it γ-*like* DNA pol. The chloroplast γ-*like* DNA pol differed from the mitochondrial DNA pol described by Litvak's group in wheat embryos.

Because, by analogy with animal DNA pols, the plant cells α-*like* and the γ-*like* enzymes were sensitive and resistant, respectively, to aphidicolin,[110] they studied the effect of aphidicolin on the synthesis of DNA in the nucleus, chloroplast and mitochondrion of cultured rice cells, as assessed by autoradiography by light and electron microscopy (see Figure 1.4). Their results[83] clearly showed that aphidicolin specifically prevented the synthesis of nuclear DNA, while it had no effect on the synthesis of organellar DNA, thus proving that the α-like DNA pol played an essential role in the replication of nuclear DNA and that this DNA pol was not involved in the replication of plastid and mitochondrial DNA. The replication of plant organellar DNA required a different DNA pol, the γ-*like* DNA pol described in chloroplast or a mitochondrial DNA pol, if different from the γ-*like* one, both of them able of strand displacement synthesis of the organellar DNAs.

Once again, by analogy with the results obtained with mammalian cells, Spadari *et al.* demonstrated by autoradiography in 1983, that aphidicolin, by inhibiting the α-*like* DNA pol, and thus nuclear DNA replication, could be used to synchronize cultured *Daucus carota* cells by inducing accumulation of cells at the G1/S boundary of the cell cycle. A single blockage with aphidicolin synchronized 95% of the cell population (the remaining non-synchronized 5% cells most probably representing the fraction of non-multiplying cells), and this was reversible, resulting in immediate and synchronous resumption of nuclear DNA synthesis following removal of the drug.[111] Obviously a small fraction of the cells was blocked in the process of DNA synthesis; if necessary these cells, which were estimated to represent <10% of total population, could be forced to follow the main stream population by an appropriate second exposure to aphidicolin.

We previously described how Hübscher and Spadari demonstrated in 1979 a role of DNA pol β in DNA repair by exploiting the unique property of brain neurons which, in adult rats, are devoid of DNA pol α and do not carry out nuclear DNA replication, thus allowing a study of UV-induced DNA repair synthesis. Similarly Spadari's group in Pavia used plant protoplasts prepared from leaves of *Nicotiana sylvestris*, which, under the experimental conditions used, did not carry out nuclear DNA replication, to study UV-induced DNA repair synthesis which also occurs in cells of higher plants.[81] They clearly demonstrated by autoradiography that in this system UV-induced DNA repair synthesis was resistant to aphidicolin and thus was not performed, at least under these physiological conditions, by the α-*like* DNA pol. As mentioned above, there was no evidence as to the existence of a low M.W. β-*like* DNA pol in plant cells and such repair synthesis could be due to a yet unknown DNA pol.

As a matter of fact plant genomes sequence has allowed in more recent years the identification in plant cells of homologs for 10 DNA pols and possibly two plastid DNA pols (see also Chapter 3).

1.9 Virus-induced DNA polymerases

Viruses are leading causes of many human diseases. Some oncogenic DNA viruses, such as polyoma and SV40, are known to induce cellular DNA synthesis upon infection and to cause a rise in various cellular DNA pols, in particular DNA pol α. But often a virus-induced DNA pol is required for viral DNA replication and this has led to isolation of viral DNA pols and to a search for antiviral compounds that might selectively inhibit the viral DNA pol and viral DNA replication. Here we shall limit the history of the discovery of viral DNA pols to some human viruses such as Herpes, Vaccinia, Hepatitis B and retroviruses for their medical interest.

1.9.1 *Herpes virus DNA polymerase*

Among human viruses, Herpes viruses (whose name derives from the Greek word *herpein* which means to creep, reflecting the creeping and spreading nature of the skin lesions they usually cause) are second only to influenza and cold viruses, in causing overt disease. Once a patient has become infected by a herpes virus, the infection remains for life because the virus remains silent ("latent") in peripheral sensory ganglia [Herpes simplex-1 (HSV-1) or 2 (HSV-2), Varicella Zoster virus (VZV)] or in monocytes or lymphocytes [Epstein-Barr virus (EBV), cytomegalovirus (CMV), Herpes lymphotropic virus, Human herpes virus-7 (HHV-7)] or in unknown target cells such as Human herpes virus-8 (HHV-8) and Kaposi's sarcoma-associated herpes virus (KSHV). Latency is often followed by subsequent reactivation and recurrence of infection.

Herpes viruses encode their own DNA-dependent DNA pols. In addition, most herpes viruses encode thymidine kinase (TK) that activates several nucleoside analogs to triphosphates that are incorporated by viral DNA pol thus inhibiting replicating viruses. Herpesviruses replicate their DNA in the cell nucleus, and Keir *et al.* in 1966 observed that the infective process in BHK or Hep-2 cells by HSV-1 was accompanied by an increase, among other enzymes, of a DNA pol activity with physical and immunological properties distinct from host cellular DNA pols. They also investigated the template specificity of the induced DNA pol in salt-fractionated crude extracts.[112] The first extensive purification and characterization, however, was reported by Weissbach *et al.* in 1973 from nuclei of HSV-1 HeLa infected cells. The enzyme had a high MW (180,000) and was optimally active at 0.15 M K_2SO_4 or 250 mM KCl, salt concentrations that inhibited the HeLa DNA pols α and β by 90%. The best primer/template was oligodG/polydC, approximately 10 times better than activated DNA, perhaps a reflection of the 67% G+C base composition of HSV-1 genome.[113] These results opened the way to the identification and isolation of a novel virus-induced DNA pol in all other members of the herpes group of viruses and to the demonstration that the HSV-induced DNA pol is required for viral replication.

1.9.2 *Vaccinia virus DNA polymerase*

Vaccinia virus, belonging to the poxvirus family and responsible for smallpox, contains a linear, double-stranded DNA genome of approximately 190 Kbp and it is unique among DNA viruses because it replicates only in the cytoplasm of the host cell. Kenneth Berns *et al.* (1969) and Citarella *et al.* (1972), in Weissbach's laboratory, were the first ones to partially purify and characterize a vaccinia-directed DNA pol from the cytoplasmic fraction of virus-infected HeLa cells.[114,115] Subsequently purified to homogeneity by Mark Chalberg and Paul Englund in 1979,[116] the enzyme is a 110 kDa single polypeptide with an active 3′→5′ proofreading exonuclease activity. Like DNA pol α and Herpesvirus DNA pol, vaccinia DNA pol is strongly inhibited by aphidicolin.[78]

1.9.3 *DNA polymerase activity in Hepatitis B particle*

Whereas the DNA pols induced during infection by Herpes or Vaccinia viruses are localized in the cell, a DNA pol activity within the virion of the Hepatitis B Dane particle was first reported by William Robinson's group in 1973.[117] They also showed that the particle contained a small circular double-stranded DNA with a MW of 1.6×10^6 daltons, which served as a primer-template for the DNA pol. The DNA pol was a specific marker for Dane particles and Robinson used the DNA pol reaction also to radiolabel the Dane particle core for a specific and sensitive radioimmunoprecipitation assay for antibodies against hepatitis B core antigen (anti-Hbc).[117–119] Hepatitis B virus replicates through reverse transcription of a pregenomic RNA by viral pol that possesses a protein-priming, reverse transcriptase, DNA pol and RNase H activities. Thus the reverse transcription activity of the viral polymerase can be the target of nucleoside analogs inhibiting viral replication and the polymerase can be used for *in vitro* screening of novel anti-HBV compounds directed against wild-type and mutants of the polymerase.

1.9.4 *Retroviruses reverse transcriptase*

Reverse transcription, the flow of genetic information from RNA to DNA, typifies retroviruses more than any other properties. Howard Temin in the 1960s was the first to suggest that some RNA viruses could replicate through a DNA intermediate, subsequently integrated into the host genome. His hypothesis was based upon the observation that replication of retroviruses was sensitive to inhibitors of DNA synthesis. As mentioned earlier, the discovery in purified virions of a novel polymerase, a RNA-dependent DNA pol (reverse transcriptase), was made in 1970 by Temin at the University of Wisconsin-Madison and independently by Baltimore at MIT. This discovery astonished the scientific world because it violated the central dogma of molecular biology, that is the flow of genetic information from DNA to RNA to proteins. The AIDS pandemic later stimulated renewed interest in reverse transcriptase as a therapeutic target.

1.10 1999–2000: Appearance of many novel specialized DNA polymerases

As soon as the human genome sequence was published in 1999, a "golden age" of DNA pols followed. More than 30 years ago it was hypothesized by Miroslav Radman that error prone DNA pols might exist in bacteria as well as in eukaryotes that could, like terminal deoxynucleotidyl transferases, insert random nucleotides in a template-independent fashion, and that DNA damage–induced mutagenesis might be a specific cellular response to genotoxic damage. This event is known as SOS response and in *E. coli* more than 30 inducible genes were identified (reviewed in Refs. 120 and 121). The isolation of mutants defective exclusively in SOS mutagenesis led to the identification of the *umu*C, *umu*D and *din*B genes in *E. coli*. These genes are now known as the two translesion DNA pols IV (*din*B) and V (*umu*C and *umu*D) (see Chapter 4).

1.10.1 *DNA polymerase ζ, the lesion extender*

Over 40 years ago Christopher Lawrence discovered that in yeast Rev 3 and Rev 7 genes are indispensable for UV mutagenesis and for mutagenesis derived from translesion DNA synthesis (TLS) through abasic sites.[120] The two subunits form the heterodimer of DNA pol ζ.[122] This enzyme belongs to the B family, but lacks a 3′→5′ exonuclease proofreading activity and is essential, because mice without the DNA pol ζ gene die during embryogenesis.[123] Its function is not to incorporate opposite a lesion but to extend from a variety of lesions (for details see Chapter 4).

1.10.2 *DNA polymerase θ*

Identification of mammalian pol Q initially arose from interest in the mus308 gene product of *Drosophila melanogaster*. Characterization of the mus308 gene showed that it encodes an unusual domain configuration, with the C-terminal part of the protein encoding a DNA pol, and the N-terminal part of the protein encoding a DNA-helicase.[124]

A cDNA encoding the DNA pol domain of a homologous human enzyme, designated DNA pol θ (pol Q) was identified by Farida Sharief *et al.*,[125] while Mineaki Seki *et al.*[126] isolated the first complete and functional human Pol Q cDNA, showing that it encodes for a 2590-amino-acid protein with both a DNA pol and helicase like domain.

The enzyme belongs to the B family pols and lacks a 3′→5′ exonuclease proofreading activity. Pol θ is capable of TLS and recently it has been suggested to be important in the repair of genomic double-strand breaks (DSBs) and to play a prominent role in nonhomologous end joining (NHEJ) (for details see Chapter 3).

1.10.3 *DNA polymerases λ and μ, two family X DNA polymerases*

In 2000 Luis Blanco reported DNA pol λ as a novel eukaryotic DNA pol with a potential role in meiosis.[127] Soon thereafter he also identified a second family X polymerase that he called DNA pol μ, homologous to TdT, that could act as a DNA mutator in eukaryotic cells.[128] These two DNA pols were subsequently studied by various laboratories. Both of these enzymes appear to have important DNA repair functions such as base excision repair, translesion DNA synthesis, double-stranded break repair and, in a physiological sense, they have important roles in the hypermutation of immunoglobulin genes, a recombination event leading to the diversity of antibodies (for details see Chapter 4).

1.10.4 *The complex Y family of DNA polymerases*

DNA pols of the Y family belong to the five subfamilies which include: (i) the *umu*C and *umu*D (pol V) branch that is only found in prokaryotes; (ii) the *din*B branch which is found in *E. coli* (pol IV), in archaea (Dbp/Dpo4) and in eukaryotes (DNA pol κ). The other three subfamilies are found exclusively in eukaryotes being (iii) DNA pol η, (iv) Rev1 and (v) DNA pol ι.

The discoveries of novel DNA pols of the Y family, were mostly reported in the year 1999–2000 (reviewed in Ref. 129). In *E. coli* the groups of Myrion Goodman, Roger Woodgate and Zvi Livneh simultaneously published in 1999 that DNA pol V is a dimer of the post-translationally modified *Umu*D protein in complex with *Umu*C.[130,131] Soon thereafter the goups of Takehiko Nohmi and Robert Fuchs identified *Din*B as a DNA pol and called it DNA pol IV.[132] Also in 1999 the groups of Fumio Hanaoka and Prakash purified a DNA pol from human Xeroderma pigmentosum variant (XP-V) cells and called it DNA pol η.[133,134] Ohmori's[135] and Errol Friedberg's[136,137] groups described the human *DINB1* gene product as DNA pol κ. Wensheng Lin *et al.*[138] identified in human cells Rev1, a homolog of the *S. cerevisiae* REV1, which also codes for a dCMP transferase that is dependent upon a DNA template. Finally, the most astonishing discovery in these years was the fact that Woodgate *et al.* identified

DNA pol ι,[139] that has such a low fidelity that it even violates the Watson-Crick base pair rule! (For details on these enzymes see Chapter 3).

1.10.5 *PrimPol*

As we mentioned earlier, a primase activity was identified in one of the four subunit of the DNA pol α complex. Recently, however, an additional DNA pol-primase, called PrimPol has been discovered.[140–142] The enzyme belongs to the archaea-eukaryotic primase superfamily. PrimPol activity is regulated by single stranded binding proteins and, differently from the DNA pol α/primase, can start DNA chains with desoxyribonucleotides. Its localization is both nuclear and mitochondrial and the enzyme appears to facilitate replication fork progression by acting as a TLS DNA pol or as a specific DNA primase reinitiating downstream lesions that block DNA synthesis. For details see Chapter 3.

1.10.6 *Present nomenclature for eukaryotic DNA polymerases*

In 2001 Burgers *et al.* proposed a novel nomenclature for eukaryotic DNA pols.[143] The current 17 DNA pols are listed in Table 1.4.

1.11 Concluding remarks, parts 1.4–1.10

With few exceptions (such as the protists *yeast* and *amoebae* which are individual, unicellular organisms), eukaryotes are complex multicellular organisms made up of many specialized cells containing DNA/chromosomes enclosed by a membrane defining the nucleus, whose presence gives the name to these organisms. Eukaryotic cells are also much larger and have a greater degree of structural organization and functional complexity than prokaryotic cells. Furthermore nearly all of them possess self-replicating organelles such as *mitochondria* (plant cells and various algae have *plastids*, among which the most prominent and well studied are the *chloroplasts* present in green parts of the plant). Both types, mitochondrion and plastid, have probably evolved from aerobic bacteria. They

Table 1.4 Nomenclature of eukaryotic DNA polymerases

Greek name	HUGO name	Class	Other names	Proposed main function
alpha	POLA	B	POL1	DNA replication
beta	POLB	X		Base excision repair
gamma	POLG	A	MIP1	Mt replication
delta	POLD1 B	B	POL3	DNA replication
epsilon	POLE	B	POL2	DNA replication
PrimPol	PRIMPOL	Archea type primase		Mt replication (initiation) Translesion synthesis Repriming at forks arrested at DNA damage
zeta	POLZ	B	REV3	Bypass synthesis
eta	POLH	Y	RAD30, XPV	Bypass synthesis
theta	POLQ	A	mus308	DNA repair
nu	POLN	A		Bypass synthesis
iota	POLI	Y	RAD30B	Bypass synthesis
kappa	POLK	Y	DinB1	Bypass synthesis
lambda	POLL	X	POL4	Base excision repair
mu	POLM	X		Non-homologous end joining
	REV1L Y	Y	REV1	Bypass synthesis
	TdT	X		Antibody diversity
Telomerase	TERT	RT		Telomere mainenance

Adapted from Burgers *et al.*[143]

retain their own bacteria-like chromosomal DNA and are self-replicating organelles that apparently divide independently of the cell, although over time most of the original genes were transferred to the host cell and therefore most of their proteins are now coded for by the nuclear DNA. Because of this they are unable to live on their own if isolated from the eukaryotic cell.

The basic mechanisms of DNA synthesis are similar in prokaryotes and eukaryotes, but DNA replication is much more complicated in eukaryotes. It starts at a specific point in the cell cycle, the beginning of the S phase and at several points in the chromosome leading to duplicate chromosomes first and then to nuclear and cell division. The surface area/volume ratio in eukaryotic cells is smaller than in prokaryotes and explains the lower metabolic rate and longer generation time in eukaryotes.

No wonder that, by analogy with multiple DNA pols (at least 5) found in *E. coli*, we have just seen that eukaryotes as well contain many DNA pols, e.g. at least 8 in yeast and at least 17 in mammalian cells. The model plant organism *Arabidopsis thaliana* could contain 10 DNA pols. To replicate and maintain genome integrity (including the organellar genome) many DNA pols are thus required, and some cooperate with multiple proteins induced/expressed in response to DNA damage and involved in the various types of error-free DNA repair mechanisms as well as in DNA damage checkpoint. Sometimes specialized DNA pols (e.g. the Y family translesion DNA pols) allow greater tolerance to damage leading to enhanced mutagenesis as well as to an increased rate of survival (see Chapters 3 and 4).

Viruses are intracellular parasites, and for their replication strategies they must either operate within limits imposed by the host cells or circumvent these limits. Several viruses, such as parvoviruses, papillomaviruses, polyomaviruses, adenoviruses and herpesviruses, enter the cell nucleus to replicate their DNA and either induce and use a cellular DNA pol (parvoviruses, papillomaviruses, polyomaviruses), or encode their own DNA pol (adenoviruses, vaccinia viruses and herpesviruses). Hepatitis B virus differs from the other DNA viruses in that it is more similar to the oncornaviruses (RNA tumor viruses) in its mode of replication.

Finally, the RNA tumor viruses (retroviruses) are an important class with an RNA genome that must be copied to DNA by an RNA-dependent DNA pol (reverse transcriptase) prior to integration in the host cell chromosomes. A member of these viruses, the *human immunodeficiency virus* (HIV) has in the last decades caused the AIDS pandemic, and its reverse transcriptase is the therapeutic target of many inhibitors.

The discovery of DNA polymerases has been a long and hard work, spanning more than 60 years, and took the best efforts of some of the

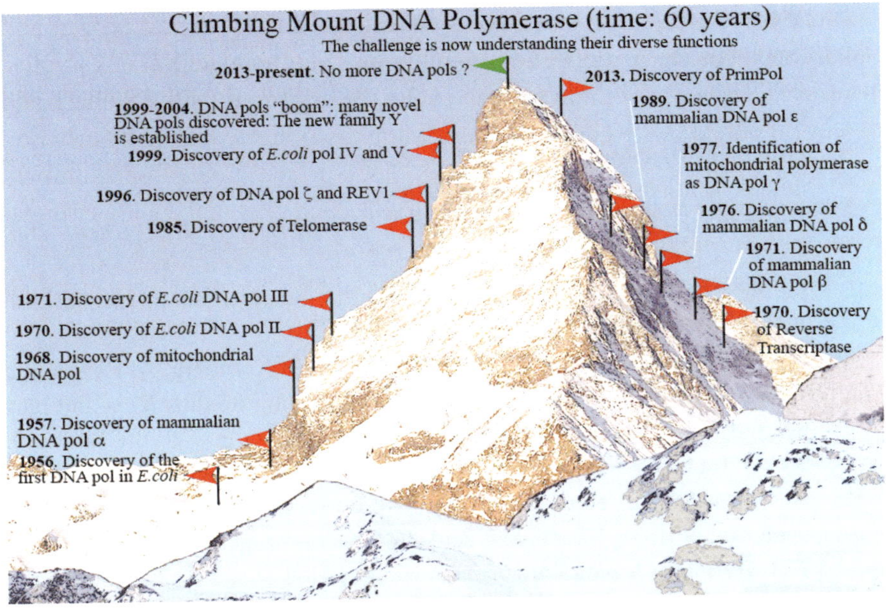

Figure 1.5 **Climbing Mount DNA polymerase**. The most significant achievements are indicated in chronological (ascending) order by the red flags. The background picture shows Mt. Cervino (Italy)/Mt. Matterhorn (Switzerland) (4,478 m/14,730 ft), in the Swiss-Italian Alps.

most talented scientists of several disciplines. As always in science, it did not follow a straight line, but rather it was a winding and climbing trail, with dead ends, crevices and pitfalls. Figure 1.5 illustrates this extraordinary effort.

References

1. Watson, J. D. & Crick, F. H. Molecular structure of nucleic acids; a structure for deoxyribose nucleic acid. *Nature* **171**, 737–738 (1953).
2. Watson, J. D. & Crick, F. H. Genetical implications of the structure of deoxyribonucleic acid. *Nature* **171**, 964–967 (1953).
3. Meselson, M. & Stahl, F. W. The Replication of DNA in Escherichia Coli. *Proc Natl Acad Sci U S A* **44**, 671–682 (1958).
4. Ycas, M. *The biological code*. (John Wiley & Sons, 1969).

5. Woese, C. R. *The genetic code*. (Harper and Row, 1967).
6. Kornberg, A. *For the love of enzymes*. (Harvard University Press, 1989).
7. Kornberg, A. & Baker, T. A. *DNA replication*. (W.H. Freeman and Company, 1992).
8. Grunberg-Manago, M., Oritz, P. J. & Ochoa, S. Enzymatic synthesis of nucleic acidlike polynucleotides. *Science* **122**, 907–910 (1955).
9. Hurwitz, J. The discovery of RNA polymerase. *J Biol Chem* **280**, 42477–42485 (2005).
10. Leloir, L. F. Two decades of research on the biosynthesis of saccharides. *Science* **172**, 1299–1302 (1971).
11. Lehman, I. R., Bessman, M. J., Simms, E. S. & Kornberg, A. Enzymatic synthesis of deoxyribonucleic acid. I. Preparation of substrates and partial purification of an enzyme from Escherichia coli. *J Biol Chem* **233**, 163–170 (1958).
12. Bessman, M. J., Lehman, I. R., Simms, E. S. & Kornberg, A. Enzymatic synthesis of deoxyribonucleic acid. II. General properties of the reaction. *The Journal of Biological Chemistry* **233**, 171–177 (1958).
13. Gellert, M. Formation of covalent circles of lambda DNA by *E. coli* extracts. *Proc Natl Acad Sci U S A* **57**, 148–155 (1967).
14. Weiss, B. & Richardson, C. C. Enzymatic breakage and joining of deoxyribonucleic acid, I. Repair of single-strand breaks in DNA by an enzyme system from Escherichia coli infected with T4 bacteriophage. *Proc Natl Acad Sci U S A* **57**, 1021–1028 (1967).
15. Olivera BM, L. I. Linkage of polynucleotide through phosphodiester bonds by an enzyme from Escherichia coli. *Proc Natl Acad Sci U S A* **57**, 1426–1433 (1967).
16. Gefter, M. L., Becker, A. & Hurwitz, J. The enzymatic repair of DNA. I. Formation of circular lambda-DNA. *Proc Natl Acad Sci U S A* **58**, 240–247 (1967).
17. Cozzarelli, N. R., Melechen, N. E., Jovin, T. M. & Kornberg, A. Polynucleotide cellulose as a substrate for a polynucleotide ligase induced by phage T4. *Biochem Biophys Res Commun* **28**, 578–586 (1967).
18. Goulian, M. & Kornberg, A. Enzymatic synthesis of DNA. 23. Synthesis of circular replicative form of phage phi-X174 DNA. *Proc Natl Acad Sci U S A* **58**, 1723–1730 (1967).
19. De Lucia, P. & Cairns, J. Isolation of an *E. coli* strain with a mutation affecting DNA polymerase. *Nature* **224**, 1164–1166 (1969).
20. Kornberg, T. & Gefter, M. L. DNA synthesis in cell-free extracts of a DNA polymerase-defective mutant. *Biochem Biophys Res Commun* **40**, 1348–1355 (1970).

21. Kornberg, T. & Gefter, M. L. Purification and DNA synthesis in cell-free extracts: properties of DNA polymerase II. *Proc Natl Acad Sci U S A* **68**, 761–764 (1971).
22. Knippers, R. DNA polymerase II. *Nature* **228**, 1050–1053 (1970).
23. Moses, R. E. & Richardson, C. C. A new DNA polymerase acitvity of Escherichia coli. II. Properties of the enzyme purified from wild-type *E. coli* and DNA-ts mutants. *Biochem Biophys Res Commun* **41**, 1565–1571 (1970).
24. Kornberg, T. & Gefter, M. L. Deoxyribonucleic acid synthesis in cell-free extracts. IV. Purification and catalytic properties of deoxyribonucleic acid polymerase III. *J Biol Chem* **247**, 5369–5375 (1972).
25. Sanger, F. *et al.* Nucleotide sequence of bacteriophage phi X174 DNA. *Nature* **265**, 687–695 (1977).
26. Sanger, F., Nicklen, S. & Coulson, A. R. DNA sequencing with chain-terminating inhibitors. *P Natl Acad Sci USA* **74**, 5463–5467 (1977).
27. Okazaki, R., Okazaki, T., Sakabe, K., Sugimoto, K. & Sugino, A. Mechanism of DNA chain growth. I. Possible discontinuity and unusual secondary structure of newly synthesized chains. *Proc Natl Acad Sci USA* **59**, 598–605 (1968).
28. Brutlag, D., Schekman, R. & Kornberg, A. A possible role for RNA polymerase in the initiation of M13 DNA synthesis. *Proc Natl Acad Sci USA* **68**, 2826–2829 (1971).
29. Greider, C. W. & Blackburn, E. H. Identification of a specific telomere terminal transferase activity in Tetrahymena extracts. *Cell* **43**, 405–413 (1985).
30. Alberts, B. M. Function of gene 32-protein, a new protein essential for the genetic recombination and replication of T4 bacteriophage DNA. *Fed Proc* **29**, 1154–1163 (1970).
31. Alberts, B. M., Amodio, F. J., Jenkins, M., Gutmann, E. D. & Ferris, F. L. Studies with DNA-cellulose chromatography. I. DNA-binding proteins from Escherichia coli. *Cold Spring Harb Symp Quant Biol* **33**, 289–305 (1968).
32. Alberts, B. M. & Frey, L. T4 bacteriophage gene 32: a structural protein in the replication and recombination of DNA. *Nature* **227**, 1313–1318 (1970).
33. Wang, J. C. Interaction between DNA and an Escherichia coli protein omega. *J Mol Biol* **55**, 523–533 (1971).
34. Wang, J. C. DNA topoisomerases: why so many? *J Biol Chem* **266**, 6659–6662 (1991).
35. Bollum, F. J. Oligodeoxyribonucleotide-primed reactions catalyzed by calf thymus polymerase. *J Biol Chem* **237**, 1945–1949 (1962).

36. Cohen, S. N., Chang, A. C., Boyer, H. W. & Helling, R. B. Construction of biologically functional bacterial plasmids in vitro. *Proc Natl Acad Sci U S A* **70**, 3240–3244 (1973).
37. Bollum, F. J. & Potter, V. R. Incorporation of thymidine into deoxyribonucleic acid by enzymes from rat tissues. *J Biol Chem* **233**, 478–482 (1958).
38. Friedkin, M., Tilson, D. & Roberts, D. Studies of deoxyribonucleic acid biosynthesis in embryonic tissues with thymidine-C14. *J Biol Chem* **220**, 627–637 (1956).
39. Friedkin, M. & Wood, H. I. Utilization of thymidine-C14 by bone marrow cells and isolated thymus nuclei. *J Biol Chem* **220**, 639–651 (1956).
40. Reichard, P. & Estborn, B. Utilization of desoxyribosides in the synthesis of polynucleotides. *J Biol Chem* **188**, 839–846 (1951).
41. Friedkin, M. & Roberts, D. The enzymatic synthesis of nucleosides. II. Thymidine and related pyrimidine nucleosides. *J Biol Chem* **207**, 257–266 (1954).
42. Lu, K. H. & Winnick, T. The correlation of growth with incorporation of radioactive metabolites into nucleic acids in embryonic tissue culture. *Exp Cell Res* **7**, 238–242 (1954).
43. Bollum, F. J. Calf thymus polymerase. *J Biol Chem* **235**, 2399–2403 (1960).
44. Yoneda, M. & Bollum, F. J. Deoxynucleotide-Polymerizing Enzymes of Calf Thymus Gland. I. Large Scale Purification of Terminal and Replicative Deoxynucleotidyl Transferases. *J Biol Chem* **240**, 3385–3391 (1965).
45. Fry, M. & Loeb L. A. *Animal cell DNA polymerases.* (CRC Press, Inc, 1986).
46. Weissbach, A., Schlabach, A., Fridlender, B. & Bolden, A. DNA polymerases from human cells. *Nat New Biol* **231**, 167–170 (1971).
47. Baril, E. F., Brown, O. E., Jenkins, M. D. & Laszlo, J. Deoxyribonucleic acid polymerase with rat liver ribosomes and smooth membranes. Purification and properties of the enzymes. *Biochemistry* **10**, 1981–1992 (1971).
48. Chang, L. M. & Bollum, F. J. Low molecular weight deoxyribonucleic acid polymerase in mammalian cells. *J Biol Chem* **246**, 5835–5837 (1971).
49. Haines, M. E., Holmes, A. M. & Johnston, I. R. Distinct cytoplasmic and nuclear DNA polymerases from rat liver. *FEBS Lett* **17**, 63–67 (1971).
50. Berger, H., Jr., Huang, R. C. & Irvin, J. L. Purification and characterization of a deoxyribonucleic acid polymerase from rat liver. *J Biol Chem* **246**, 7275–7283 (1971).
51. Chang, L. M. Phylogeny of DNA polymerase-beta. *Science* **191**, 1183–1185 (1976).

52. Chang, L. M. & Bollum, F. J. Antigenic relationships in mammalian DNA polymerase. *Science* **175**, 1116–1117 (1972).
53. Hecht, N. B. Interconvertibility of mouse DNA polymerase activities derived from the nucleus and cytoplasm. *Biochim Biophys Acta* **312**, 471–483 (1973).
54. Lazarus, L. H. & Kitron, N. Letter: Cytoplasmic DNA polymerase: polymeric forms and their conversion into an active monomer resembling nuclear DNA polymerase. *J Mol Biol* **81**, 529–534 (1973).
55. Spadari, S., Muller, R. & Weissbach, A. The dissimilitude of the low and high molecular weight deoxyribonucleic acid-dependent deoxyribonucleic acid polymerases of HeLa cells. *J Biol Chem* **249**, 2991–2992 (1974).
56. Smith, R. G., Abrell, J. W., Lewis, B. J. & Gallo, R. C. Serological analysis of human deoxyribonucleic acid polymerases. Preparation and properties of antiserum to deoxyribonucleic acid polymerase I from human lymphoid cells. *J Biol Chem* **250**, 1702–1709 (1975).
57. Brun, G. M., Assairi, L. M. & Chapeville, F. Immunological relationships between chick embryo polymerases. *J Biol Chem* **250**, 7320–7323 (1975).
58. Weissbach, A., Baltimore, D., Bollum, F., Gallo, R. & Korn, D. Nomenclature of eukaryotic DNA polymerases. *Science* **190**, 401–402 (1975).
59. Temin, H. M. & Mizutani, S. RNA-dependent DNA polymerase in virions of Rous sarcoma virus. *Nature* **226**, 1211–1213 (1970).
60. Baltimore, D. RNA-dependent DNA polymerase in virions of RNA tumour viruses. *Nature* **226**, 1209–1211 (1970).
61. Chang, L. M. & Bollum, F. J. Low molecular weight deoxyribonucleic acid polymerase from rabbit bone marrow. *Biochemistry* **11**, 1264–1272 (1972).
62. Fridlender, B., Fry, M., Bolden, A. & Weissbach, A. A new synthetic RNA-dependent DNA polymerase from human tissue culture cells (HeLa-fibroblast-synthetic oligonucleotides-template-purified enzymes). *Proc Natl Acad Sci U S A* **69**, 452–455 (1972).
63. Spadari, S. & Weissbach, A. HeLa cell R-deoxyribonucleic acid polymerases. Separation and characterization of two enzymatic activities. *J Biol Chem* **249**, 5809–5815 (1974).
64. Chang, L. M., Brown, M. & Bollum, F. J. Induction of DNA polymerase in mouse L cells. *J Mol Biol* **74**, 1–8 (1973).
65. Chiu, R. W. & Baril, E. F. Nuclear DNA polymerases and the HeLa cell cycle. *J Biol Chem* **250**, 7951–7957 (1975).

66. Baril, E. F., Jenkins, M. D., Brown, O. E., Laszlo, J. & Morris, H. P. DNA polymerases I and II in regenerating rat liver and Morris hepatomas. *Cancer Res* **33**, 1187–1193 (1973).
67. Mayer, R. J., Smith, R. G. & Gallo, R. C. DNA-metabolizing enzymes in normal human lymphoid cells. VI. Induction of DNA polymerases alpha, beta, and gamma following stimulation with phytohemagglutinin. *Blood* **46**, 509–518 (1975).
68. Bertazzoni, U. *et al.* Variations of DNA polymerase-alpha and -beta during prolonged stimulation of human lymphocytes. *Proc Natl Acad Sci U S A* **73**, 785–789 (1976).
69. Spadari, S. & Weissbach, A. RNA-primed DNA synthesis: specific catalysis by HeLa cell DNA polymerase alpha. *Proc Natl Acad Sci U S A* **72**, 503–507 (1975).
70. Hubscher, U., Kuenzle, C. C. & Spadari, S. Variation of DNA polymerases-alpha, -beta. and -gamma during perinatal tissue growth and differentiation. *Nucleic Acids Res* **4**, 2917–2929 (1977).
71. Spadari, S., Focher, F. & Hubscher, U. Developmental activity profile of DNA polymerases delta and alpha in rat neurons suggests a coordinated *in vivo* function. *In Vivo* **2**, 317–320 (1988).
72. Bolden, A., Noy, G. P. & Weissbach, A. DNA polymerase of mitochondria is a gamma-polymerase. *J Biol Chem* **252**, 3351–3356 (1977).
73. Hubscher, U., Kuenzle, C. C. & Spadari, S. Identity of DNA polymerase gamma from synaptosomal mitochondria and rat-brain nuclei. *Eur J Biochem* **81**, 249–258 (1977).
74. Hubscher, U., Kuenzle, C. C., Limacher, W., Scherrer, P. & Spadari, S. Functions of DNA polymerases alpha, beta, and gamma in neurons during development. *Cold Spring Harb Symp Quant Biol* **43 Pt 1**, 625–629 (1979).
75. Bertazzoni, U., Scovassi, A. I. & Brun, G. M. Chick-embryo DNA polymerase gamma. Identity of gamma-polymerases purified from nuclei and mitochondria. *Eur J Biochem* **81**, 237–248 (1977).
76. Hubscher, U., Kuenzle, C. C. & Spadari, S. Functional roles of DNA polymerases beta and gamma. *Proc Natl Acad Sci U S A* **76**, 2316–2320 (1979).
77. Bucknall, R. A., Moores, H., Simms, R. & Hesp, B. Antiviral effects of aphidicolin, a new antibiotic produced by Cephalosporium aphidicola. *Antimicrob Agents Chemother* **4**, 294–298 (1973).

78. Pedrali-Noy, G. & Spadari, S. Effect of aphidicolin on viral and human DNA polymerases. *Biochem Biophys Res Commun* **88**, 1194–1202 (1979).
79. Pedrali-Noy, G. *et al.* Synchronization of HeLa cell cultures by inhibition of DNA polymerase alpha with aphidicolin. *Nucleic Acids Res* **8**, 377–387 (1980).
80. Geuskens, M., Hardt, N., Pedrali-Noy, G. & Spadari, S. An autoradiographic demonstration of nuclear DNA replication by DNA polymerase alpha and of mitochondrial DNA synthesis by DNA polymerase gamma. *Nucleic Acids Res* **9**, 1599–1613 (1981).
81. Sala, F. *et al.* DNA repair synthesis in plant protoplasts is aphidicolin-resistant. *FEBS Lett* **138**, 213–217 (1982).
82. Spadari, S., S. F., Pedrali-Noy, G. Aphidicolin: a specific inhibitor of nuclear DNA replication in eukaryotes. *Trends in Biochemical Sciences* **7**, 29–32 (1982).
83. Sala, F. *et al.* Functional roles of the plant alpha-like and gamma-like DNA polymerases. *FEBS Lett* **124**, 112–118 (1981).
84. Waser, J., Hubscher, U., Kuenzle, C. C. & Spadari, S. DNA polymerase beta from brain neurons is a repair enzyme. *Eur J Biochem* **97**, 361–368 (1979).
85. Hardt, N., Pedrali-Noy, G., Focher, F. & Spadari, S. Aphidicolin does not inhibit DNA repair synthesis in ultraviolet-irradiated HeLa cells. A radioautographic study. *Biochem J* **199**, 453–455 (1981).
86. Byrnes, J. J., Downey, K. M., Black, V. L. & So, A. G. A new mammalian DNA polymerase with 3′ to 5′ exonuclease activity: DNA polymerase delta. *Biochemistry* **15**, 2817–2823. (1976).
87. Tan, C. K., Castillo, C., So, A. G. & Downey, K. M. An auxiliary protein for DNA polymerase-delta from fetal calf thymus. *J Biol Chem* **261**, 12310–12316 (1986).
88. Focher, F. *et al.* Calf thymus DNA polymerase delta independent of proliferating cell nuclear antigen (PCNA). *Nucleic Acids Res* **17**, 1805–1821 (1989).
89. Syvaoja, J. & Linn, S. Characterization of a large form of DNA polymerase delta from HeLa cells that is insensitive to proliferating cell nuclear antigen. *J Biol Chem* **264**, 2489–2497 (1989).
90. Wintersberger, E. Occurrence of a DNA-polymerase in isolated yeast mitochondria. *Biochem Biophys Res Commun* **25**, 1–7 (1966).

91. Wintersberger, U. & Wintersberger, E. Studies on deoxyribonucleic acid polymerases from yeast. 2. Partial purification and characterization of mitochondrial DNA polymerase from wild type and respiration-deficient yeast cells. *Eur J Biochem* **13**, 20–27 (1970).
92. Eckstein, H., Paduch, V. & Hilz, H. Synchronized yeast cells. 3. DNA synthesis and DNA polymerase after inhibition of cell division by x-rays. *Eur J Biochem* **3**, 224–231 (1967).
93. Chang, L. M. DNA polymerases from bakers' yeast. *J Biol Chem* **252**, 1873–1880 (1977).
94. Conaway, R. C. & Lehman, I. R. A DNA primase activity associated with DNA polymerase alpha from Drosophila melanogaster embryos. *Proc Natl Acad Sci U S A* **79**, 2523–2527 (1982).
95. Plevani, P., Badaracco, G., Augl, C. & Chang, L. M. DNA polymerase I and DNA primase complex in yeast. *J Biol Chem* **259**, 7532–7539 (1984).
96. Johnson, L. M., Snyder, M., Chang, L. M., Davis, R. W. & Campbell, J. L. Isolation of the gene encoding yeast DNA polymerase I. *Cell* **43**, 369–377 (1985).
97. Bauer, G. A., Heller, H. M. & Burgers, P. M. DNA polymerase III from Saccharomyces cerevisiae. I. Purification and characterization. *J Biol Chem* **263**, 917–924 (1988).
98. Budd, M. E., Sitney, K. C. & Campbell, J. L. Purification of DNA polymerase II, a distinct DNA polymerase, from Saccharomyces cerevisiae. *J Biol Chem* **264**, 6557–6565 (1989).
99. Sitney, K. C., Budd, M. E. & Campbell, J. L. DNA polymerase III, a second essential DNA polymerase, is encoded by the S. cerevisiae CDC2 gene. *Cell* **56**, 599–605 (1989).
100. Morrison, A. *et al.* REV3, a *Saccharomyces cerevisiae* gene whose function is required for induced mutagenesis, is predicted to encode a nonessential DNA polymerase. *J Bacteriol* **171**, 5659–5667 (1989).
101. Burgers, P. M. *et al.* Revised nomenclature for eukaryotic DNA polymerases. *Eur J Biochem* **191**, 617–618 (1990).
102. Castroviejo, M., Tarrago-Litvak, L. & Litvak, S. Partial purification and characterization of two cytoplasmic DNA polymerases from ungerminated wheat. *Nucleic Acids Res* **2**, 2077–2090 (1975).
103. Sahai Srivastava, B. I. & Grace, J. T., Jr. A 7S DNA polymerase in the cytoplasmic fraction from higher plant. *Life Sci* **14**, 1947–1954 (1974).
104. Tarrago-Litvak, L., Castroviejo, M. & Litvak, S. Studies on a DNA polymerse gamma-like enzyme from wheat embryos. *FEBS Lett* **59**, 125–130 (1975).

105. Gardner, J. M. & Kado, C. I. High molecular weight deoxyribonucleic acid polymerase from crown gall tumor cells of periwinkle (Vinca rosea). *Biochemistry* **15**, 688–697 (1976).
106. Mory, Y. Y., Chen, D. & Sarid, S. De Novo Biosynthesis of Deoxyribonucleic Acid Polymerase during Wheat Embryo Germination. *Plant Physiol* **55**, 437–442 (1975).
107. McLennan, A. G. & Keir, H. M. The deoxyribonucleic acid polymerases of non-vertebrate eukaryotes. *Biochem Soc Symp*, 55–73 (1977).
108. Amileni, A., Sala, F., Cella, R. & Spadari, S. The major DNA polymerase in cultured plant cells: partial purification and correlation with cell multiplication. *Planta* **146**, 521–527 (1979).
109. Sala, F., Amileni, A. R., Parisi, B. & Spadari, S. A gamma-like DNA polymerase in spinach chloroplasts. *Eur J Biochem* **112**, 211–217 (1980).
110. Sala, F. *et al.* Specific and reversible inhibition by aphidicolin in the alpha-like DNA polymerase of plant cells. *FEBS Lett* **117**, 93–98 (1980).
111. Sala F., G. M., Nielsen, E., Magnien, E., Dvreux, M., Pedrali-Noy, G. & Spadari, S. Synchronization of nuclear DNA synthesis in cultured Daucus carota L. cells by aphidicolin. *FEBS Letters* **153**, 204–208 (1983).
112. Keir, H. M., Subak-Sharpe, H., Shedden, W. I., Watson, D. H. & Wildy, P. Immunological evidence for a specific DNA polymerase produced after infection by herpes simplex virus. *Virology* **30**, 154–157 (1966).
113. Weissbach, A., Hong, S. C., Aucker, J. & Muller, R. Characterization of herpes simplex virus-induced deoxyribonucleic acid polymerase. *J Biol Chem* **248**, 6270–6277 (1973).
114. Berns, K. I., Silverman, C. & Weissbach, A. Separation of a new deoxyribonucleic acid polymerase from vaccinia-infected HeLa cells. *J Virol* **4**, 15–23 (1969).
115. Citarella, R. V., Muller, R., Schlabach, A. & Weissbach, A. Studies on vaccinia virus-directed deoxyribonucleic acid polymerase. *J Virol* **10**, 721–729 (1972).
116. Challberg, M. D. & Englund, P. T. Purification and properties of the deoxyribonucleic acid polymerase induced by vaccinia virus. *J Biol Chem* **254**, 7812–7819 (1979).
117. Kaplan, P. M., Greenman, R. L., Gerin, J. L., Purcell, R. H. & Robinson, W. S. DNA polymerase associated with human hepatitis B antigen. *J Virol* **12**, 995–1005 (1973).
118. Robinson, W. S., Clayton, D. A. & Greenman, R. L. DNA of a human hepatitis B virus candidate. *J Virol* **14**, 384–391 (1974).

119. Robinson, W. S. & Greenman, R. L. DNA polymerase in the core of the human hepatitis B virus candidate. *J Virol* **13**, 1231–1236 (1974).
120. Lawrence, C. W. & Christensen, R. UV mutagenesis in radiation-sensitive strains of yeast. *Genetics* **82**, 207–232. (1976).
121. Friedberg, E. C., Wagner, R. & Radman, M. Specialized DNA polymerases, cellular survival, and the genesis of mutations. *Science* **296**, 1627–1630. (2002).
122. Lawrence, C. W. Cellular functions of DNA polymerase zeta and Rev1 protein. *Adv Protein Chem* **69**, 167–203 (2004).
123. Wittschieben, J. *et al.* Disruption of the developmentally regulated Rev3I gene causes embryonic lethality. *Curr. Biol.* **10**, 1217–1220 (2000).
124. Harris, P. *et al.* Molecular cloning of Drosophila mus308, a gene involved in DNA cross-link repair with homology to prokaryotic DNA polymerase I genes. *Mol. Cell. Biol.* **16**, 5764–5771 (1996).
125. Sharief, F. S., Vojta, P. J., Ropp, P. A. & Copeland, W. C. Cloning and chromosomal mapping of the human DNA polymerase theta (POLQ), the eighth human DNA polymerase. *Genomics* **59**, 90–96. (1999).
126. Seki, M., Marini, F. & Wood, R. D. POLQ (Pol theta), a DNA polymerase and DNA-dependent ATPase in human cells. *Nucleic Acids Res* **31**, 6117–6126 (2003).
127. Garcia-Diaz, M. *et al.* DNA polymerase lambda (Pol lambda), a novel eukaryotic DNA polymerase with a potential role in meiosis. *J Mol Biol* **301**, 851–867 (2000).
128. Dominguez, O. *et al.* DNA polymerase mu (Pol mu), homologous to TdT, could act as a DNA mutator in eukaryotic cells. *Embo J* **19**, 1731–1742 (2000).
129. Prakash, S., Johnson, R. E. & Prakash, L. Eukaryotic translesion synthesis DNA polymerases: specificity of structure and function. *Annu Rev Biochem* **74**, 317–353 (2005).
130. Reuven, N. B., G., A., Maor-Shoshani, A. & Livneh, Z. The mutagenesis protein UmuC is a DNA polymerase activated by UmuD', RecA ans SSB and is specialized in translesion replication. *J. Biol. Chem.* **274**, 31763–31766 (1999).
131. Tang, M. *et al.* UmuD'(2)C is an error-prone DNA polymerase, Escherichia coli pol V. *Proc Natl Acad Sci U S A* **96**, 8919–8924. (1999).
132. Wagner, J. *et al.* The dinB gene encodes a nocel E. coil DNA polymerase, DNA pol IV, involved in mutagenesis. *Mol. Cell* **4**, 281–286 (1999).
133. Masutani, C. *et al.* The XPV (xeroderma pigmantosum variant) gene encodes human DNA polymerase eta. *Nature* **399**, 700–704 (1999).

134. Johnson, R. E., Kontratick, C. M., Prokash, S. & Prakash, L. hRad30 mutations in the variant form of Xeroderma pigmentosum. *Science* **284**, 263–265 (1999).
135. Ogi, T., Kato, T., Jr., Kato, T. & Ohmori, H. Mutation enhancement by DINB1, a mammalian homologue of the Escherichia coli mutagenesis protein dinB. *Genes Cells* **4**, 607–618 (1999).
136. Gerlach, V. L. *et al.* Human and mounse homologs of the Escherichia coli DinB (DNA polmyerase IV), members of the UmuC/DinB family. *Proc Natl Acad Sci U S A* **96**, 11922–11927 (1999).
137. Gerlach, V. L. *et al.* Human DNA polymerase kappa: a novel DNA polymerase of unknown biological function encoded by the DINB1 gene. *Cold Spring Harb Symp Quant Biol* **65**, 41–49 (2000).
138. Lin, W., Xin, H., Zhang, Y., Yuan, F. & Wang, Z. The human Rev1 gene codes for a DNA template dependent dCMP transferase. *Nucl Acids Res* **27**, 4468–4475 (1999).
139. Tissier, A., McDonald, J. P., Frank, E. G. & Woodgate, R. poliota, a remarkably error-prone human DNA polymerase. *Genes Dev* **14**, 1642–1650. (2000).
140. Wan, L. *et al.* hPrimpol1/CCDC111 is a human DNA primase-polymerase required for the maintenance of genome integrity. *Embo Reports* **14**, 1104–1112 (2013).
141. Garcia-Gomez, S. *et al.* PrimPol, an Archaic Primase/Polymerase Operating in Human Cells. *Molecular cell* **52**, 541–553 (2013).
142. Bianchi, J. *et al.* PrimPol Bypasses UV Photoproducts during Eukaryotic Chromosomal DNA Replication. *Molecular Cell* **52**, 566–573 (2013).
143. Burgers, P. M. *et al.* Eukaryotic DNA polymerases: proposal for a revised nomenclature. *J Biol Chem* **276**, 43487–43490 (2001).

Chapter 2

DNA Polymerases: General Aspects

2.1 Synthesis and maintenance of DNA in nature need DNA polymerases

In all three kingdoms of life (bacteria, archaea, and eukaryotes) DNA constitutes the genetic material. Even many viruses consist of DNA. Universally in any organism in nature, a DNA molecule is built up of adenine (A) pairing to thymine (T) and cytosine (C) pairing to guanine (G). As initially suggested by Watson and Crick in 1953, by this base pairing rule the mother DNA strand encodes the instruction for the synthesis of the new daughter DNA strand in DNA replication during the cell division. The mother strand also serves as a copy to replace the damaged strand in different DNA repair events.[1] Ever since the first discovery in 1956, the enzymes that catalyze the synthesis of DNA are called DNA polymerases (pols)[2,3] (for details see Chapter 1). DNA replication and the various DNA repair events all rely on DNA synthesis carried out by those DNA pols (Figure 2.1).

Nature has evolved a system where DNA can be faithfully transmitted from one generation to the next by replication. However, a few mistakes performed in a DNA replication cycle might be critical to guarantee changes that possibly allow the evolution of species. A DNA molecule in a single mammalian cell has a length of 2 m, contains about 6×10^9 bases and faces up to 100,000 alterations per day. Such alterations include depurination, depyrimidation, oxidation, non-enzymatic methylation, bulky

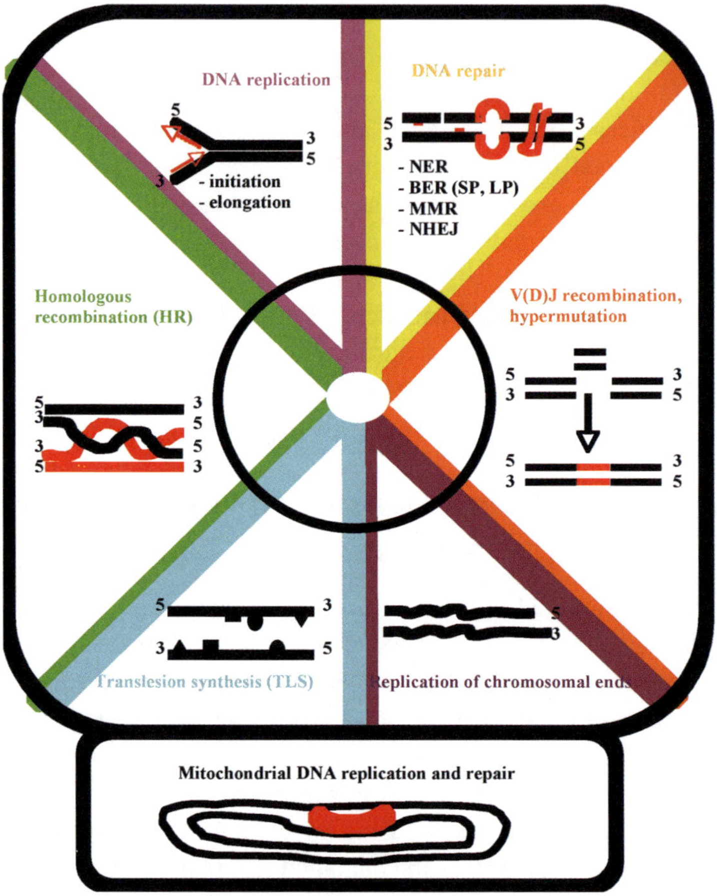

Figure 2.1 **Cellular events that need DNA polymerases**. **NER**: nucleotide excision repair; **BER**: base excision repair; **SP**: short patch; **LP**: long patch; **MMR**: mismatch repair; **NHEJ**; non-homologous end joining; **HR**: homologous recombination; **TLS**: translesion synthesis. For more details see text.

adduct formation, DNA deletions, DNA expansions, recombinations and breaks of the phosphodiester backbone, all of which can lead to single-stranded or even double-stranded breaks. The enormous and difficult joint task of the various DNA repair mechanisms is to keep these alterations at a "forgivable" level so that the continuity of the cellular function is guaranteed at any time. Whenever mistakes accumulate, the chances increase that those alterations stably establish themselves in the DNA. If DNA repair is not keeping up in time, the consequences of such alterations in the DNA sequence can be cell death or changes of the cellular physiology. This programmed cell death, called apoptosis, is a physiological event to eliminate useless cells that contain altered DNA coding for proteins with reduced, enhanced or ablated functions. However, if those cells cannot undergo apoptosis, even more mutations can accumulate. These can lead to a gain of unlimited replication potential and uncontrolled cell division, a process that can eventually give rise to cancer. Still, as deleterious as mutations may seem, an intermediate level of accumulating DNA mutations might provide a driving force for the evolution of species.

Depending on the organism, DNA replication relies on two to three different DNA pols[4,5] (for details see also Chapter 3). DNA repair events need DNA pols capable of short-gap filling as in short-patch base excision repair (SP-BER) or DNA pols that can synthesize longer patches of DNA [in long patch (LP-) BER,[6] nucleotide excision repair (NER),[7] mismatch repair (MMR)[8,9] and non-homologous end-joining[4,10,11]]. Moreover, long patch DNA synthesis is also required for homologous recombination and physiological processes occurring in the context of immunoglobulin recombination.[12] As indicated above, the DNA pols also have to face the many alterations continuously being generated in the DNA. Specifically selected for those tasks a variety of DNA pols that act as translesion DNA pols has evolved.[13,14] Translesion synthesis (TLS) can either be faithful (error free) or mutagenic (error prone).

DNA pols have a unique polarity and synthesize DNA only in the 5′ to 3′ direction (see below). This leads to a fundamental problem to fully replicate the chromosomal ends, the telomeres, giving rise to a phenomenon called telomere shortening. A specialized DNA pol, called telomerase, can fulfill this function and prevent the shortening of telomeres.[15] Finally, DNA pols are not only simple working horses to synthesize DNA but they can also be involved in control mechanisms such as checkpoint control during the cell cycle.[16]

2.2 The DNA polymerase reaction

The basic reaction mechanism of DNA pols is universal (Figure 2.2). A DNA pol needs (i) the four bases in their activated form as deoxyribonucleoside triphosphates (dATP, dTTP, dCTP and dGTP) as substrate,

(a) **Components**

1. **Template (DNA)**
2. **Primer (DNA or RNA)**
3. **Four deoxyribonucleoside triphosphates (dATP, dGTP, dCTP, dTTP)**
4. **Divalent cation, generally $MgCl_2$**
5. **DNA polymerase**

(b)

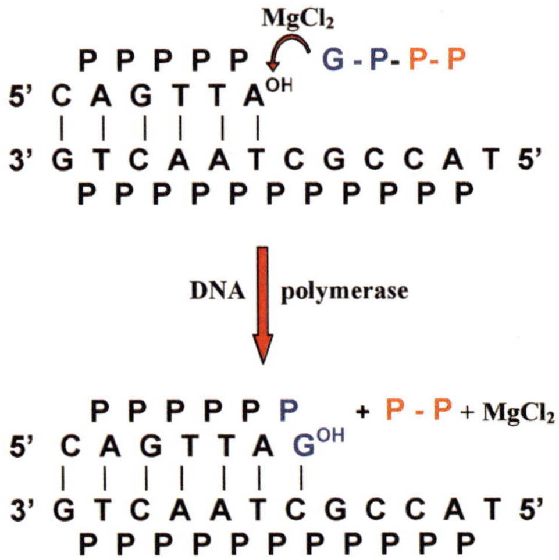

Figure 2.2 **The DNA polymerase reaction**. For details see text.

(ii) a DNA template that instructs the DNA pol which bases to incorporate (A to T and C to G and vice versa), (iii) a primer containing a 3′-OH group at its end which is hybridized to a template (there are only a few exceptions to this, e.g. the so called protein priming in bacteriophage ϕ29 and in adenovirus), (iv) the divalent cation magnesium as cofactor, and (v) in some cases other proteins called DNA pol auxiliary proteins (see below for further details).

DNA pols incorporate a base by the following five-step mechanism:[17,18] first, a correct hydrogen bonding between the incoming dNTP and the templating base is established (A to T and G to C and vice versa); second, water is excluded from the active site; third, the geometric selection in the active site takes place; fourth, the dNTP binding affinity leads to an induced conformational change in the active site, and fifth, the phosphodiester bond is formed between the last base of the primer and the new nucleotide by the release of pyrophosphate.

DNA pols, as everything else in nature, can make mistakes and occasionally incorporate wrong bases. To reduce the amount of those mistakes many, but not all DNA pols, contain a second activity, the 3′→5′ exonuclease.[19] This so-called proofreading exonuclease mechanism removes non-properly paired bases. Such incorrectly base paired nucleotides can either derive from a wrong pairing of two normal nucleotides or be caused by altered or missing bases that prevented the DNA pol active site from choosing the correct nucleotide to be incorporated. If an incorrect base pair is overlooked by the DNA pol, the so called post-replication mismatch repair (MMR) system can remove the incorrectly incorporated base (for details see later).[20]

How accurate is then the overall DNA synthesis conducted by DNA pols? While spontaneous correct base pairing in the absence of a DNA pol can occur at a frequency of 10:1 correct vs. incorrect incorporation events, DNA pols display frequencies of correct incorporation that range from 10:1 to 100,000:1. The reasons why DNA pols have different fidelities and what are the advantages of having low-fidelity DNA pols will be discussed in later chapters. By the intrinsic 3′→5′ exonuclease, DNA pols can improve their fidelities up to 1,000,000:1. However, since the higher eukaryotic genomes, including humans, contain 6×10^9 base pairs, a mistake every 1,000,000 base pairs would lead to the occurrence of 6×10^3 mistakes per cell division, a number which is unacceptable for the preservation of the integrity of genomic information. This problem is taken care of by DNA pol auxiliary proteins such as proliferating cell nuclear antigen (PCNA),[21] replication factor C (RF-C)[22] and replication protein A (RP-A),[23] which improve the processivity and the fidelity of DNA pols by a great extent (see DNA pol holoenzymes below). Finally, the above mentioned post-replication MMR system can discriminate between the old and newly synthesized DNA strand and thus remove the incorrectly incorporated nucleotide.[20] The auxiliary proteins and the post-replicative mismatch repair system together contribute to reduce the error rate of

Table 2.1 Fidelity of DNA synthesis

Step	Fidelity
Spontaneous base pairing without DNA polymerase	10^{-1}
A "sloppy" DNA polymerase	10^{-1} to 10^{-4}
A faithful DNA polymerase with 3′→5′ exonuclease	10^{-5} to 10^{-6}
A faithful DNA polymerase and auxiliary proteins	10^{-7} to 10^{-8}
A faithful DNA polymerase, auxiliary proteins and the post-replication mismatch repair system	10^{-9} to 10^{-10}

replicative DNA pols to around $10^{-10} – 10^{-11}$. In summary these four steps, DNA pols, their 3′→5′ exonuclease activity, DNA pol auxiliary proteins and the MMR system, can account for the overall DNA replication fidelity required in nature (Table 2.1).

2.3 The universal right-hand structure of a DNA polymerase

So far, the crystal structures that have been obtained for many DNA pols from all kingdoms of life have revealed a quite common structure.[24,25] The overall architecture of a DNA pol resembles a human right hand and consists of three domains: palm, fingers and thumb (Figure 2.3). Upon binding of the DNA pol to the template/primer and the dNTP, conformational changes in the DNA pol as well as in the DNA itself lead to the assembly of a DNA pol active site. These changes allow a catalytically competent DNA pol to exert its function. The finger domain interacts with the incoming dNTP and single-stranded DNA. The thumb on the other hand binds to double-stranded DNA. This double-stranded DNA is either a primer hybridized to a template or represents the newly synthesized DNA with its templating strand. The palm domain harbors the catalytic residues that bind magnesium ions, which are required for the phosphoryl transfer reaction (Figure 2.2). The phosphoryl transfer reaction includes the attachment of the dNMP to the 3′-OH group of the primer or the growing DNA chain and the release of pyrophosphate. The

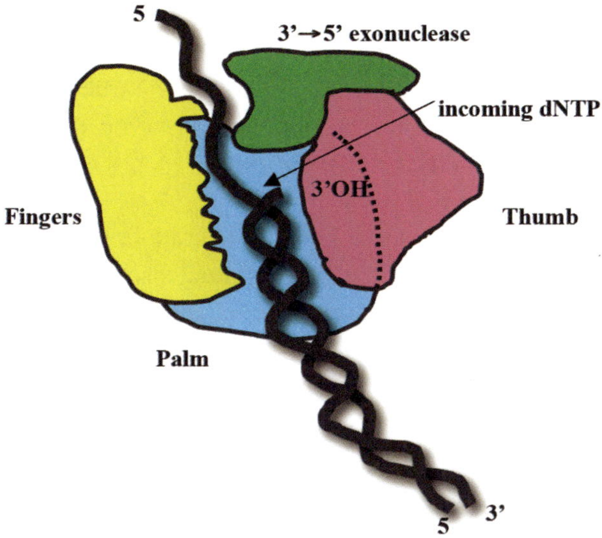

Figure 2.3 **Structure of a DNA polymerase**. Schematic representation of a typical DNA pol containing a 3′→5′ exonuclease proofreading activity. Yellow: fingers; pink: thumb; blue: palm; green 3′→5′ exonuclease domain.

palm domain, where the catalytic reaction occurs, is extremely conserved in most DNA pols analyzed so far. Exceptions are these: (i) the pol β-like nucleotidyltransferase superfamily X such as DNA pol β,[26] (ii) the DNA pols III from *E. coli* and *Thermus aquaticus*[27] and the DNA pol X from *Deinococcus radiodurans*.[28] Despite these differences, all DNA pols utilize the same two-metal (mostly magnesium)-ion mechanism for dNTP addition. In addition to promote the chemical step, sequential binding of the metal ions causes changes in the relative orientation of the base and triphosphate moieties of the incoming nucleotide, which favor correct base pairing.[29] Time-resolved X-ray crystallography also revealed the presence of a third metal ion, during the polymerization reaction. It has been hypothesized that this metal comes into play after phosphodiester bond formation and its role is likely to stabilize the transition state, facilitating pyrophosphate release.[30] However, more recent structural studies with DNA Pol η, unequivocally showed that formation of the ternary complex with both the DNA and dNTP substrate coordinated by the first two metal ions, creates a new binding site for the third metal, which is essential for catalysis. Binding of the third metal has been shown to constitute the true rate-limiting step for catalysis.[31] In contrast to the widely conserved palm domain, the fingers and the thumbs differ in the various DNA pol families (see below and Chapter 3).

As already mentioned, some DNA pols contain an associated 3′→5′ exonuclease activity (Figure 2.3). The exonuclease domains of the DNA pol are structurally conserved in all DNA pols tested so far.[19] These DNA pols contain two active sites — one for polymerization and one for exonucleolytic proofreading. The proofreading DNA pols must have a mechanism to balance between DNA synthesis and proofreading exonuclease action. It is likely that the structure of the DNA at the primer terminus and a few bases upstream of the active site, is responsible for the switch between "going forward" (DNA synthesis) and "going back" (proofreading). It is thought that a non-properly base paired primer terminus becomes single-stranded and is frayed out to bind to the exonuclease domain, while the correctly base paired primer terminus forms perfectly double-stranded DNA that preferentially binds to the DNA pol active site. It seems that when a DNA pol is incorporating nucleotides, the exonuclease domain is in closed conformation that does not allow the binding of single-stranded DNA. However, upon incorporation of a wrong nucleotide, the single-stranded DNA primer end allows the formation of an open exonuclease domain conformation ready for proofreading.

2.4 The eight DNA polymerase families and their functions: An overview

So far, eight different DNA pol families have been grouped based on sequence homologies. They are called family A, B, C, D, X, Y, RT, and AEP (Table 2.2).[4] None of the three kingdoms of life contains all the eight families. Prokaryotes contain five, namely A, B, C, X, and Y, archaea harbor five (B, D, X, Y, and AEP), and eukaryotes are composed of the six families A, B, X, Y, RT, and AEP. Moreover, the DNA pols from a

Table 2.2 The eight DNA polymerase families

Family	Viral	Bacteria	Archaea	Eukaryotes
A	T3 DNA pol	DNA pol I *E. coli*		DNA pol γ
	T5 DNA pol	DNA pol I *T. aquaticus*		DNA pol θ
	T4 DNA pol			DNA pol ν
	T7 DNA pol			
B	T6 DNA pol	DNA pol II *E. coli*	DNA pol BI	DNA pol α
	RB69 DNA pol		DNA pol BII	DNA pol δ
	Adeno[a] DNA pol		DNA pol BIII	DNA pol ε
	HSV-1[b] DNA pol			DNA pol ζ/Rev3
	Vaccinia DNA pol			
	Phi29 DNA pol			
C		DNA pol III *E. coli*		
		DNA pol III *B. subtilis*		
		DNA E *B. subtilis*		
		DNA pol III *T. aquaticus*		
D			DNA pol D	
X	ASFV[c] DNA pol	DNA pol X *D. radiodurans*	DNA pol X *M. mazei*	DNA pol β (pol IV S.c.[d])
		DNA pol X *B.subtilis*	DNA pol X *M. thermauto-trophicus*	DNA pol λ (pol LSP S. p[e])
				DNA pol μ

(*Continued*)

Table 2.2 (*Continued*)

Family	Viral	Bacteria	Archaea	Eukaryotes
		DNA pol X *L. monocytogenes*	DNA pol X *T. volcanium*	
		DNA pol X *S. saprolyticus*	DNA pol X *F. acidarmanus*	TdT[f]
		DNA pol X *S. aureus*		
		DNA pol X *D. reducens*		
		DNA pol X *A. aeolicus*		
		DNA pol X *T. thermophilus*		
		DNA pol X *T. denitrificans*		
		DNA pol X *T. aquaticus*		
Y		DNA pol IV *E .coli*	Dpo4 DNA pol[g]	DNA pol η
		DNA pol V *E. coli*	Dbh DNA pol[g]	DNA pol κ
				DNA pol i
				Rev1
RT	RT[h]			Telomerase
AEPi				PrimPol

a: Adenovirus; b: Herpes simplex virus; c: African swine fever virus; d: *Saccharomyces cerevisiae;* e: *Schizosaccaromyces pombe;* f: Terminal deoxyribonucleotidyl transferase; g: From *Sulfolobus solfataricus;* h: Reverse transcriptases from retroviruses and lentiviruses; i. Archaeo-eukaryotic primase superfamily variety of viruses of prokaryotes and eukaryotes belong to the three families A, B, and X.

As noted above, the overall structures of DNA pols are very similar. It appears that the different DNA pol families have evolved in evolution as an adaptation of an organism to certain survival strategies. Viruses contain 1–2 different DNA pols (the African swine fever virus codes for an own X family repair DNA pol in addition to its replicative enzyme, see also below),[32] while bacteria and archaea, as simple single cell organisms, code for 4–5 different DNA pols. In contrast to prokaryotes, eukaryotic cells possess two- to fourfold more distinct DNA pols, a fact that might reflect the higher complexity of DNA maintenance and replication when a nucleus is present. However, it is remarkable that all those eukaryotic DNA pols, with the exception of telomerase (RT family), belong to the same DNA pol families that already exist in prokaryotes. This means that, apart from the telomerase, no new DNA pol families have evolved with the development of eukaryotic cells. The telomerase obviously evolved due to the need of eukaryotes to solve the replication problem at the end of their chromosomes. Thus, the strategy of the eukaryotes was rather to multiply DNA pols within a family. In eukaryotes there are 3 family A (DNA pol γ, DNA pol θ and DNA pol ν); 5 family B (DNA pol α, DNA pol δ, DNA pol ε, DNA pol ζ and Rev3); 4 family X (DNA pol β, DNA pol λ, DNA pol μ and TdT); 4 family Y members (DNA pol η, DNA pol κ, DNA pol ι, and Rev1) and one of the archeo-eukaryotic primase superfamily (PrimPol). This multiplication within the families asks for an explanation, which we will try to provide in the following paragraphs.

In order to understand the reason for a multiplication of family members, we have to consider first their functional tasks. For many decades the scientific community believed that each single DNA pol had a unique and particular function. In higher eukaryotes, DNA pol α was considered as the replicative enzyme, DNA pol β as the repair enzyme and DNA pol γ as the mitochondrial replicase (for the rationale see Chapter 1). Eventhough this statement is still true today, the overall picture concerning the functional tasks of those and all the recently discovered DNA pols is much more complex and far from being solved in detail. What follows is an attempt to group proposed DNA pol functions in viruses, bacteria, archaea, and eukaryotes:

(i) *Viruses* (Table 2.3): (DNA viruses either rely on cellular DNA pols or have their own replicative DNA pols (e.g. T4 and T7 in bacteria and adenovirus, herpes simplex virus and vaccinia virus in eukaryotes). The African swine fever virus, the causative agent of African swine fever, is a large double-stranded DNA virus, which replicates in the cytoplasm of infected cells. It is the only member of the Asfarviridae family, and is an exception to other viruses in the context of DNA pols since it contains a pol X DNA pol, likely with properties in DNA repair. Such a repair DNA pol allows better survival of the virus under conditions in the infected cell that might otherwise lead to genomic instability.
(ii) *Bacteria* (Table 2.3): Bacteria have sophisticated replication and repair machineries with at least one replicative DNA pol, belonging to the C family (pol III) and five DNA pols for repair (A family DNA pol I, B family DNA pol II, C family DNA pol III, and Y family DNA pols IV and V). It is the repair task that obviously became important for the evolution of bacteria. For BER, NER, MMR and recombinational repair they rely on DNA pol I, DNA pol II and DNA pol III. Moreover, translesion synthesis over damaged DNA requires DNA pol IV and DNA pol V (Table 2.4). For replication, on the other hand, it appears that the family C pol III forms a triple DNA pol replisome containing three times the same DNA pol (see also holoenzyme below).
(iii) *Archaea* (Table 2.4): For DNA replication, archaea appear to have DNA pols of two different families: B and D. This is the first time in evolution that an organism relies on DNA pols belonging to two different families for replication. Archaea have to survive harsh conditions such as high temperatures, high salt, high sulfur, and high metals just to mention a few. For repair they rely on family X and B and for TLS on family Y DNA pols.
(iv) *Eukaryotes* (Table 2.5): From Table 2.5 it appears that eukaryotes contain at least 17 different DNA pols. As already indicated, with the exception of telomerase, no new families have evolved in comparison with prokaryotes. In contrast, eukaryotic cells even miss the C and the D families present in prokaryotes and archaea. However, to compensate,

Table 2.3 Functions of viral and bacterial DNA polymerases[a]

Family	DNA polymerase	Main function
A	T3 DNA pol	replication
	T5 DNA pol	replication
	T4 DNA pol	replication
	DNA pol I *E. coli*	replication, repair
	DNA pol I *T. aquaticus*	replication
B	T6 DNA pol	replication
	DNA RB69 pol	replication
	Adeno[b] DNA pol	replication
	HSV1[c] DNA pol	replication
	Vaccinia DNA pol	replication
	Phi29 DNA pol	replication
	DNA pol II E. coli	repair, replication?
C	DNA pol III *E. coli*	replication
	DNA pol III *B. subtilis*	replication
	DNA pol E *B. subtilis*	replication
	DNA pol III *T. aquaticus*	replication
X	ASFV[d] DNA pol	repair
	DNA pol X *D. radiodurans*	repair (double-stranded breaks)
	DNA pol X *B. subtilis*	repair
	DNA pol X *L. moncytogenes*	repair
	DNA pol X *S. saprolyticus*	repair
	DNA pol X *S. aureus*	repair
	DNA pol X *D. reducens*	repair
	DNA pol X *A. aeolicus*	repair
	DNA pol X *T. thermophilus*	repair
	DNA pol X *T. denitrificans*	repair
	DNA pol X *T. aquaticus*	repair
Y	DNA pol IV *E.coli*	translesion synthesis
	DNA pol V *E. coli*	translesion synthesis
RT	reverse transcriptase	DNA synthesis from RNA

a: For more details and references see Chapter 3; b: Adenovirus; c: Herpes simplex virus; d: African swine fever virus

Table 2.4 Functions of the archaeal DNA polymerases

Family	DNA polymerase	Main function
B	DNA pol BI	Replication, repair?
	DNA pol BII	Replication, repair?
	DNA pol BIII	Repair (BER), replication?
D	DNA pol D	replication
X	DNA pol X *M. mazei*	repair
	DNA pol X *M. thermautotrophicus*	repair
	DNA pol X *T. volcanium*	repair
	DNA pol X *F. acidarmanus*	repair
Y	Dpo4 DNA pol[a]	translesion synthesis
	Dbh DNA pol[a]	translesion synthesis

a: From *Sulfolobus solfataricus*

the four families A, B, X and Y in eukaryotes each contain 3–4 different DNA pols. Studies in many laboratories can be summarized by saying that redundant functions of DNA pols seem to be an important survival strategy for eukaryotes (see also Chapter 4). Moreover, new DNA synthesis events evolved, such as: (i) mitochondrial DNA replication and repair, (ii) physiological repair in immunoglobulin recombination and in antigen receptor diversity generation, and (iii) in sister chromatid cohesion. In eukaryotic DNA replication three different DNA pols are required (DNA pols α, δ and ε). Mitochondrial replication requires DNA pol γ for elongation and PrimPol for initiation. The latter seems also required for TLS and repriming of arrested forks in eukaryotes. DNA repair events each require several DNA pols such as DNA pols β, λ, θ, δ and ε in BER, DNA pols δ, ε, κ and η in NER, DNA pols δ in MMR, DNA pols λ and μ in DSBR and DNA pols λ, μ and TdT in immunoglobulin recombination. Often in BER, NER and MMR, the cell borrows the replication enzymes DNA pols δ and ε for repair. To be able to cope with modified, frequently miscoding nucleotides that are produced by DNA damage, TLS has been established. It is a synthesis event aimed at decreasing the genetic alterations after DNA damage.

Table 2.5 Functions of the eukaryotic DNA polymerases

Family	DNA polymerase	Main function
A	DNA pol γ	replication in mitochondria
	DNA pol θ	replication of cross links, base excision repair, hypermutation of immunoglobulin genes
	DNA pol ν	repair
B	DNA pol α	replication (priming)
	DNA pol δ	replication (lagging strand), repair
	DNA pol ε	replication (leading strand), repair
	DNA pol ζ/Rev3	translesion synthesis (extension)
X	DNA pol β (pol IV S.c.[a])	base excision repair
	DNA pol λ (pol LSP in S.p.[b])	base excision repair, double-strand break repair, immunoglobulin recombinational repair, translesion synthesis
	DNA pol μ	immunoglobulin recombinational repair
	TdT[c]	antigen receptor diversity
Y	DNA pol η	translesion synthesis (incorporation)
	DNA pol κ	translesion synthesis (incorporation, extension)
	DNA pol ι	translesion synthesis (incorporation)
	Rev1	translesion synthesis (incorporation)
RT	Telomerase	replication of chromosome ends
AEP	PrimPol	replication mitochondria; translesion synthesis; repriming at forks arrested at DNA damage

a: *Saccharomyces cerevisiae;* b: *Schizosaccaromyces pombe;* c: Terminal deoxynucleotidyl transferase.

For this purpose, eukaryotes have evolved an entire set of specialized DNA pols from the three families Y, X and B. While the Y family DNA pols are responsible for incorporating nucleotides opposite certain lesions (DNA pols η, κ, ι, and Rev1), the B family DNA pol ζ/Rev3 can extend from the often mis-incorporated nucleotides opposite such lesions. 8-oxo-G is a frequent lesion caused by reactive oxygen species (ROS) and poses a big problem to the genomic integrity, as replicative DNA pols preferentially incorporate a wrong A instead of the correct C opposite the lesion. This preferential misincorporation is due to the fact that an A pairing to 8-oxo-G forms a Hoogsteen base pair (which is, in contrast to the Watson-Crick base pairing, an incorrect way to base pair two nucleotides) that creates no single-stranded DNA that could activate the closed exonuclease conformation (see also Chapter 4). Therefore, the mispairing of A to 8-oxo-G goes unnoticed by the proofreading functions of replicative DNA pols and can lead to the generation of GC to TA transversion mutations. However, the X family DNA pol λ has been shown to act as repair enzyme for this lesion by faithfully incorporating a C opposite an 8-oxo-G, instead of the wrong A that is preferentially incorporated by the replicative DNA pols. Thus, DNA pol λ seems to play a role in preventing GC to TA transversions that occur due to 8-oxo-G.

In summary, the DNA pol number increased concomitantly with the evolution towards more complex cellular structures and their functions are closely linked to the complexity of those organisms. In the following chapter we will gain more insight into the fascinating areas of expertise of all the individual DNA pols and an attempt is made to give a global picture how they work in concert in a living eukaryotic cell (Chapter 4).

2.5 DNA polymerase holoenzymes

More than 30 years ago the laboratory of the late Kornberg, a pioneer in the field of DNA pols, discovered that the replicative DNA pol III was inefficient in DNA replication compared to DNA pol I *in vitro*.[33] The subsequent isolation of an efficient replication complex from crude *E. coli* extracts soon indicated that a more complex form of DNA pol III was present in the cell. From this finding it was realized that DNA pol III functions in the context of a multipolypeptide complex, which was thereupon called DNA pol III holoenzyme.[33] DNA pol III holoenzyme contains 10 different polypeptides (Table 2.6).[34,35] These proteins are grouped into the three major functional units: (i) the DNA pol III core consisting of three subunits, (ii) the sliding clamp consisting of a homodimeric ring, and (iii) the clamp loader/polymerase connector complex consisting of six different subunits (one subunit, τ, being a dimer). A fourth functional unit is

Table 2.6 Components of DNA polymerase holoenzymes[a]

Components	*E. coli* (M_r, *k*Da)	Human (M_r, *k*Da)	Function
DNA polymerase	*DNA pol III core (166)*	*DNA pol ε (350.3)*	replication (leading strand)
	α (129.9)	p260 (261.5)	
	ε (27.5)	p59 (59.5)	
	θ (8.6)	p17 (17.0)	
		p12 (12.3)	
DNA polymerase	*DNA pol III core (166)*	*DNA pol δ (238.7)*	replication (lagging strand)
	α (129.9)	p125 (123.6)	
	ε (27.5)	p66 (51.4)	
	θ (8.6)	p50 (51.3)	
		p12 (12.4)	
Sliding clamp	*Dimer (81.2)* β-subunit (40.6)	*Trimer (86.1)* PCNA (28.7)	processivity factor for DNA polymerases
Clamp loader	*γ/τ complex (291.1)*	*RF-C (314.9)*	pentameric clamp loader
	γ/τ_2 (47.5/71.1)	p140 (128.2)	
	δ (38.7)	p40 (39.7)	
	δ′ (36.9)	p38 (38.5)	
	χ (16.6)	p37 (39.2)	
	ψ (15.2)	p36 (40.6)	

a: Only the two examples of *E. coli* and human are exemplified.

the single-stranded binding protein (SSB), a necessary component for a DNA pol III holoenzyme to efficiently synthesize DNA. This unit is mentioned here since there is a lot of evidence in the literature that SSB, as well as its eukaryotic counterpart replication protein A (RP-A), physically interact with the holoenzymes from *E. coli* and eukaryotes, respectively. The DNA pol III holoenzyme contains two copies of the DNA pol III core which are connected together by the clamp loader. Both DNA pol III core enzymes themselves are attached to a homodimeric ring like protein called the β-clamp. In such a conformation, the dimeric holoenzyme can act in a coordinated way at the replisome. In summary the DNA pol III holoenzyme is a complex formed by 17 polypeptides: two DNA pol III core enzymes consisting of 6, two β-clamps consisting of 4 and the clamp loader/connector complex consisting of 7 polypeptides.

In eukaryotes, as exemplified by human enzymes, the situation is much more complex and the picture is less clear than in *E. coli*. The functional units of the human replication complex are: (i) the three DNA pols α, δ and ε core enzymes, each containing 4 subunits,[36] (ii) the sliding clamp called proliferating cell nuclear antigen (PCNA) existing as a homotrimeric ring[21,37] and (iii) the clamp loader replication factor C (RF-C) formed by 5 different subunits.[22,38] How the three DNA pols act together is not known in detail. Figure 2.4 documents a simplified view of

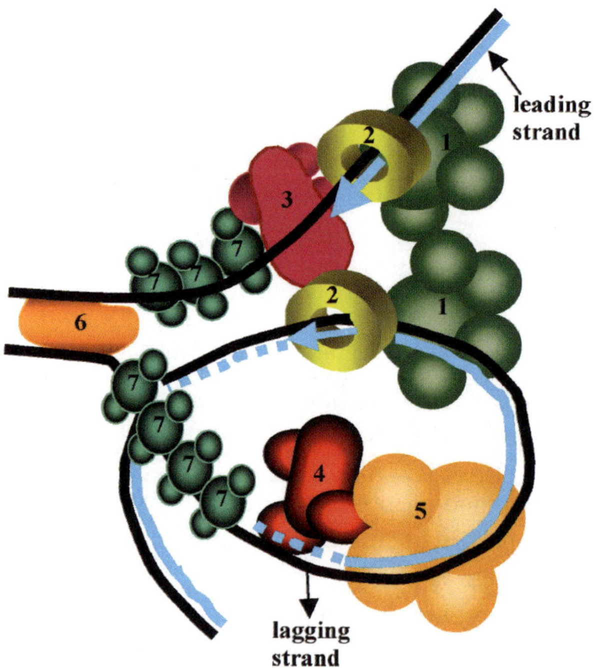

Figure 2.4 **Dynamics at the eukaryotic DNA replication fork**. DNA forms a loop and dimerisation of the two polymerases via a putative dimerisation factor is proposed, thus rendering the same directionality for the leading and the lagging DNA strands. 1: Replication factor C (RF-C); 2: proliferating cell nuclear antigen (PCNA); 3: DNA pol ε; 4: DNA pol α; 5: DNA pol δ; 6: DNA helicase; 7: replication protein A (RP-A); light blue square: initiator RNA and DNA. For more details see text. Adapted from Toueille and Hübscher.[41]

how they might act dynamically at the DNA replication fork. The model of Alberts *et al.* (1980) proposes that the DNA forms a loop and the two polymerases dimerize via a putative dimerization factor, thus rendering the same directionality for the leading and the lagging DNA strands.[39] After the complex events of initiation of DNA replication (for a review and references therein see Ref. 40), the DNA at the origin of replication eventually becomes single-stranded. Starting from this point, the DNA replication fork is formed by recruitment of the diverse DNA replication factors including the DNA pols, the clamp and the clamp loader. The single-stranded DNA is immediately covered by RP-A, which also participates in unwinding of DNA. DNA pol α/primase is then recruited thanks to its interaction with RP-A and Cdc45 (see below for other proteins interacting with DNA pols) and initiates *de novo* synthesis by first synthesizing an RNA fragment of about ten bases, succeeded by a 30-nucleotide DNA fragment. This initiation step and the following events occur during synthesis at the site of each Okazaki fragment on the lagging strand, but only once for the leading strand. After the initiation step, an ATP-dependent polymerase switch from DNA pol α to DNA pol ε (at the leading strand) and δ (at the lagging strand) occurs, for which hydrolysis of ATP by RF-C is required. PCNA is simultaneously loaded by RF-C at the primer end. DNA pol ε bound to PCNA is then able to synthesize long stretches of DNA. When DNA pol δ encounters the 5′ end of the previous Okazaki fragment, can perform strand displacement DNA synthesis. The flap structure that is generated by this displacement is eventually a substrate for combined cleavage by flap endonuclease 1 (Fen 1) and the Dna2 endonucleases, for whom the single-stranded binding protein RP-A, PCNA and RF-C can act as modulators. The resulting nick is finally sealed by DNA ligase I (for reviews and references therein see Refs. 41 and 42).

2.6 DNA polymerases, ring-like clamps and clamp loaders

Proliferating cell nuclear antigen (PCNA) was originally identified as a DNA sliding clamp for replicative DNA pols and as an essential component of the eukaryotic chromosomal DNA replisome.[21] Beyond this function, PCNA can interact with multiple partners, which are involved in several metabolic pathways, such as Okazaki fragment processing, DNA repair (BER, NER, MMR), translesion DNA synthesis (TLS), DNA methylation, chromatin remodelling and cell cycle regulation.[37] PCNA in mammalian cells thus appears to play a key role in controlling a variety of reactions through the coordination and organization of different partners. From these data two major questions emerged: how do these proteins access PCNA in a coordinated manner and how does PCNA temporally and spatially organise their functions? Here we only expand on the functional interactions of PCNA with DNA pols. For other PCNA functions the reader might consult a specific book about PCNA (*Proliferating Cell Nuclear Antigen* (PCNA), Lee H. ed., 2006 ISBN 81-308-0096-9) or a recent review.[43]

PCNA belongs to the so called DNA sliding clamp family, which includes also the *E. coli* DNA pol III β-subunit and the bacteriophage T4 gene45 protein. These proteins perform the essential functions of providing replicative DNA pols (III in *E. coli* and ε/δ in eukaryotes) with the high processivity required to duplicate an entire genome. Crystallographic studies of the β-subunit and PCNA revealed that both these proteins adopt superimposable ring-like structures in solution, with a central hole sufficiently large to accommodate the double helix of DNA, even though they show almost no sequence similarity.[44] In eukaryotes, three identical PCNA monomers, each comprising two similar domains, are joined in a head-to-tail arrangement to form a homotrimer (Figure 2.5). This is composed of six repeating domains and exhibits six-fold symmetry. The resulting ring has two non-equivalent surfaces: an outer surface composed of β-sheets and a layer of α-helices rich in basic residues lining the inner side of the hole. Those α-helices are positioned perpendicularly to the phosphate backbone of DNA. Owing to this unique structure, PCNA is topologically linked to the double helix by encircling it, but still able to freely slide along the DNA lattice by virtue of the α-helices lining the inner channel. This is believed to happen because both the inner α-helices and the DNA backbone are negatively charged and thus repelling each other. PCNA and its homologs increase the processivity of a DNA pol by engaging in protein–protein interactions with their outer surface and tethering it to the DNA. Thus, those ring-like structures prevent the DNA pol from dissociating while advancing along the template DNA. Hence they were also named sliding clamps.

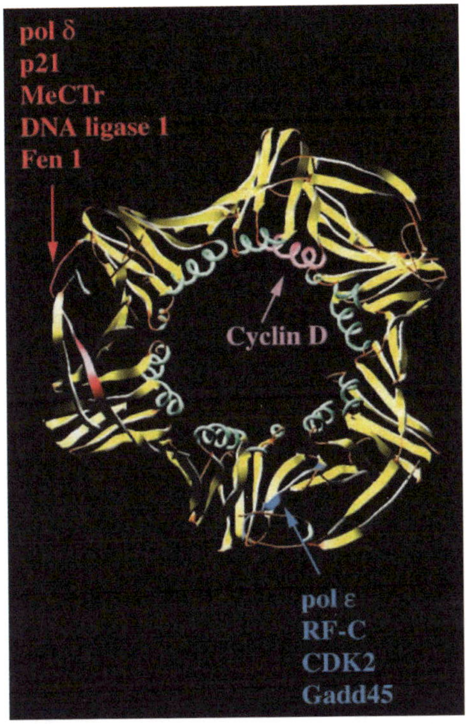

Figure 2.5 **Structure of human PCNA**. Note that several proteins have overlapping binding sites. For more details see Chapter 4. Human DNA pols in different DNA transactions. Reproduced from Jonsson and Hübscher.[45]

From Table 2.7 it is evident that PCNA indeed interacts with 10 different DNA pols from the families B, X and Y. As discussed above, by interacting with the leading strand replicase DNA pol ε and the lagging strand replicase DNA pol δ, PCNA provides processivity to these enzymes. PCNA's interaction with the BER enzymes DNA pols β and λ allows them to perform accurate repair synthesis. The translesion DNA pols ζ, η, κ, i, and Rev1 all interact with PCNA as well.

For these interactions to take place, posttranslational modifications of both PCNA and DNA pols are necessary. They will be described in more detail in Chapters 3 and 4.

The ring like structure prevents PCNA from binding to and encircling DNA on its own. A protein complex called replication factor C

Table 2.7 Functional consequences of the clamp interactions with different DNA polymerases[a]

Family	DNA polymerase (DNA pol)	Clamp[b]	
		PCNA (function for DNA pol)	Rad9/Rad1Hus1 (function for DNA pol)
B	DNA pol α	—	DNA pol α recruits the 9-1-1 complex to DNA
B	DNA pol δ	replication lagging strand	—
B	DNA pol ε	replication leading strand	—
B	DNA pol ζ	translesion DNA synthesis (extender), regulates pol ζ dependent mutagenesis	mutagenesis, recruitment to double-stranded breaks
X	DNA pol β	base excision repair	stimulation of DNA pol β
X	DNA pol λ	base excision repair, DSBR, translesion DNA synthesis	—
X	TdT	DNA damage tolerance	—
Y	DNA pol η	translesion DNA synthesis	—
Y	DNA pol i	translesion DNA synthesis	—
Y	DNA pol κ	translesion DNA synthesis	DNA pol κ recruits the 9-1-1 complex to DNA
Y	Rev 1	translesion DNA synthesis	recruitment to double-strand breaks

a: For detailed reviews of other PCNA interacting proteins see Refs. 21 and 37.
b: Posttranslational modifications of the clamps will be discussed in Chapter 4.

(RF-C) has been found to load PCNA onto DNA. RF-C is a complex of five polypeptides with distinct functions and belongs to the AAA+ machine of ATPases.[22]

The loading mechanism is believed to work as follows[46,47] (Figure 2.6): (A) RF-C binds ATP and forms a binary complex with PCNA; (B) this leads to a conformational change in RF-C with the consequence that the PCNA ring is opened; (C) in a next step this binary complex, having a higher affinity to DNA, encircles the DNA whereupon the PCNA ring is closed again and thus encircling the DNA; (D) by hydrolyzing ATP to

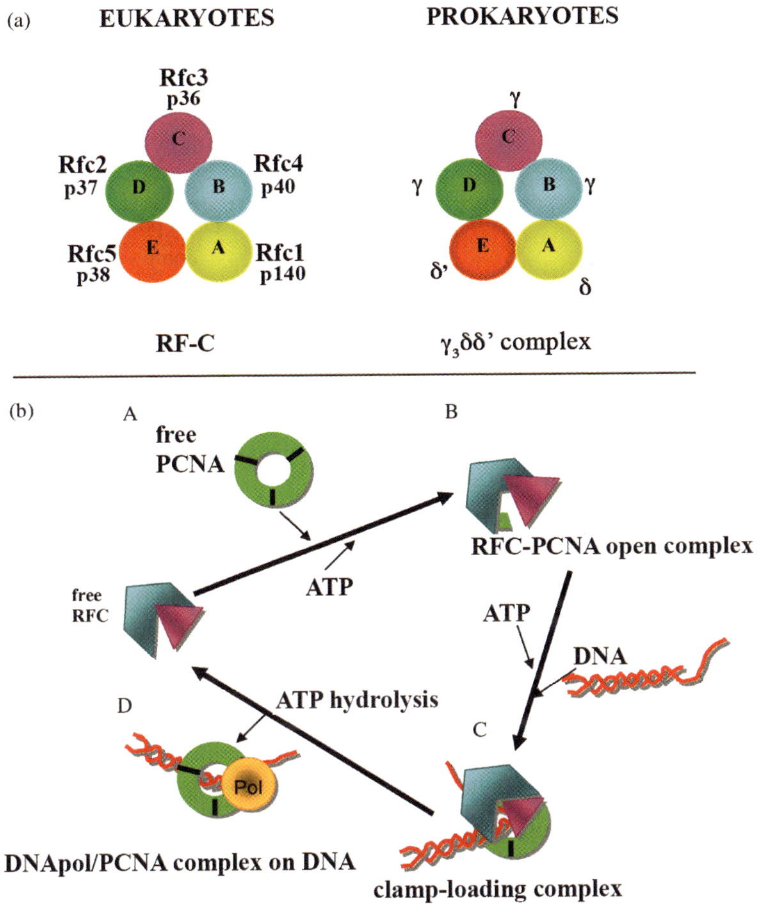

Figure 2.6 **PCNA is loaded onto DNA by RF-C**. **(a).** Subunit compositions of eukaryotic RF-C and the bacterial γ-complex. Rfc1-5 refers to yeast *S. cerevisiae* while p140 and p36, 37, 38 and 40 refer to the mammalian subunits. **(b).** (**A**) the ATP bound RF-C forms a complex with the ring closed PCNA; (**B**) a conformational change in RF-C opens the PCNA ring; (**C**) this open complex has a higher affinity to DNA. The PCNA ring is closed; (**D**) finally upon hydrolysis of ATP, RF-C dissociates from the complex, the DNA pol binds to PCNA and can thus synthesize DNA. The cycle is then repeated.

ADP and P_i, RF-C leaves PCNA and the DNA, allowing the DNA pol to bind to PCNA. It is likely that the DNA pols δ and ε help displacing RF-C while binding to PCNA. After dissociation from PCNA, RF-C binds another ATP molecule and is ready for loading the next PCNA molecule.

2.7 DNA polymerases, alternative clamps and clamp loaders

Alternative clamps:[48] Genotoxic stress activates checkpoint-signaling pathways that halt cell cycle progression, regulate DNA repair and trigger apoptosis. Studies in yeast and humans suggest that Rad17, Rad9, Rad1, Hus1 play key roles in checkpoint activation. Three of these proteins Rad9, Hus1, and Rad1 form a heterotrimeric complex (called the 9-1-1 complex), which resembles a PCNA-like sliding clamp, whereas Rad17 is part of a clamp-loading complex that is related to the PCNA clamp loader, RF-C. In response to DNA damage, the 9-1-1 complex is loaded onto DNA in a Rad17-RF-$C_{2\text{-}5}$-dependent manner. The crystal structure of the 9-1-1 complex has been solved and suggested a toroidal heterotrimeric structure similar to the homotrimeric structure of PCNA.[49,50] The DNA-bound 9-1-1 complex then facilitates ATR-mediated phosphorylation and activation of Chk1, a protein kinase that regulates S-phase progression, G2/M arrest, and replication fork stabilization. In addition, the 9-1-1 complex also directly participates in DNA repair by interacting with and/or stimulating several components of different repair pathways including BER, NER, DSBR and TLS.

The 9-1-1 complex interacts with at least five different DNA pols from the three families B, X and Y (Table 2.7). The first description was the interaction of the 9-1-1 complex with the BER enzyme DNA pol β,[51] thus stimulating its activity, increasing its affinity for the primer-template and enhancing its DNA strand displacement synthesis. The interaction of the 9-1-1 complex with DNA pol ζ regulates its mutagenicity and helps to control this DNA pol to properly extend from a lesion.[52] A similar effect has been described by the interaction of the 9-1-1 complex with Rev1, a deoxycytidyltransferase interacting with many TLS DNA pols.[53] DNA pol α, together with the protein TopBP1 (see below), can directly recruit the 9-1-1 complex to stalled replication forks.[54]

Alternative clamp loaders:[41] In addition to the "classical" RF-C protein involved in DNA replication and the checkpoint clamp-loader Rad17-RF-$C_{2\text{-}5}$, the four small subunits of RF-C are part of other alternative clamp-loaders (Figure 2.7). First, the sister chromatid cohesion factor,

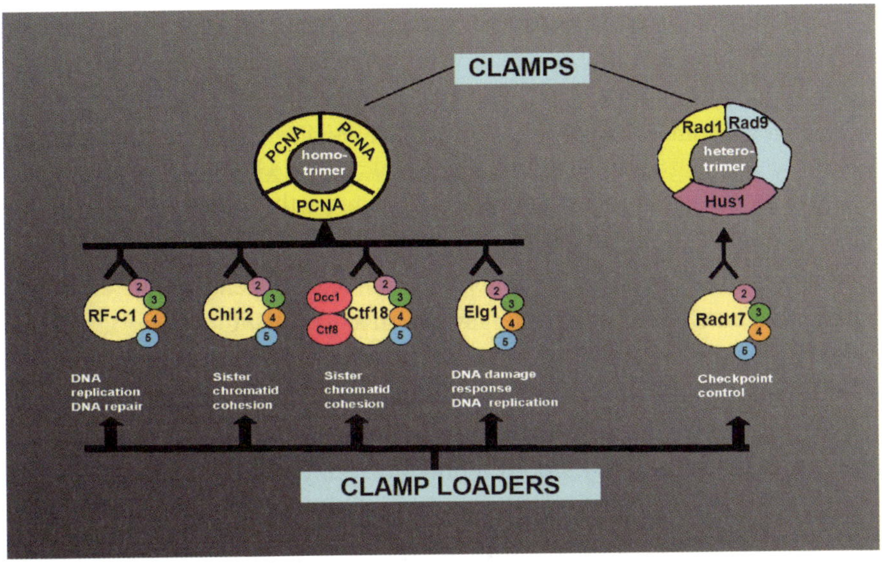

Figure 2.7 **Clamps and clamp loaders**. Under normal conditions e.g. during replication the RF-C heteropentameric clamp loader (RF-$C_{1\text{-}5}$) loads PCNA in an ATP dependent manner. In other cellular metabolic events such as sister chromatid cohesion, DNA damage response and checkpoint control, alternative clamp loaders (consisting of the four small subunits RF-$C_{2\text{-}5}$ and an additional protein) can either load PCNA or the alternative clamp, the 9-1-1 complex. For details see text. Reproduced from Toueille and Hübscher.[41]

Chl12, exhibits sequence similarity to the large subunit of RF-C. The complex formed by association of Chl12 with the RF-$C_{2\text{-}5}$ subunits is structurally similar to RF-C and was shown to be able to load PCNA in an ATP dependent manner onto circular DNA. The Chl12-RF-$C_{2\text{-}5}$ complex can stimulate DNA pol δ activity. Second, another set of proteins involved in sister chromatid cohesion, Ctf18, Dcc1 and Ctf8 can also associate with the four small RF-C subunits. Similarly to the previous complex, this complex is also a clamp-loader for PCNA, and can substitute for the replicative RF-C in a DNA pol δ catalyzed DNA replication assay. Furthermore Ctf18 RF-$C_{2\text{-}5}$ can unload PCNA in the presence of RP-A and can stimulate DNA pol η. Third, the Elg1 protein shares sequence homology with the large subunit of the clamp loader RF-C and forms a complex with the RF-$C_{2\text{-}5}$ subunits.

Genetic data from yeast studies indicated that Elg1 has a role in DNA damage response and in maintenance of the genomic integrity, as well as in DNA replication. Yeast Elg1 interacts physically and genetically with PCNA, in a manner that depends on PCNA modification, and exibits preferentially affinity for SUMOylated PCNA.[55] Moreover, Elg1-RF-$C_{2\text{-}5}$ contributes to genome stability via telomere length regulation through a putative replication-mediated pathway. The telomeric function of Elg1 was shown to be independent of recombination and dependent on an active telomerase and DNA pol α suggesting that Elg1-$RFC_{2\text{-}5}$ plays a role in checkpoint pathway and DNA replication.

Finally, the checkpoint clamp-loader Rad17-RF-$C_{2\text{-}5}$, when phosphorylated, preferentially associates with the site of ongoing replication and interacts with the leading strand replication enzyme DNA pol ε.

2.8 Replicative DNA polymerases and interacting proteins

DNA pols, in whatever context they act, have to interact with many other proteins. As an example we introduce some selected proteins that interact with the three replicative DNA pols α, δ and ε. The eukaryotic replicative DNA helicase, termed CMG consists of 11 subunits, the so called Cd45-MCM-GINS proteins. First, DNA pol α, the initiating replicase, interacts with the Mcm 2–7 components of CMG.[56] It furthermore binds to the GINS proteins,[57,58] and Cdc45[59] which acts as connectors of DNA pol α to the replication machinery. A physical interaction of DNA pol α with Mcm10 and DNA has been described.[60] This interaction with Mcm10 recruits DNA pol α to chromatin and the interaction with And1/Ctf4 stabilizes DNA pol α at the replication fork.[61] A key role for Ctf4 has been described for coupling the Mcm 2–7 helicase to the initiating DNA pol α.[62] Structural studies have shown that Ctf4 forms a homotrimer. The N-ter of DNA pol α p180 and the Sld5 subunit of GINS physically interact with each protomer. It has ben suggested that two molecules of DNA pol α and one molecule of the GINS complex associate to the Ctf4 trimer.[63] These interactions are also important in coordinating DNA damage response events at the replication fork.[64]

Second, DNA pol ε, the leading strand replicase, forms a functional 15-subunit holoenzyme for eukaryotic leading-strand DNA replication.[65] Furthermore, it interacts with Mrc1 (ortholog of metazoan claspin) that links replication and the S phase checkpoint[66] and it can interact with Cdc45, another connector protein.[59] Interaction with Dpb11 (in *S. cerevisiae*)/Cut5 (in *S. pombe*)/TopBP1 (in human) recruits DNA pols α and ε to DNA.[67] TopBP1 and DNA pol α together recruit the 9-1-1 complex to DNA when replication forks are stalled after DNA damage.[54] Third, DNA pol δ, the lagging strand replicase, interacts with Werner and Bloom DNA helicases[68,69] that help to unwind secondary structures in the DNA. Finally, INO80 interacts with all three DNA pol thus regulating replisome function and stability.[70]

2.9 DNA polymerases and the single-stranded DNA binding replication protein A

DNA replication, DNA repair and DNA recombination all expose single-stranded DNA that is extremely prone to genotoxic attacks. Replication protein A (RP-A) protects single-stranded DNA from nucleolytic attack, removes hairpins and blocks DNA reannealing until the above mentioned pathways are terminated.[23,71]

Many DNA repair studies showed that RP-A plays a major role in coordinating the repair mechanisms and is therefore more than a simple "protector" for single-stranded DNA. Essential roles in BER, NER, DSBR and MMR were identified for RP-A. Moreover, RP-A can sense DNA damage and activate checkpoints.[72] Several studies suggested that in this step RP-A is a key player. Indeed, RP-A coated single-stranded DNA is a common structure generated at the sites of DNA damage. The ATR checkpoint kinase is recruited to the sites of single-stranded DNA by binding of its regulatory partner, ATRIP, to the RP-A protein. Furthermore, experiments in yeast suggested that RP-A single-stranded DNA filaments allow the sensing of stalled replication forks via recruitment of Ddc1 (homologue to the human Rad9) and Ddc2 (homologue to the human ATRIP). This recruitment leads to checkpoint activation and to stabilization of the stalled replication fork and confirmed the crucial role of RP-A in the DNA damage-sensing step of the checkpoint response. Phosphorylation and SUMOylation of the RPA complex have been shown to control specific aspects of DNA damage signaling and repair by modulating the interaction between RPA and its partners.[73] Likewise, it has been recently shown that RPA is ubiquitinated in chromatin in response to replication fork stalling, and that ubiquitination is mediated by the E3 ligase RFWD3. Furthermore, RFDW promotes replication fork restart and homologous recombination at stalled forks.[74]

RP-A contains three subunits with Mr of 70,000 (RP-A1), 32,000 (RP-A2) and 14,000 (RP-A3). An alternative form of RP-A, containing a different 32 kDa subunit called RP-A4, has recently been described. This form may prevent cellular proliferation by inhibition of DNA replication and might play a role in maintaining genetic stability in quiescent cells.[75] RPA has also been shown to regulate the activity of the recent discovered PrimPol.[76]

Studies carried out in our laboratories suggest important roles of RP-A in helping DNA pols to act faithfully: (i) RP-A can act as a "fidelity clamp" for DNA pol α;[77] (ii) RP-A, together with PCNA and Fen 1 can modulate the strand displacement activity of DNA pol δ in Okazaki fragment synthesis, which promotes nt processing;[78] (iii) RP-A, together with PCNA, can suppress completely the deoxyribonucleotidyltransferase activity of DNA pol λ, while stimulating its polymerase activity;[79] (iv) RP-A, but not PCNA, selectively prevents the misincorporation of an incorrect nucleotide by DNA pol λ, without affecting the correct incorporation;[80] (v) RP-A and PCNA allowed correct incorporation of C opposite a template 8-oxo-G 1200- fold more efficiently than the incorrect A by DNA pol λ, and 68-fold by DNA pol η, respectively. These findings suggested for the first time a novel accurate mechanism to reduce the deleterious consequences of oxidative damage and, in addition, point to an important role for RP-A and PCNA in determining a functional hierarchy among different DNA pols in lesion bypass;[81] (vi) RP-A and PCNA enhance the fidelity of translesion synthesis by DNA pol λ over 2-hydroxy-adenine (2-OH-A) in a repeated sequence;[82] (vii) RP-A and PCNA act as molecular switches to activate the DNA pol λ-dependent highly efficient and faithful repair of A:8-oxo-G mismatches in human cells, while they inhibit the less faithful DNA pol β thus coordinating the DNA pol selection in 8-oxo-guanine repair.[83]

2.10 Chapter summary

In this Chapter we summarized that synthesis of DNA in nature needs robust enzymes called DNA pols. The DNA pols are used for all DNA transactions (DNA replication, the different DNA repair events, DNA recombination and physiological recombination in the immune system) in a cell. The DNA pol reaction is complex and its structure resembles a right hand, consisting of fingers, a palm and a thumb. Nature has brought forward eight different DNA pol families and higher eukaryotic cells contain a total of 15 DNA pols, plus Rev1 and the telomerase. They harbor different, mostly redundant functions. Replicative DNA pols act in large complexes called holoenzymes. They consist, besides the DNA pol core, of a clamp and a clamp loader. In eukaryotes the clamp is called proliferating cell nuclear antigen (PCNA) and the clamp loader replication factor C (RF-C). The fact that PCNA interacts with 10 different DNA pols suggests a central regulatory task for this important protein. Additionally, alternative clamps and several alternative clamp loaders exist. The fact that many of these are involved in checkpoint sensing suggests an important role for them in guiding the appropriate DNA pol under certain conditions to the right place to work. Replicative DNA pols interact with many proteins to build a stable replisome. Finally, the single-stranded binding protein-replication protein A (RP-A) does more than bind and protect DNA. It has additional functions in checkpoint regulation. The investigations of RP-A interacting with different DNA pols have recently revealed key roles for this very important protein in various DNA transactions. Table 2.8 documents a gradient of fidelity of different mammalian DNA pols in cellular events such as DNA replication, DNA repair and V(D)J recombination while Figure 2.8 documents that the number of different DNA pols increases in evolution from viruses to mammals. Figure 2.9 highlights the different subunit composition of DNA pols across the eukaryotic families.

Table 2.8 Redundant functions of DNA polymerases

DNA replication	**DNA repair**	**V(D)J recombination**
Pols α, γ, δ and ε	Pol β, η, κ, λ, ζ, θ, ν and Rev1	Pol ι, λ, and μ
Fidelity: 10^{-4}–10^{-6}	Fidelity: 10^{-2}–10^{-4}	Fidelity: 10^{0}–10^{-2}

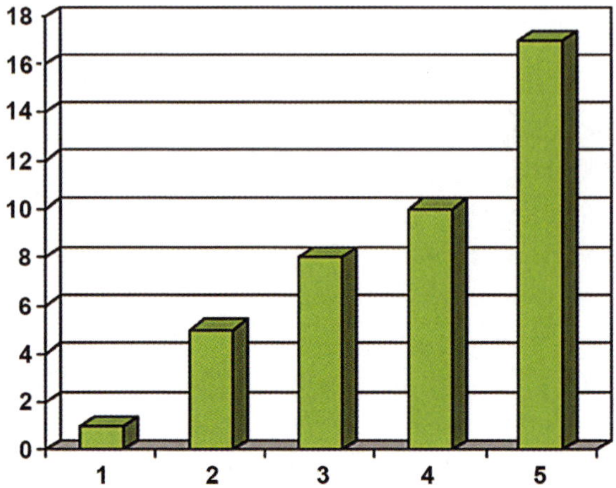

Figure 2.8 **The number of different DNA polymerases increases in evolution from viruses to mammals**. 1:Viruses; 2: prokaryotes; 3: eukaryotes (single cell); 4: eukaryotes (plants); 5: eukaryotes (mammals).

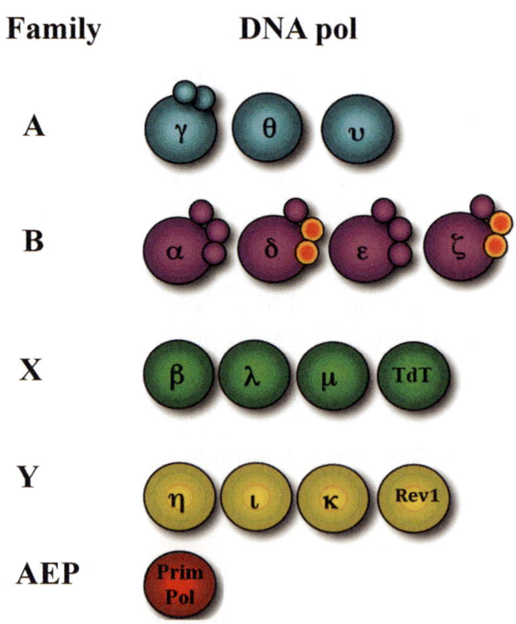

Figure 2.9 **Subunit compositions of the mammalian DNA polymerases**. For further details see Chapter 4. The small subunit of DNA pol γ is a homodimer. DNA pol ζ can form a complex with two small subunits of DNA pol δ. Telomerase is not shown.

References

1. Watson, J. D. & Crick, F. H. Genetical implications of the structure of deoxyribonucleic acid. *Nature* **171**, 964–967 (1953).
2. Bessman, M. J., Kornberg, A., Lehman, I. R. & Simms, E. S. Enzymic synthesis of deoxyribonucleic acid. *Biochim Biophys Acta* **21**, 197–198 (1956).
3. Kornberg, A. Enzymatic synthesis of deoxyribonucleic acid. *Harvey Lect* **53**, 83–112 (1957).
4. Hubscher, U., Maga, G. & Spadari, S. Eukaryotic DNA polymerases. *Annu Rev Biochem* **71**, 133–163 (2002).
5. Hübscher, U. & Shevelev, I. *Replication fork.* 1632–1635 (Springer Verlag, 2006).
6. Wallace, S. S. Base excision repair: A critical player in many games. *DNA Repair* **19**, 14–26 (2014).
7. Spivak, G. Nucleotide excision repair in humans. *DNA Repair* **36**, 13–18 (2015).
8. Fishel, R. Mismatch Repair. *Journal of Biological Chemistry* **290**, 26395–26403 (2015).
9. Kunkel, T. A. & Erie, D. A. Eukaryotic Mismatch Repair in Relation to DNA Replication. *Annu Rev Genet* **49**, 291–313 (2015).
10. Rossi, M. L., Purohit, V., Brandt, P. D. & Bambara, R. A. Lagging strand replication proteins in genome stability and DNA repair. *Chem Rev* **106**, 453–473 (2006).
11. Ceccaldi, R., Rondinelli, B. & D'Andrea, A. D. Repair Pathway Choices and Consequences at the Double-Strand Break. *Trends Cell Biol* **26**, 52–64 (2016).
12. Gearhart, P. J. & Wood, R. D. Emerging links between hypermutation of antibody genes and DNA polymerases. *Nat Rev Immunol* **1**, 187–192. (2001).
13. Prakash, S., Johnson, R. E. & Prakash, L. Eukaryotic translesion synthesis DNA polymerases: specificity of structure and function. *Annu Rev Biochem* **74**, 317–353 (2005).
14. Acharya, N., Brahma, A., Haracska, L., Prakash, L. & Prakash, S. Mutations in the ubiquitin binding UBZ motif of DNA polymerase eta do not impair its function in translesion synthesis during replication. *Mol Cell Biol* **27**, 7266–7272 (2007).
15. Kelleher, C., Teixeira, M. T., Forstemann, K. & Lingner, J. Telomerase: biochemical considerations for enzyme and substrate. *Trends Biochem Sci* **27**, 572–579 (2002).

16. Wang, S. W., Toda, T., MacCallum, R., Harris, A. L. & Norbury, C. Cid1, a fission yeast protein required for S-M ckeckpoint control when DNA polymerase δ or ε is inactivated. *Mol. Cell. Biol.* **20**, 3234–3244 (2000).
17. Kunkel, T. A. & Bebenek, K. DNA replication fidelity. *Annu Rev Biochem* **69**, 497–529 (2000).
18. Kunkel, T. A. DNA replication fidelity. *J Biol Chem* **279**, 16895–16898 (2004).
19. Shevelev, I. V. & Hubscher, U. The 3′ 5′ exonucleases. *Nat Rev Mol Cell Biol* **3**, 364–376 (2002).
20. Jiricny, J. The multifaceted mismatch-repair system. *Nat Rev Mol Cell Biol* **7**, 335–346 (2006).
21. Maga, G. & Hubscher, U. Proliferating cell nuclear antigen (PCNA): a dancer with many partners. *J Cell Sci* **116**, 3051–3060 (2003).
22. Davey, M. J., Jeruzalmi, D., Kuriyan, J. & O'Donnell, M. Motors and switches: AAA+ machines within the replisome. *Nat Rev Mol Cell Biol* **3**, 826–835 (2002).
23. Fanning, E., Klimovich, V. & Nager, A. R. A dynamic model for replication protein A (RPA) function in DNA processing pathways. *Nucleic Acids Res* **34**, 4126–4137 (2006).
24. Joyce, C. M. & Steitz, T. A. Function and structure relationships in DNA polymerases. *Annu Rev Biochem* **63**, 777–822. (1994).
25. Steitz, T. A. DNA polymerases: structural diversity and common mechanisms. *J. Biol. Chem.* **274**, 17395–17398 (1999).
26. Sawaya, M. R., Pelletier, H., Kumar, A., Wilson, S. H. & Kraut, J. Crystal structure of rat DNA polymerase β: evidence for a common polymerase mechanism. *Science* **264**, 1930–1935 (1994).
27. Bailey, S., Wing, R. A. & Steitz, T. A. The structure of T. aquaticus DNA polymerase III is distinct from eukaryotic replicative DNA polymerases. *Cell* **126**, 893–904 (2006).
28. Leulliot, N. *et al.* The Family X DNA Polymerase from Deinococcus radiodurans Adopts a Non-standard Extended Conformation. *J Biol Chem* **284**, 11992–11999 (2009).
29. Freudenthal, B. D., Beard, W. A. & Wilson, S. H. Structures of dNTP Intermediate States during DNA Polymerase Active Site Assembly. *Structure* **20**, 1829–1837 (2012).
30. Nakamura, T., Zhao, Y., Yamagata, Y., Hua, Y. J. & Yang, W. Watching DNA polymerase eta make a phosphodiester bond. *Nature* **487**, 196–201 (2012).
31. Gao, Y. & Yang, W. Capture of a third Mg2+ is essential for catalyzing DNA synthesis. *Science* **352**, 1334–1337 (2016).

32. Garcia-Escudero, R., Garcia-Diaz, M., Salas, M. L., Blanco, L. & Salas, J. DNA polymerase X of African swine fever virus: insertion fidelity on gapped DNA substrates and AP lyase activity support a role in base excision repair of viral DNA. *J Mol Biol* **326**, 1403–1412 (2003).
33. McHenry, C. & Kornberg, A. DNA polymerase III holoenzyme of Escherichia coli. Purification and resolution into subunits. *J Biol Chem* **252**, 6478–6484 (1977).
34. Johnson, A. & O'Donnell, M. Cellular DNA replicases: components and dynamics at the replication fork. *Annu Rev Biochem* **74**, 283–315 (2005).
35. O'Donnell, M. Replisome architecture and dynamics in Escherichia coli. *J Biol Chem* **281**, 10653–10656 (2006).
36. Garg, P. & Burgers, P. M. DNA polymerases that propagate the eukaryotic DNA replication fork. *Crit Rev Biochem Mol Biol* **40**, 115–128 (2005).
37. Moldovan, G. L., Pfander, B. & Jentsch, S. PCNA, the Maestro of the Replication Fork. *Cell* **129**, 665–679 (2007).
38. Mossi, R. & Hubscher, U. Clamping down on clamps and clamp loaders — the eukaryotic replication factor C. *Eur J Biochem* **254**, 209–216 (1998).
39. Sinha, N. K., Morris, C. F. & Alberts, B. M. Efficient in vitro replication of double-stranded DNA templates by a purified T4 bacteriophage replication system. *J Biol Chem* **255**, 4290–4293 (1980).
40. Sclafani, R. A. & Holzen, T. M. Cell cycle regulation of DNA replication. *Annu Rev Genet* **41**, 237–280 (2007).
41. Toueille, M. & Hubscher, U. Regulation of the DNA replication fork: a way to fight genomic instability. *Chromosoma* **113**, 113–125 (2004).
42. Georgescu, R. E. *et al.* Mechanism of asymmetric polymerase assembly at the eukaryotic replication fork. *Nat Struct Mol Biol* **21**, 664–670 (2014).
43. Georgescu, R., Langston, L. & O'Donnell, M. A proposal: Evolution of PCNA's role as a marker of newly replicated DNA. *DNA Repair* **29**, 4–15 (2015).
44. Hingorani, M. M. & O'Donnell, M. Sliding clamps: a (tail)ored fit. *Curr Biol* **10**, R25–29 (2000).
45. Jonsson, Z. O. & Hubscher, U. Proliferating cell nuclear antigen: more than a clamp for DNA polymerases. *Bioessays* **19**, 967–975 (1997).
46. Bowman, G. D., O'Donnell, M. & Kuriyan, J. Structural analysis of a eukaryotic sliding DNA clamp-clamp loader complex. *Nature* **429**, 724–730 (2004).
47. Trakselis, M. A. & Bell, S. D. Molecular biology: the loader of the rings. *Nature* **429**, 708–709 (2004).

48. Sancar, A., Lindsey-Boltz, L. A., Unsal-Kaccmaz, K. & Linn, S. Molecular mechanisms of mammalian DNA repair and the DNA damage checkpoints. *Annu Rev Biochem* **73**, 39–85. (2004).
49. Dore, A. S., Kilkenny, M. L., Rzechorzek, N. J. & Pearl, L. H. Crystal structure of the rad9-rad1-hus1 DNA damage checkpoint complex — implications for clamp loading and regulation. *Mol Cell* **34**, 735–745 (2009).
50. Xu, M. *et al.* Structure and functional implications of the human rad9-hus1-rad1 cell cycle checkpoint complex. *J Biol Chem* **284**, 20457–20461 (2009).
51. Toueille, M. *et al.* The human Rad9/Rad1/Hus1 damage sensor clamp interacts with DNA polymerase beta and increases its DNA substrate utilisation efficiency: implications for DNA repair. *Nucleic Acids Res* **32**, 3316–3324 (2004).
52. Sabbioneda, S. *et al.* The 9-1-1 checkpoint clamp physically interacts with polzeta and is partially required for spontaneous polzeta-dependent mutagenesis in Saccharomyces cerevisiae. *J Biol Chem* **280**, 38657–38665 (2005).
53. Waters, L. S. *et al.* Eukaryotic translesion polymerases and their roles and regulation in DNA damage tolerance. *Microbiol Mol Biol Rev* **73**, 134–154 (2009).
54. Yan, S. & Michael, W. M. TopBP1 and DNA polymerase-alpha directly recruit the 9-1-1 complex to stalled DNA replication forks. *J Cell Biol* **184**, 793–804 (2009).
55. Parnas, O. *et al.* Elg1, an alternative subunit of the RFC clamp loader, preferentially interacts with SUMOylated PCNA. *Embo Journal* **29**, 2611–2622 (2010).
56. Forsburg, S. L. Eukaryotic MCM proteins: beyond replication initiation. *Microbiol Mol Biol Rev* **68**, 109–131 (2004).
57. De Falco, M. *et al.* The human GINS complex binds to and specifically stimulates human DNA polymerase alpha-primase. *EMBO Rep* **8**, 99–103 (2007).
58. Tercero, J. A., Labib, K. & Diffley, J. F. DNA synthesis at individual replication forks requires the essential initiation factor Cdc45p. *EMBO J* **19**, 2082–2093 (2000).
59. Zou, L. & Stillman, B. Assembly of a complex containing Cdc45p, replication protein A, and Mcm2p at replication origins controlled by S-phase cyclin-dependent kinases and Cdc7p-Dbf4p kinase. *Mol Cell Biol* **20**, 3086–3096 (2000).
60. Warren, E. M., Huang, H., Fanning, E., Chazin, W. J. & Eichman, B. F. Physical interactions between MCM10, DNA, AND DNA polymerase {alpha}. *J Biol Chem* (2009).

61. Zhu, W. *et al.* Mcm10 and And-1/CTF4 recruit DNA polymerase alpha to chromatin for initiation of DNA replication. *Genes Dev* **21**, 2288–2299 (2007).
62. Gambus, A. *et al.* A key role for Ctf4 in coupling the MCM2-7 helicase to DNA polymerase alpha within the eukaryotic replisome. *EMBO J* **28**, 2992–3004 (2009).
63. Simon, A. C. *et al.* A Ctf4 trimer couples the CMG helicase to DNA polymerase alpha in the eukaryotic replisome. *Nature* **510**, 293-+ (2014).
64. Fumasoni, M., Zwicky, K., Vanoli, F., Lopes, M. & Branzei, D. Error-Free DNA Damage Tolerance and Sister Chromatid Proximity during DNA Replication Rely on the Pol alpha/Primase/Ctf4 Complex. *Molecular Cell* **57**, 812–823 (2015).
65. Langston, L. D. *et al.* CMG helicase and DNA polymerase epsilon form a functional 15-subunit holoenzyme for eukaryotic leading-strand DNA replication. *P Natl Acad Sci USA* **111**, 15390–15395 (2014).
66. Lou, H. *et al.* Mrc1 and DNA polymerase epsilon function together in linking DNA replication and the S phase checkpoint. *Mol Cell* **32**, 106–117 (2008).
67. Masumoto, H., Sugino, A. & Araki, H. Dpb11 controls the association between DNA polymerases alpha and epsilon and the autonomously replicating sequence region of budding yeast. *Mol Cell Biol* **20**, 2809–2817 (2000).
68. Kamath-Loeb, A. S., Johansson, E., Burgers, P. M. & Loeb, L. A. Functional interaction between the Werner Syndrome protein and DNA polymerase delta. *Proc Natl Acad Sci USA* **97**, 4603–4608 (2000).
69. Selak, N. *et al.* The Bloom's syndrome helicase (BLM) interacts physically and functionally with p12, the smallest subunit of human DNA polymerase delta. *Nucleic Acids Res* **36**, 5166–5179 (2008).
70. Papamichos-Chronakis, M. & Peterson, C. L. The Ino80 chromatin-remodeling enzyme regulates replisome function and stability. *Nat Struct Mol Biol* **15**, 338–345 (2008).
71. Wold, M. S. Replication protein A: a heterotrimeric, single-stranded DNA-binding protein required for eukaryotic DNA metabolism. *Annu Rev Biochem* **66**, 61–92 (1997).
72. Zou, L. & Elledge, S. J. Sensing DNA damage through ATRIP recognition of RPA-ssDNA complexes. *Science* **300**, 1542–1548. (2003).
73. Marechal, A. & Zou, L. RPA-coated single-stranded DNA as a platform for post-translational modifications in the DNA damage response. *Cell Res* **25**, 9–23 (2015).

74. Elia, A. E. H. *et al.* RFWD3-Dependent Ubiquitination of RPA Regulates Repair at Stalled Replication Forks. *Molecular Cell* **60**, 280–293 (2015).
75. Mason, A. C. *et al.* An alternative form of replication protein a prevents viral replication in vitro. *J Biol Chem* **284**, 5324–5331 (2009).
76. Wan, L. *et al.* hPrimpol1/CCDC111 is a human DNA primase-polymerase required for the maintenance of genome integrity. *Embo Reports* **14**, 1104–1112 (2013).
77. Maga, G., Frouin, I., Spadari, S. & Hubscher, U. Replication protein A as a "fidelity clamp" for DNA polymerase alpha. *J Biol Chem* **276**, 18235–18242 (2001).
78. Maga, G. *et al.* Okazaki fragment processing: modulation of the strand displacement activity of DNA polymerase δ by the concerted action of replication protein A, proliferating cell nuclear antigen and flap endonuclease 1. *Proc. Natl. Acad. Sci. USA,* **98**, 14298–14303 (2001).
79. Maga, G. *et al.* DNA elongation by the human DNA polymerase lambda polymerase and terminal transferase activities are differentially coordinated by proliferating cell nuclear antigen and replication protein A. *J Biol Chem* **280**, 1971–1981 (2005).
80. Maga, G., Shevelev, I., Villani, G., Spadari, S. & Hubscher, U. Human replication protein A can suppress the intrinsic in vitro mutator phenotype of human DNA polymerase lambda. *Nucleic Acids Res* **34**, 1405–1415 (2006).
81. Maga, G. *et al.* 8-oxo-guanine bypass by human DNA polymerases in the presence of auxiliary proteins. *Nature* **447**, 606–608 (2007).
82. Crespan, E., Hubscher, U. & Maga, G. Error-free bypass of 2-hydroxyadenine by human DNA polymerase lambda with Proliferating Cell Nuclear Antigen and Replication Protein A in different sequence contexts. *Nucleic Acids Res* **35**, 5173–5181 (2007).
83. Maga, G. *et al.* Replication protein A and proliferating cell nuclear antigen coordinate DNA polymerase selection in 8-oxo-guanine repair. *Proc Natl Acad Sci USA* **105**, 20689–20694 (2008).

Chapter 3

Human DNA Polymerases: From Structure to Function

3.1 The high number of specialized pathways in eukaryotic cells requires a plethora of specialized DNA synthesizing enzymes

Based on sequence homology and structural similarities, DNA polymerases (pols) have been grouped into eight different families: A, B, C, D, X, Y, AEP and reverse transcriptase (RT).[1] The evolution of DNA pols from prokaryotic organisms to metazoan, can be correlated to the increasing levels of complexity, from single-cell organisms to the highly organized hierarchical development of tissues and organs during embryogenesis occurring in vertebrates and plants. Very likely the replicative DNA pols of prokaryotes and eukaryotes evolved from two independent lineages.[2] Eukaryotic-type B family enzymes probably evolved from ancient RNA-dependent RNA polymerases and are present in both Archaea and eukaryotes. Prokaryotic-type C family enzymes, instead, evolved later in the history of life, likely from a template-independent nucleotidyltransferase such as poly(A) polymerase. However, regardless of their phylogenetic relationships, the universally conserved function of DNA pols among all DNA-based life forms (i.e. all cellular-based organisms and several viruses) is their ability to copy the genetic information stored in the double-helix, both for its duplication (DNA replication) or for the maintenance of its integrity (DNA repair). As a consequence, a minimal set of replicative and repair DNA pols is present in all eukaryotic forms (Table 3.1).

The transition from prokaryotic to eukaryotic cells, marked by the inner compartmentalization of cellular functions through the development of membrane-limited subcellular compartments (nucleus) or the endosymbiotic acquisition of organelles (mitochondria, plastids), brought up the need of additional specialized functions. For example, the development of a cellular cycle, made of temporally and spatially defined different phases, required the confinement of the DNA replication and repair machineries within the appropriate phases as well as their integration in the cell cycle regulatory circuits.[3–8]

The organelles of endosymbiotic origins such as mitochondria and plastids,[9] possess their own DNA and replicative machineries, distinct from those responsible for the replication of nuclear DNA, even though encoded by the nuclear genome.[10,11] Thus, as a consequence of both their larger genomes and their more complex regulation, eukaryotic cells possess a higher number of replicative and repair DNA pols (including organelle-specific enzymes) than prokaryotes (Table 3.1). An example of this diversification is the structure of the replisome. At the prokaryotic DNA replication fork a dimer of DNA pol III holoenzyme performs both leading and lagging strand DNA synthesis, whereas RNA priming is made by the primase activity associated with the replicative helicase. The eukaryotic replication fork, on the other hand, seems to employ three different DNA pols[12–17]: DNA pol α/primase initiates DNA synthesis on both strands, DNA pol ε synthesizes the leading strand, whereas DNA pol δ works at the lagging strand. RNA priming is catalyzed by a heterodimeric DNA primase which is an intrinsic component of the DNA pol α four subunit complex.

Such a division of labor at the eukaryotic DNA replication fork could allow for several layers of regulation, each DNA pol being a sensor and a transducer of signals for the cell cycle regulatory circuits.

However, an additional DNA pol-primase called PrimPol has been recently discovered.[18–20] PrimPol activity is regulated by single-stranded

Table 3.1 DNA polymerases in diverse eukaryotic organisms

DNA polymerase					
Saccaromyces cerevisiae	*Arabidopsis thaliana*	*Homo sapiens*	Subunits[a]	Family	Main Function
DNA pol α/primase	DNA pol α/primase	DNA pol α/primase	p180 p68 p58(Pri2) p49(Pri1)	B	DNA replication (initiation)
		PrimPol	p64	Archea-type Primase (AEP)	Mitochondrial replication (initiation) Translesion synthesis Repriming at forks arrested at DNA damage
DNA pol δ	DNA pol δ	DNA pol δ	p125 p66 (PolD2) p50 (PolD3) p12[b](PolD4)	B	DNA replication (lagging strand)
DNA pol ε	DNA pol ε	DNA pol ε	p261 p59 p17 p12	B	DNA replication (leading strand)

(Continued)

Table 3.1 (*Continued*)

DNA polymerase					
Saccaromyces cerevisiae	***Arabidopsis thaliana***	***Homo sapiens***	**Subunits**[a]	**Family**	**Main Function**
DNA pol ζ	DNA pol ζ	DNA pol ζ	p350(Rev3L) p24(Rev7) p66 (PolD2) p50 (PolD3)	B	TLS[c] (extender)
DNA pol γ	DNA pol γ-like	DNA pol γ	p140 p55	A	Mitochondria/plastid DNA replication
—	DNA pol θ	DNA pol θ	p250	A	DNA repair
—	—	DNA pol ν	p100	A	DNA repair
DNA pol IV	DNA pol λ	DNA pol λ	p70	X	DNA repair/V(D)J recombination
—	—	DNA pol β	p39	X	DNA repair
—	—	DNA pol μ	p55[d]	X	DNA repair/V(D)J recombination
—	—	TdT	p58	X	DNA repair/V(D)J recombination
DNA pol η	DNA pol η	DNA pol η	p79	Y	TLS (inserter)
Rev1	Rev1	Rev1	p138	Y	TLS (inserter)
—	—	DNA pol ι	p80	Y	TLS (inserter)
—	DNA pol κ	DNA pol κ	p99	Y	TLS (inserter/extender)

[a]The molecular weights refer to the human proteins. [b]the orthologue of p12 has not been found yet in *S.cerevisiae*. [c]TLS, translesion DNA synthesis. [d]p55 forms a homodimer

DNA binding proteins and, differently from the DNA pol α-primase, can start DNA chains with deoxynucleotides. Its localization is both nuclear and mitochondrial and the enzyme appears to facilitate replication fork progression by acting as a translesion (TLS) DNA polymerase or as a specific DNA primase reinitiating downstream of lesions that block DNA synthesis (see also Section 3.5.5).

The assignment of DNA pols to specific strands during DNA replication in eukaryotic cells has been studied in yeast by using mutant versions of DNA pols δ and ε with specific errors signature (for reviews see Refs. 21 and 22). The studies clearly indicated that DNA pol ε replicated the leading strand and DNA pol δ replicated the lagging strand. Subsequent *in vitro* studies with reconstituted eukaryotic proteins have confirmed the assignment of DNA pol ε to the leading strand and DNA pol δ to the lagging strand.[23] Indeed, DNA pol δ appears to be enriched on the lagging strand and DNA pol ε on the leading strand *in vivo*.[24]

Nevertheless, a recent article by Robert Johnson *et al.*[25] raises the entire question of strand assignments again and concludes that DNA pol δ replicates both leading and lagging strands. The authors attribute to different mismatch repair mechanisms on the leading and lagging strand the apparent lack of contribution of DNA pol δ to leading strand replication (see also Ref. 26 for a review).

A third possibility has been offered in a very recent article.[27] Here the authors also find that DNA pol δ can play an important role in establishing leading strand synthesis, but the scenario they propose is different. They suggest that, at the replication fork and after DNA helicase activation of the origin, DNA pol α synthesizes the primer for the leading strand. Then DNA pol δ, together with PCNA, will take over the 3′ end primer generated by DNA pol α. However DNA pol δ synthesis will be relatively inefficient and it will be taken over by a more efficient DNA pol ε-CMG complex, which will promote the maximal rate of leading strand synthesis.

Clearly more investigation is required to completely unravel the roles of different pols at the eukaryotic replication fork. Of particular importance, to this aim, is to note that all the studies dealing with this issue have been performed with yeast strains or proteins, suggesting caution about interpreting what is really going on in human cells.

The transition from unicellular eukaryotes (such as yeast) to metazoan is reflected by a further specialization of DNA pol functions. The requirement for tissue- or even cell-specific events, hierarchically organized within a rigid temporal program during embryogenesis, as well as the division of labor among different tissues and organs in the adult organism, determined the need of highly regulated and specialized enzymes. As a consequence, all the four DNA pol families typical of eukaryotic cells (A, B, X and Y) were expanded in metazoan through a series of duplication/diversification events, leading to the acquisition of new members with specialized functions (Table 3.1). An example of such expansion is the V(D)J recombination process, an essential component of the immune system. In vertebrate cells, at least three DNA pols from the family X, DNA pol λ, DNA pol μ and Terminal-deoxynucleotidyl Transferase (TdT), are involved in V(D)J recombination at specific steps of the immunoglobulin repertoire development.[28–30] In all other eukaryotes, including multicellular organisms such as plants, with tissues and organs but without a vertebrate-like immune system, only one member of family X (the ortholog of DNA pol λ) is present (Table 3.1). This suggests that in vertebrate cells, the X family of DNA pols underwent a large expansion as a consequence of immune system development.

3.2 Eukaryotic DNA polymerase structure: The "right hand" of the cell

3.2.1 *Common features*

Albeit DNA pols from different families are quite dissimilar in terms of their primary sequence, they fold into a conformation resembling a human right hand composed of three distinct domains designated as palm, thumb, and fingers[31] (see also Chapter 2). The palm subdomain, which contains the catalytically important residues, can be superimposed among members of the A, B, Y and RT families. This subdomain consists of four- to six-stranded β-sheets flanked by two α-helices. Comparison with the structure of the bacteriophage RB69 DNA pol (Figure 3.1) showed that the most conserved residues are located 10 Å from the active site, into three regions that emanate from the palm, the fingers, and the thumb and converge at the catalytic site, forming a continuous conserved surface.

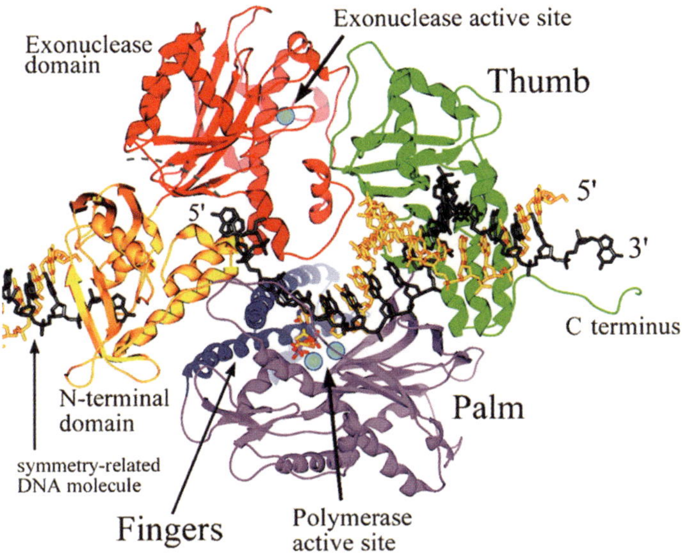

Figure 3.1 **Structure of the B family DNA polymerase from bacteriophage RB69**. The different domains are highlighted in different colors. The right-hand analogy includes the fingers, palm and thumb subdomain, while the exonuclease domain is positioned on top of the palm with its catalytic site facing the polymerase active site. (Reproduced from Ref. 32 with permission)

Unlike the palm, the other two subdomains, thumb and fingers, are unrelated among the different families, but they function similarly using analogous secondary structural elements. The fingers are involved in correctly positioning the template and the incoming complementary dNTP, and the thumb is important in DNA binding and processivity. All DNA pols catalyze the 5′→3′ addition of nucleotides in a template-dependent manner, following a universally conserved reaction mechanism.[33] The phosphoryl transfer reaction is catalyzed by a two-metal-ion mechanism. Two Mg^{2+} ions form a pentacoordinated transition state with the phosphate groups of the incoming nucleotide, through interaction with conserved carboxylate residues in region I and region II. Another common feature of all DNA pols is the obligatory sequential ordered reaction mechanism, whereby the DNA substrate binds first, followed by the dNTP. The sequential association of each substrate with the enzyme is accompanied by a rearrangement of the relative positions of regions within the structural domains. Of particular significance is the concerted movement of the finger subdomains that rotate toward the palm to switch from an "open" to a "closed" conformation, forming the binding pocket for the incoming dNTP. This transition sets the enzyme in its catalytically competent state, properly aligning the active site's carboxylate residues and the α-phosphate of the incoming dNTP for nucleophilic attack and phosphodiester bond formation. In most DNA pols, the rate of the open-to-close transition is slower than the rate of chemical bond formation, thus representing the limiting step of the overall DNA pol reaction.

3.2.2 *Specific features of the different families*

Family A. Family A DNA pol θ is the largest human pol, with a single catalytic subunit of 2590 aa.[34] Recent structural studies have uncovered unique features for this enzyme.[35–37] Three domains characterize DNA pol θ: a superfamily 2 helicase domain at its N-ter, a family A DNA pol domain at its C-ter and a connection domain. Indeed, DNA pol θ is the only human DNA pol which also contains a helicase domain. With respect to the prototypic family A pols, three unique insertions are present in DNA pol θ: insert 1 in the thumb and inserts 2 and 3 in the palm subdomains. The crystal structure of the pol domain showed that insert 2 has a role in mediating the open-to-close transition occurring upon binding the incoming nucleotide, by interacting with the finger domain. In particular, DNA pol θ was shown to "grasp" the primer terminus, thanks to the Arg2254 side chain of insert 2, which contacts the 3′ terminal phosphate of the primer and the O helix of the thumb subdomain, in its close conformation. Since mutations in this domain abolish the ability of DNA pol θ to bypass AP sites and other lesions, it appears that insert 2 is important to stabilize the binding of the enzyme to aberrant primer termini. DNA pol θ possesses also an exonuclease domain, which is, however, not functional. Within this domain, two additional unique insertions (loop 1 and loop 2) are present, which may take interactions with the connecting and/or helicase domains. The structure of the helicase domain of DNA pol θ has been also solved. It comprises five subdomains: two core helicase domains (D1 and D2) and three globular domains: a winged helix domain (aa 514–612), a helical domain (aa 613–789) and a helix-hairpin-helix domain (aa 790–891). The N-ter D1 (aa 1–289) and C-ter D2 (aa 290–513) RecA helicase domains, contain all the canonical helicase motifs (I-VI) of superfamily 2 helicases. The structures in complex with either a non-hyrolyzable ATP analog or with ADP, show a similar mode of nucleotide binding with other helicases. However, comparison of the apo- and nucleotide bound structures failed to reveal large conformational changes as are shown to take place in the case of other helicases. The roles of the other three domains are unclear. Based on sequence and structural homologies with the close homolog Hel308 helicase, they are suggested to be involved in protein–protein or protein–DNA interactions. It has to be noted that, so far, the helicase domain of DNA pol θ was shown to have DNA-dependent ATPase activity, but not DNA unwinding capability. Thus, whether it works as a proper helicase in the context of the full length protein still remains an open question.

Family A DNA pol ν is smaller than DNA pol γ or θ, comprising only 900 aa. A truncated form (aa 192–863) has been crystallized in complex with DNA.[38] DNA pol ν shows three insertions analogous to those found in DNA pol θ. Inserts 1 and 2 are located at the tip and base of the thumb subdomain, while Insert 3 is located in the palm. It also possesses an inactive exonuclease domain (aa 192–416). One striking feature of the structure of the binary complex with DNA is a partially open conformation, where the fingers domain is rotated towards the DNA. In this conformation, an incoming nucleotide may reach the active site, but its base pairing with the templating base is blocked. Thus, additional conformational changes are required to fully open the catalytic site and then subsequently adopt the catalytically competent closed conformation. This unusual open conformation is stabilized by specific interactions among different residues, most notably a hydrogen bond between Tyr 682 and Arg 692. In addition, Lys 679 located near the nascent base pair, may potentially form hydrogen bonds with both a template guanine and an incoming TTP, thus providing an explanation for the preference of T incorporation opposite G, characteristic of DNA pol ν. Another unique feature of DNA pol ν is the extended β-hairpin of Insert 2 (aa 592–606), which creates a cavity absent in other family A DNA pols. This exposes to the solvent the primer strand upstream of the terminal base (residues –2 to –4), while it is usually buried in the DNA binding groove in other DNA pols. This feature suggested that

DNA pol ν may be particularly prone to primer-slippage events. For example, when facing a blocking lesion, DNA pol ν may loop out the primer, exposing the base at position –1 to be used as template for translesion synthesis, thus effectively bypassing the need for using a damaged base as template. Primed loop-out may also be a mechanism for DNA trinucleotide-repeat expansion.

Family B. As listed in Table 3.1, the eukaryotic replicative enzymes DNA pols α, δ and ε consist each of multiple subunits, with the largest one being the one responsible for the catalytic activity. Both from sequence and structural analysis, it appears that the catalytic subunits of eukaryotic DNA pols are composed of a central domain that is evolutionarily very conserved.[1] On the basis of the DNA pol α sequence, six highly similar regions termed I–VI have been identified, whose relative position along the primary sequence is conserved. From the N- to the C-terminus the order is: IV, II, VI, III, I, and V. The highly conserved catalytic site of family B DNA pols is located in the palm, and contains three invariant aspartic residues, two in region I (-YG**D**T**D**S- motif) and one in region II (-**D**xxSLYPS- motif). The -SLYPS- motif is also important for deoxyribonucleoside triphosphate (dNTP) binding. These motifs are absolutely conserved in DNA pol α and δ subfamilies, whereas DNA pol ε has considerably diverged from the consensus, so that the region I motif became -EL**D**T**D**G- and region II has become -**D**xxAMYPN-. The large subunits of DNA pol α, δ, and ε are representative of three distinct subfamilies within the B-type DNA pols. DNA pol δ is the most conserved, with 90% identity between human and mouse and 49% between human and yeast. In comparison, the identity between human and yeast DNA pol α and DNA pol ε is 35% and 39%, respectively. The N-terminal part of DNA pol δ (amino acid residues 1–305), albeit generally poorly conserved, contains three regions of high homology: a nuclear targeting signal (NTS) and the so-called NT-1 and NT-2 regions. The C-terminal part (amino acids 850–1105) is more conserved, with three nearly identical regions termed CT-1 to 3 and a zinc-finger domain (ZnF2), which is 89% identical between human and yeast DNA pol δ. C-terminal zinc-finger domains are also present in α- and ε-like DNA pols. Both the N- and C-terminal parts of DNA pol ε show about 25% identity between human and yeast. However, the C-terminal region of DNA pol ε contains about 1000 extra amino acids found only in the DNA pol ε subfamily members, characterized by an highly acidic region (residues 1918–1948) and a zinc-finger domain (residues 2125–2222). Crystal structures were solved[39] for a truncated form of the catalytic subunit of yeast DNA pol α (aa 349–1258), either alone or in complex with an RNA/DNA hybrid (binary complex) or with the nucleic acid and an incoming nucleotide (ternary complex). The p180 subunit is organized in the typical thumb, palm and fingers subdomains. In addition it bears an exonuclease domain, which is, however, enzymatically inactive due to mutations in the four catalytic aminoacids. Moreover, a β-hairpin, which is conserved in most of the B-family exonuclease domains, in DNA pol α adopts an α-helix conformation (aa 667–693). Its function seems to stabilize the unpaired bases at positions –2 and –3 of the template strand. These features explain why DNA pol α is devoided of proofreading exonuclease activity. Superimposition of the apoenzyme with the binary and ternary complexes, revealed extensive motions not only in the fingers and thumb subdomains, as registered for the majority of DNA pols, but also in the palm. It has been proposed that these conformational changes of the palm, together with thumb motions, may help DNA pol α to disengage from the RNA/DNA primer, allowing its substitution with another replicative DNA pol. The DNA pols δ and ε catalytic subunits both contain a 3′→5′ proofreading exonuclease domain at their N terminus. A truncated form of DNA pol δ large subunit (aa 67–985) has been crystallized in complex with DNA and nucleotide[40] and found similar to the other known crystal structures of exonuclease proficient family B members.[32,41–44] In these structures, the exonuclease domain is folded around a central β-sheet that contains the active site and, together with the palm domain, creates a ring-shaped structure with a central hole, where the template/primer duplex DNA is positioned. As typical of B-family DNA pols, the three catalytic carboxylates are located in the palm subdomain. Interestingly, in the structure of DNA pol δ, besides the two canonical catalytic metals, a third metal ion has been detected coordinating the γ-phosphate of the incoming nucleotide and the carboxylate of Glu802. The catalytic subunit of DNA pol δ possesses an exonuclease domain, which is separated from the pol active site by 45Å. A β-hairpin segment of the exonuclease domain, contacts the major groove of the nascent DNA, acting as a wedge between ds DNA and the ss 5′ terminal end of the template strand. The role of this β-hairpin is to promote the switching of a mismatched primer terminus from the pol to the exonuclease site. This arrangement is similar to what has been observed for RB69 and T4 DNA pols. The crystal structures of two N-ter catalytic fragments of yeast DNA pol ε (aa 1–1228 and 1–1187, respectively), have been reported, in complex with DNA and nucleotide[45,46] and revealed a new domain that is able to encircle the nascent double stranded DNA and that was never observed before in the B-family DNA pols. This domain, consisting of three β-strands and two α-helices, contains a non-catalytic metal binding site, takes close contacts with the phosphate backbone of the DNA template and it is implicated in increasing the processivity of DNA pol ε. Hence, it has been named P (processivity) domain. This domain has been also proposed to regulate the switching between the exonucleolytic and the polymerization modes. In DNA pol δ this is achieved by a contact between the nascent DNA and a β-hairpin element, which helps discriminating between ss DNA (exo mode) and ds DNA (pol mode). However, this element in DNA pol ε is too short to contact the nascent DNA.

The catalytic mechanism of the exonuclease activity involves a metal ion phosphoryl transfer, analogous to the one responsible for polymerization. The 3′→5′ exonuclease activity allows the removal of misincorporated nucleotides, ensuring the high fidelity of DNA synthesis required for faithful genome replication. During DNA synthesis, DNA pols δ and ε repetitively shuttle between a polymerizing and an editing mode. A mismatched base pair prevents the fingers from rotating toward the palm to bind the incoming dNTP. This leaves the 3′ mismatched end available for binding to the exonuclease active site, which removes the wrong nucleotide. During the switch between polymerizing and editing modes, the DNA moves toward the exonuclease active site with a rotation in the double-helix axis. This movement is aided by the tip of the thumb subdomain, which holds contact with the DNA during the movement, guiding it on a path between the two sites. Thus, the balance between polymerase and exonuclease is regulated by a competition for the 3′ end of the primer between the respective active sites. The low-resolution structure of the DNA pol ε holoenzyme from *S. cerevisiae* has been solved by cryo-electron microscopy.[47] This study revealed an extended shape for the DNA pol ε holoenzyme, where the double stranded region of the DNA template is accommodated in a large cleft residing in the "tail" constituted by the three small subunits, whereas the single-stranded region and the primer/template junction is bent while protruding into the active site. The tail seems to be connected to the catalytic subunit through a flexible link, which might help to keep the DNA in the right orientation.

Family X. The eukaryotic family X is composed of four members: DNA pols λ, β, μ and TdT.[48] They all possess a C-terminal β-like domain, comprising the catalytic site. In addition, all members except DNA pol β contain an extended N-terminus, with BRCT-, NLS- and Proline-rich domains. Crystal structures are available for all members of this family, giving plenty of information on their structure–function relationships.[49] The first X-ray crystal structure of the ternary complex of a DNA pol with its DNA substrate and an incoming nucleotide bound in the active site of a family X DNA pol was solved for mammalian DNA pol β (Figure 3.2).[50]

Although the topology of the palm subdomains of family X DNA pols is not homologous to those of other families, the geometry of the active

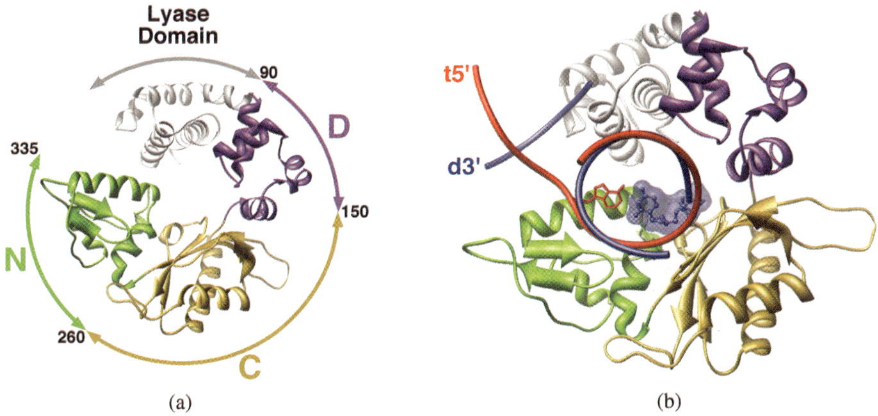

Figure 3.2 **Structure of mammalian DNA pol β.** (a). Shown are the polymerase (colored) and the N-terminal lyase (grey) domains. D, C and N refers to the polymerase subdomains. Numbers indicate the aminoacid residues for the subdomain boundaries. (b). Top view of DNA pol β bound to a one-nucleotide gapped DNA and an incoming dNTP. Red is the template strand, solid blue is the primer strand, shaded blue is the dNTP. The template (t) and downstream (d) ends are indicated (t5′ and d3′).(Reproduced from Ref. 50 with permission.)

site including the position of the substrates (metals, dNTP, and DNA) can be functionally aligned. Three conserved aspartic acid residues (D190, D192, and D256) in the palm subdomain of DNA pol β define the active site and coordinate two Mg^{2+} ions necessary for catalysis. One metal ion is coordinated by the oxygen atoms from all three phosphates of the incoming nucleotide, and by two aspartate residues. The other metal ion interacts with the α-phosphate, the three aspartate residues, and the 3′-OH of the terminal deoxyribose sugar of the primer strand. The catalytic mechanism proceeds via in-line displacement of a proton from the 3′-OH which is transferred to the nearby D256 residue of DNA pol β. Nucleophilic attack of the α-phosphate generates a bipyramidal pentacoordinated transition state. This transition state species leads to inversion of the α-phosphate stereochemistry and release of the pyrophosphate group. The active site then returns to the pre-catalytic state, ready for the next round of catalysis. The chemistry step in DNA pol β is dependent upon large-scale subdomain motions. The incoming nucleotide triggers movement of the carboxyl-terminal thumb subdomain of DNA pol β from its "open" conformation to a "close" one, so that the α-helix N of the thumb subdomain can directly interact with the nascent base pair. The side chain of R283 from α-helix N shifts to interact with the DNA minor groove. Such minor groove interactions help DNA pol β to assess the proper geometry of the incoming nucleotide and the primer terminus. In addition, the residue R258 in the open conformation bonds with catalytic aspartate D192, preventing its interaction with the metal ions. Upon binding of the correct nucleotide, movement of the aromatic residue F272 disrupts the interaction between D192 and R258, allowing D192 to re-orient and coordinate the active site metals. The crystal structure of the catalytic domain of DNA pol λ complexed with a gapped DNA substrate[51] showed that the architecture of the active site is substantially different from the other DNA pols.[52] The interactions between DNA pol λ and the template strand are fewer than for family B DNA pols, and involve only the 3′ terminal base pair. The N-terminal 8 kDa lyase domain makes contacts with the downstream 5′-phosphate. Crystal structures of the DNA pol λ ternary complex with DNA and dNTP, showed that the active site is in a "closed" conformation even prior to dNTP binding. Binding of the nucleotide shifts the template strand and causes the repositioning of a β-hairpin in the palm subdomain, a loop in the thumb subdomain and of three key residues: Y505, N513 and R517. These side chains establish hydrogen bonds with the 3′-terminal base pair and with the incoming nucleotide. Thus, DNA pol λ takes much less interactions with both the template strand and the incoming nucleotide than family A or B DNA pols. Structures of DNA pol λ in complex with a mismatched (G:G) primer/template,[53] showed that the templating G adopted a *syn* conformation, with little or no distortion of either the amino acid sidechains or the DNA backbone. These data indicated that DNA pol λ is able to easily accommodate even a bulky purine:purine mismatch in its active site, providing a rationale for its TLS ability. DNA pol λ shows a very high rate of –1 frameshifts generation. A structure has been solved of DNA pol λ in complex with a frameshift intermediate.[54] The DNA contained an extrahelical template nucleotide upstream the active site; however, even this highly distorted primer did not cause significant rearrangement of DNA pol λ active site. Further crystal structures revealed that, when bound to a 2-nt gap, pol λ scrunches the template and binds the additional uncopied template base in an extrahelical position within a binding pocket that comprises three conserved aminoacids.[55] Thus, DNA pol λ seems to have specialized for the elongation of aberrant primer ends, such as those arising from the presence of lesions, or of primer/template pairs with minimal homology, such as those generated during NHEJ.

Family X comprises the only known teminal deoxynucleotidyl transferases of mammalian cells: TdT, DNA pol λ and DNA pol μ. While DNA pols λ and μ are also capable of template-dependent DNA polymerization, TdT is a strictly template-independent polymerase. This feature can be explained at the structural level[56] by the presence of an extended loop between the β-strands 3 and 4, called Loop I, which occupies the site where normally the primer/template junction is bound, occluding the space for the template strand, so that only a single-stranded DNA can be bound. Loop I is also present in DNA pol μ, and is of a similar length of that of TdT, but has higher conformational flexibility, explaining the limited template-dependent capability of DNA pol μ. Interestingly the equivalent regions in DNA pol β and λ are much shorter and thus less able to interfere with binding of the template strand. A second important determinant to template-independent activity is a histidine residue present in the

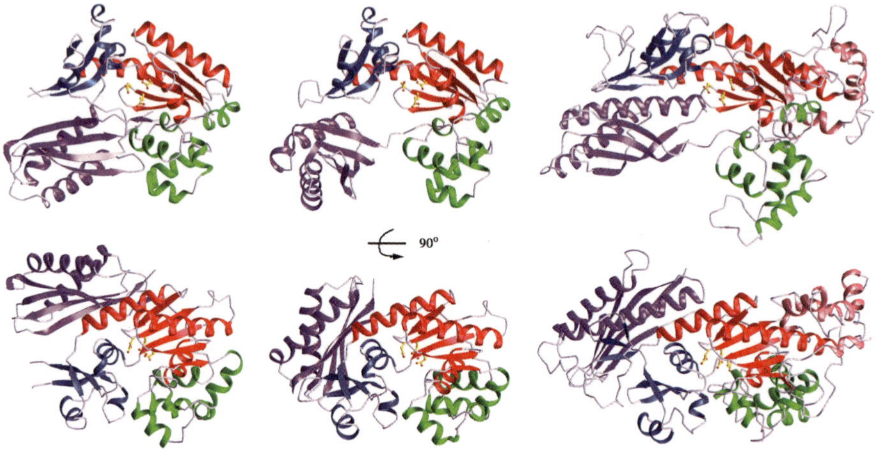

Figure 3.3 **Structures of archaeal and eukaryotic family Y DNA polymerases**. The different domains are highlighted in different colors. Red, palm; blue, fingers; green, thumb; purple, little finger. (Reproduced from Ref. 57 with permission.)

active sites of DNA pol μ (H329) and TdT (H342) but absent in DNA pols β and λ.

Family Y. The Y-family DNA pols η, ι and κ share five conserved motifs (motifs I–V). Motifs I (I/V-D-M/L/F) and III (A/L-SIDE-V/A-F/Y) are responsible for catalysis and coordination of the metal ions. Motif II (Y-x-A-R/K) is involved, together with motif IV, in dNTP binding. Finally, motifs IV and V contain hairpin-helix-hairpin domains. Crystal structures[57–59] of the polymerase domain of archaeal (Dbh and Dpo4) and eukaryotic Y-family members (REV1, DNA pol η, ι, and κ) have been solved (Figure 3.3). They show that the first 250–350 N-terminal residues including the five conserved motifs constitute the catalytic core of these DNA pols, folding in the thumb, palm, and fingers subdomains found in all known DNA and RNA pols. Despite a lack of apparent sequence homology, the palm domain of the A-, B-, and Y-family DNA pols, as well as RT's, are highly conserved. In addition to the palm, DNA pol η contains a small α-helical "wrist" domain. The thumb and fingers are smaller in Y-family DNA pols than in other families. The thumb domain is constituted by a small and stubby α-helical structure, whereas the fingers are shorter than in family B DNA pols and composed of a mixed α + β structure. Intriguingly, the fingers of Y-family DNA pols lack the O α-helix, which is present in high-fidelity DNA pols and has been implicated in selectivity against mismatched nucleotides. At the C-terminus, Y-family DNA pols possess a unique domain of about 100 residues, called the little finger (LF) domain. A common characteristic of these enzymes is the presence of a more spacious active site, with little contacts between the enzyme and the DNA template or the incoming dNTP. This may explain their ability to accommodate bulky DNA lesions which are normally excluded by the tighter active sites of high-fidelity DNA pols.

3.3 Eukaryotic DNA polymerases accessory subunits

Among eukaryotic DNA pols, the three replicative enzymes DNA pols α, δ, ε, the translesion DNA pol ζ of family B and the mitochondrial replicase DNA pol γ from family A are heteromultimeric enzymes consisting of a large (>100 kDa) subunit and up to three smaller subunits (Table 3.1 and Figure 2.8 in Chapter 2). The large subunit contains the DNA polymerase activity and, in the case of DNA pols γ, δ, and ε, also the 3′→5′ proofreading exonuclease. Here we summarize some structural features of the small subunits.

DNA primase subunits (*p58, p49*). DNA pol α is the only family B enzyme showing an additional catalytic activity associated to two of its small subunits. The p58/p49 subunits of the DNA pol α four subunits complex, in fact, are responsible for its associated DNA-dependent RNA pol activity, commonly referred to as DNA primase. Eukaryotes and archaea often possess a DNA primase physically associated with a DNA pol, whereas other organisms using a DNA priming mechanism such as prokaryotes and some DNA viruses, have an independent DNA primase or a multifunctional helicase/primase protein. The structure of the archaeal pRN1, Pfu and human DNA pol-primases have been solved,[60–62] providing support for a common evolutionary ancestor of eukaryotic and archaeal primases (Figure 3.4).

Interestingly, the DNA pol α-associated primase is both structurally and evolutionarily distinct from family B DNA pols.[63] In fact, the

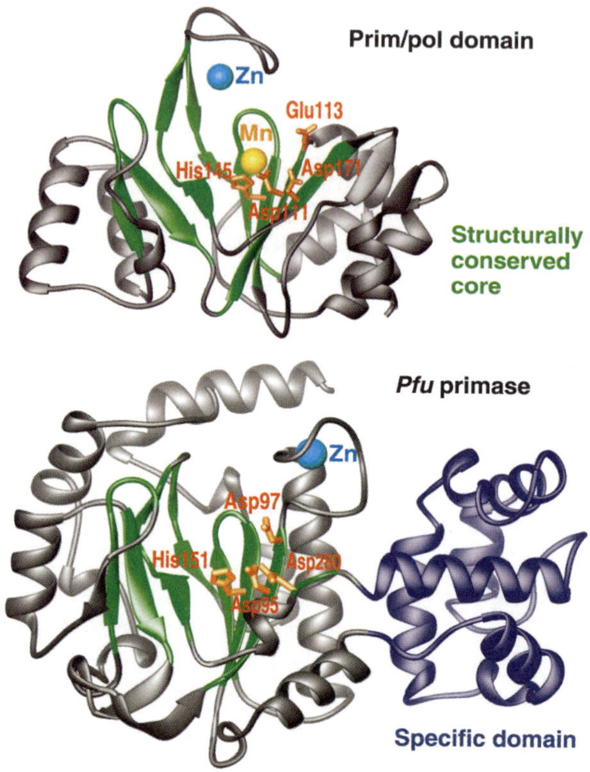

Figure 3.4 **Structures of archaeal DNA primases**. Comparison of the primase-polymerase domain of pRN1 (top) with the archaeal Pfu primase (bottom). Catalytic domains are in silver. Equivalent aminoacids are in orange. Pfu species-specific domain is in blue. (Reproduced from Ref. 60 with permission.)

primases appear related to enzymes cleaving a DNA strand using an internal tyrosine residue in the protein to accept the 5′ end of the cleaved DNA, such as DNA topoisomerases. One hypothesis is that an active site tyrosine might have originally provided to the ancestral protein the hydroxyl group required for the priming reaction, whereas a more sophisticated *de novo* polymerase activity was acquired by the priming protein later in evolution, as a consequence of the development of more complex replisomes. The conserved catalytic core of the heterodimeric archaeo-eukaryotic primase (AEP) family consists of two units packed against each other, bearing six conserved strands (β1–6) and four helices (α1–4). The N-terminal unit contains an AEP-specific α-helix/β-sheet repetitive unit, whereas the C-terminal unit contains an RNA-recognition motif related to the one previously reported in the palm domain of DNA replication and repair DNA pols. The active sites residues (a -DxD- motif in β3, a histidine in β5 and an aspartate in β6) are lying in the space between the two units. In addition, a conserved four-cysteine cluster is present at the C-terminus of the large DNA primase subunit.[64] These four cysteines seem to coordinate an $[Fe_2\text{-}S_2]$ iron-sulfur cluster of the type already reported in several DNA glycosylases, endonucleases and helicases, whose precise functional role is still not completely clear.

The recent crystal structures of the full-length human primase reveal the precise organization of the enzyme, the relative positions of its functional domains and the mode of its interaction with modelled DNA and RNA.[61,62] The structures indicate that dramatic conformational changes in primase are necessary to accomplish the initiation and then the elongation of RNA synthesis. The presence of a long linker between the N-and C terminal domains of p55 provides the structural basis for the bulk of the enzyme conformational flexibility. The three-dimensional structure of the human dimeric primase in complex with the C-ter domain of catalytic subunit of DNA pol α, was also solved. The primase active site resides in the p49 subunit. Nucleotide incorporation is mediated by three catalytic Asp residues 109, 111 and 306. The triphosphate moiety of the incoming nucleotide interacts with the -SGRRG- motif (aa 160–166), while the sugar and base pack against the -HLLK- motif (aa 315–318). The p58 subunit interacts with p49 through an extensively hydrophobic interface. It folds in two subdomains: a larger globular one and a smaller α/β mixed segment. The latter is directly engaging p49, while the larger globular domain reaches over the small subunit's active site, in order to position its Fe-S domain which is essential for initiation, but not for elongation. The DNA pol α catalytic subunit is tethered to the primase, through interactions between its C-ter and a hydrophobic ledge at the N-ter of p58. The main interactions are seen between the -SEVNLSKLF- motif (aa 1447–1455) of DNA pol α and the -ENIS- domain (aa 35–38) and the helix α3 of p49. The structure of the dimeric primase in complex with the DNA pol α p180 and p70 subunits has been recently solved. It shows that the linkers connecting p180 and the C-ter of p58 with the rest of the primosome are endowed with enough flexibility to mediate the different structural rearrangements required for sequential RNA primer synthesis by the primase and its elongation by DNA pol α. In particular, the C-ter domain of p58 packs close to p49 during initiation, then moves away during elongation and, after incorporation of 9 ribonucleotides, mediates the transfer of the RNA primer to the DNA pol α active site. The catalytic site of DNA pol α p180 is blocked by p70 during RNA primer synthesis. Subsequently, movement of the p58 C-ter linker pushes p180 away from p70, freeing its active site and allowing RNA primer binding. A structure of p58 in complex with a RNA/DNA hybrid revealed that this subunit has a high affinity for the primer and it is indeed responsible for holding the 5′ end of the nascent RNA primer throughout the elongation by the p49 and during the transfer to the DNA pol α p180 subunit.[62,65,66]

DNA polymerase γ small subunit (p55). The mitochondrial replicase DNA pol γ is a heterotrimer composed of one large subunit and a dimer of the small p55 subunit. The structure of the small (p55) subunit has been solved.[67,68] The protein shows a remarkable overall structural similarity with the glycyl-tRNA synthetase protein, but the sites essential for catalysis are not conserved. In the solved structure, p55 crystallized as a dimer, with each monomer consisting of three distinguishable domains. Domain 1 is formed by a seven-stranded β sheet, whereas domain 2 contains three strands which, together with the same structural elements of the facing monomer, constitute a six-stranded parallel β sheet which defines the monomer–monomer interface. Domain 3 contains a five-stranded β sheet. Another prominent feature is a four-helix bundle, formed by α-helices D and E in the domain 2 of each monomer. The dimer interface, comprising domains 1 and 2, contains the binding sites for two metal ions, whereas domain 3 contains the region interacting with the DNA pol γ catalytic subunit. The p55 subunit confers processivity to the catalytic subunit of pol γ.[69]

The B subunits superfamily. The second largest subunits of the family B heterotetrameric enzymes DNA pols α, δ and ε are generally designated as B subunits. Multiple sequence analysis of the B subunits of different species revealed extensive sequence similarities, identifying 12 conserved boxes (I-XII).[70] Boxes I-III define an oligonucleotide binding (OB)-fold domain which is separated by a short proline-rich region from the C-terminal calcineurin-like phosphoesterase domain spanning boxes IV to XII. Secondary structure predictions also suggest that all the B subunits share a common fold. The solved structure of the N-terminal part of the DNA pol ε B subunit,[71] revealed a predominant α-helical structure with significant similarity to the α-helical C-terminal domains of the ATPase associated with various cellular activity (AAA+) proteins, a superfamily of ATP-hydrolyzing enzymes. The AAA+ C-terminal domain forms a α-helical bundle, which forms the lid of the ATP binding site and is involved in the functional regulation of the ATPase activity. However, no catalytic activities have been found for any of the B subunits so far. Phylogenetic analysis of the B subunits superfamily members suggested two distinct lineages: the first comprises DNA pol α p70 and DNA pol ε p59, the second comprises DNA pol δ p66 and also some archaeal proteins. One possibility would be that the B subunits originated from a single ancestor before the archaea–eukaryotes separation, with two subsequent gene-duplication events leading to the separate lineages. Alternatively, two B subunits might have existed already prior to the separation between the two kingdoms, and one of them could have been lost during archaea evolution, while a duplication in eukaryotes gave rise to the DNA pol α and ε subunits.

Structural studies have helped to clarify the molecular characteristics of the DNA pol α B subunit (p70).[72] The p70 subunit is organized into two domains: an oligonucleotide/oligosaccharide binding (OB) domain at its N-ter and a phosphoesterase (PDE) domain at its C-ter. This structure, along with sequence homology, suggests that the B subunit evolved from a calcineurin-like phosphoesterase. The two domains interact with an extended interface at the C-ter of p180. The structure of the p50 subunit of pol δ in complex with the N-ter portion of p66 has also been solved. The p50 subunit showed both the PDE and OB domains found in all the B subunits of replicative B-family DNA pols. The N-ter 105aa fragment of p66 interacting with p50 adopts a winged helix-turn-helix fold. Association between the two subunits is stabilized by hydrophobic and electrostatic interactions, as well as by 37 hydrogen bonds.

3.4 Eukaryotic DNA polymerase fidelity: Structural and functional aspects

Fidelity is the ability of a DNA pol to discriminate between a complementary base pair and a mismatch.[73] The current models predict that fidelity arises from the combined effects: the size of the nascent base pair and the plasticity (or tightness) of the active site, which affect the nature of different kinetic checkpoints along the reaction pathway.[74–76] Overall, both kinetic and structural data are in agreement with the so called "induced fit" model, where fidelity is the result of a complex series of small-scale conformational changes occurring upon substrate binding, during stabilization of the ternary complex (pre-chemistry) and in phosphodiester bond formation (chemistry). These steps involve interaction of the enzyme with different chemical entities including water molecules, salts and metal ions. The microscopic rates of these small-scale motions can vary among different enzymes, thus resulting in different free energy profiles and, ultimately, in different combinations of rate limiting steps for correct vs. incorrect incorporation.[77] In addition, both the hydrogen bonding pattern between the two bases of the terminal base-pair and the interactions between the DNA pol side chains and the DNA minor groove in the proximity of the template-primer junction, play an important role in the discrimination process. However, eukaryotic DNA pols exhibit an extraordinary wide range of fidelity values. As summarized in Table 3.2, the error rates for single-base substitutions due to DNA pols can vary from 10^{-1} to $>10^{-6}$, depending on the polymerase, the mispair, and the local sequence context. Such a wide variation of fidelity values correlates with the existing structural differences among the DNA pols of the different families (see Section 4.2), which, in turn, have evolved to fulfil specific functions. In particular, the discovery of the highly error-prone Y-family TLS enzymes called for a broader definition of fidelity. This "unified theory" of DNA pol fidelity states that fidelity can be defined as the ability of a given DNA pol to generate correct base pairs irrespectively from the presence of chemical modifications on the bases (either templating or of the incoming nucleotide). This implies that the substrate-induced conformational changes and the

Table 3.2 Biochemical properties of eukaryotic DNA polymerases

DNA pol	Template[a]	Metal	Catalytic efficiency (k_{cat}/K_m, M^{-1} s^{-1})		Error rate ($\times 10^{-5}$)[a]		TSL
			3′OH	**dNTPs**	**exo[−b]**	**exo[+b]**	**Lesion (Effect)**
α	gap	Mg^{2+}	4.3 × 106	8.4 × 105	10	n.a	**T-T dimer** (blocking) **8-oxo-G** (miscoding) **AP site** (blocking)
δ	p/t	Mg^{2+}	3.3 × 10^{8}[c]	0.37 × 10^{6}[c]	10	0.1	**T-T dimer** (blocking) **8-oxo-G** (miscoding) **AP site** (miscoding)
ε	p/t	Mg^{2+}	2.1 × 10^{8}	0.5 × 10^{6}	20	0.5	**T-T dimer** (blocking) **8-oxo-G** (miscoding) **AP site** (blocking miscoding)
ζ	mismatch	Mg^{2+}	n.a.[g]	1.4 × 10^{5}	100	n.a.	**T-T dimer** (blocking miscoding) **8-oxo-G** (blocking)
γ	DNA/ RNA hybrid	Mg^{2+}	4.5 × 10^{9}	5 × 10^{7}	10	0.1	**T-T dimer** (miscoding) **8-oxo-G** (miscoding) **AP site** (blocking)
θ	p/t	Mg^{2+}	2.3–4.2 × 10^{6}	0.4 × 10^{6}	1000[d]	n.a	**T-T dimer** (blocking) **8-oxo-G** (n.a.)
ν	p/t	Mg^{2+}	n.a.	1.6–5.1 × 10^{6}	350[e]	n.a	**5S Thymine glycole** (coding)
β	gap	Mn^{2+}	5 × 10^{8}	1.7 × 10^{6}	10	n.a	**T-T dimer** (miscoding) **8-oxo-G** (miscoding) **AP site** (miscoding)
λ	gap	Mn^{2+} Mg^{2+}	5 × 10^{5}	1.2–2.7 × 10^{6}	100	n.a	**T-T dimer** (n.a.) **8-oxo-G** (miscoding) **AP site** (miscoding)
μ	gap	Mn^{2+} Mg^{2+}	n.a.	6.4 × 10^{4}	10	n.a	**T-T dimer** (coding[f]) **8-oxo-G** (miscoding) **AP site** (miscoding)

(Continued)

Table 3.2 (*Continued*)

DNA pol	Template[a]	Metal	Catalytic efficiency (k_{cat}/K_m, M^{-1} s^{-1}) 3′OH	dNTPs	Error rate ($\times 10^{-5}$)[a] exo[-b]	exo[+b]	TSL Lesion (Effect)
η	p/t	Mg^{2+}	6×10^6	4.3×10^5	3500	n.a	**T-T dimer** (coding) **8-oxo-G** (miscoding) **AP site** (blocking)
ι	p/t	Mg^{2+}	1.3×10^7	1.2×10^5	72000[e]	n.a	**T-T dimer** (blocking) **8-oxo-G** (miscoding) **AP site** (blocking)
κ	p/t	Mg^{2+}	n.a.	1.4×10^4	1000	n.a	**T-T dimer** (blocking) **8-oxo-G** (miscoding) **AP site** (blocking)

a. p/t, primer-template. Errors are base substitution rates. b. For the enzymes with intrinsic 3′→5′ exonuclease activity, the indicated exo⁻ values are those for the correspondent exo⁻ mutant proteins, where available. c. In the presence of its auxiliary factor PCNA. d. For the preferred G:T/T:T mismatches. e. For the preferred T:G mismatch. f. coding, neither miscoding nor blocking. Template information is retained. g. n.a., not available.

network of hydrogen bonding between the bases of the nascent base pair, are all influenced by the particular nature of the DNA pol's active site and by the physico-chemical features of the pairing bases. Thus, for example TLS DNA pols should not be viewed as generally low-fidelity enzymes, but rather as DNA pols specialized in catalyzing error-free bypass of a certain set of DNA lesions.

As shown in Table 3.2, replicative and repair DNA pols show high and moderate fidelity, respectively. This reflects the need for replicative enzymes to catalyze hundreds of thousands of incorporation events, whereas the activity of repair enzymes is limited to very short stretches of DNA (typically 1–10 nucleotide incorporation events). Indeed, for longer DNA tracts to be synthesized (for example such those arising in nucleotide excision repair, mismatch repair or recombination) replicative DNA pols are usually employed. The high fidelity of replicative DNA pols of the A and B families is further enhanced by the presence of intrinsic

3′→5′ exonuclease or "proofreading" activities (Table 3.2). Exonucleases catalyze the excision of nucleoside monophosphates (dNMPs) from 3′- or 5′-DNA termini.[78] To be qualified as a "proofreader" an exonuclease needs to satisfy precise criteria: it has to act unidirectionally starting from the 3′ end of a DNA strand; it needs to be non-processive in its DNA degradation; it has to act on single-stranded DNA as the optimal substrate; it must show preferential excision on a mispaired primer terminus. A non-processive or distributive manner of nucleotide hydrolysis is required in order to allow for the immediate subsequent elongation of primers during error correction. At the opposite extreme of the spectrum of fidelity variation stands the DNA pol family Y. While the differences in fidelity between replicative and repair enzymes are moderate, the eukaryotic TLS enzymes DNA pol η, ι and κ, can display up to 10^5–10^6- fold increased error rates with respect to replicative enzymes (Table 3.2). The high "infidelity" of TLS DNA pols is a direct consequence of their ability to support DNA synthesis past lesions that cannot be negotiated by the high-fidelity replicative DNA pols. Structural data offered the possibility to understand the molecular basis for the great tolerance towards various kinds of DNA damage showed by TLS DNA pols.[79,80] As already mentioned in Section 3.2, the thumb and fingers subdomains are smaller in Y-family DNA pols than in the other families. The thumb domain is constituted by a small and stubby α-helical structure, whereas the fingers are shorter than in family B DNA pols and composed of a mixed α + β structure. Intriguingly, the fingers of Y-family DNA pols lack the O α-helix, which is present in high fidelity DNA pols and has been implicated in selectivity against mismatched nucleotides. In addition, a common characteristic of these enzymes is the presence of a more spacious active site, with very little contacts between the enzyme and either the DNA template and the incoming dNTP. This may explain their ability to accommodate even bulky DNA lesions, which are normally excluded by the tighter active sites of high-fidelity DNA pols. Family Y DNA pol ι incorporates G 10-fold better than the correct A opposite a template T.[81] The molecular basis for this relaxed nucleotide selectivity lies in the fact that DNA pol ι favours Hoogsteen base pairing, rather than the classical Watson-Crick.[82] In the ternary structure of DNA pol ι, the templating A is switched to its *syn*-conformation through interaction with L62 and Q59 in the fingers subdomain, making hydrogen bonds between its N7, N6 atoms and the N3 and O4 of the incoming dTTP in *anti*-conformation. Hoogsteen base pair can be formed also between *syn*-guanine and *anti*-thymine, thus explaining the nucleotide incorporation properties of DNA pol ι. Another Y-family member, DNA pol κ, is very proficient in extending mismatched primers. Its active site is narrower than that of DNA pol η. In addition, this pol shows a unique domain at its N-ter, called N-clasp, which encircles the primer terminus. These structural features confer higher stability to DNA pol κ when bound to a mismatched primer terminus, thus allowing its extension. The Y-family enzyme Rev1 can be considered as a G template-specific DNA pol, since its preferred substrates are a templating G and an incoming dCTP nucleotide. The molecular basis of such peculiar properties has been elucidated by the resolution of the ternary complex of Rev1 with DNA and an incoming dCTP.[83] In both yeast and human ternary complexes, the incoming dCTP and the templating G do not pair with each other. The template G is evicted from the DNA and bound to the specific G-loop subdomain of Rev1. Similarly, the incoming dCTP is preferentially sampled among the four possible dNTPs, thanks to its selective pairing with an Arg residue in the N-digit domain of the enzyme. These striking structural features allow easy and error-free bypass of bulky N^2-guanosine adducts.[84,85]

In conclusion, comparison of replicative DNA pols with the specialized TLS enzymes of the Y-family suggests that the latter are more tolerant towards lesions in the DNA template thanks to a more open active site, in combination with the ability to use non-Watson-Crick base pairing. These structural features, in turn, determine the very low ability of these enzymes to discriminate against mismatched nucleotides, thus explaining their low fidelity.

3.5 Biochemical and functional properties of the different eukaryotic DNA polymerases

3.5.1 *Family A DNA polymerases*

The eukaryotic members of family A, whose prototypic enzyme is *E. coli* DNA pol I, are prokaryotic-related DNA pols specialized in the replication of the small circular DNA of organelles (mitochondria, plastids), termed DNA pol γ-like enzymes. However, in vertebrate cells, this family comprises two additional enzymes whose functions are still not completely known, DNA pol θ and ν.

DNA polymerase γ. The DNA pol γ-like enzymes are conserved in all eukaryotic cells.[86] They are named after the animal cell mitochondrial DNA replicase, DNA pol γ,[87] but related enzymes are also present in plants for the replication of plastid DNA[88–91] and in Apicomplex protozoa for the duplication of plastid-like elements.[92,93] While replication and repair of the nuclear genomes in eukaryotic cells is accomplished by at least 14 different DNA pols (Table 3.1), the duplication and maintenance of the genetic information stored within the small organelles genomes solely rely on DNA pol γ-like enzymes,[94] and, as recently identified, by the prim/pol enzyme.[19] A notable exception being the trypanosome *T. brucei* kinetoplasts, which possess several plastid DNA pols unrelated to DNA pol γ.[95] In animal cells, the average mitochondrial (mt) DNA content present in each cell represents 1% of the total DNA content. Accordingly, DNA pol γ accounts for about 1–5% of the total cellular DNA pol activity. Homology modeling allowed the reconstruction of the three-dimensional structure of the p140 catalytic subunit of the human enzyme[94] (Figure 3.5). In Chapter 5 we will discuss the processive human mitochondrial DNA synthesis and the disease related DNA pol mutations as elaborated by the structure of the DNA pol γ.

DNA pol γ is an heterotrimer composed of a single large subunit and a dimeric small subunit.[96,97] According to cDNA sequence,[98] the predicted molecular weights for the large catalytic subunit range from 115 kDa (*S. pombe*) to 143 kDa (*S. cerevisiae*), with the human protein being 140 kDa. Primary sequence alignments of the large subunits from different organisms, reveal the DNA pol family A highly conserved motifs A, B and C, as well as the exonuclease consensus motifs I, II and III. In addition, six DNA pol γ-specific elements (1–6) were recognized. Elements 5 and 6 are localized at the C-terminus of the DNA pol domain, flanking motif C, whereas elements 1–4 are in the long linker region between the exonuclease and the polymerase domains. This region spans 337 aa in yeast and 482 aa in humans, being twice as big as the corresponding linker regions of the other family A DNA pols. Mutations in the elements 1–4 have been

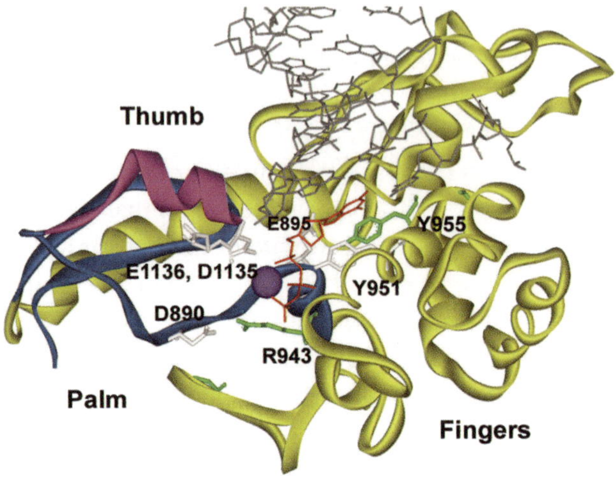

Figure 3.5 **Structural model of the human DNA polymerase γ large subunit**. The thumb, fingers, and palm subdomains are magenta, yellow, and blue, respectively. Primer and template DNA strands are gray, and the incoming ddGTP is red. The purple sphere is an Mg^{2+} ion. (Reproduced from Ref. 94 with permission.)

shown to affect DNA binding and processivity, as well as the interaction with the small subunit. The 55 kDa small subunit of human DNA pol γ (p55) forms a stable complex with the p140 large subunit in solution. Electron microscopy analysis of the DNA pol γ holoenzyme, showed that one p55 subunit interacts more closely with the p140 subunit, while the other small subunit is more exposed to the solvent.[99] Crystal structure of the DNA pol γ holoenzyme, showed that it is an heterotrimer, with the large catalytic subunit associated to a dimer of the small subunit. However, the p140 subunit contacts only with the proximal p55 within the dimer, mainly through hydrophobic interactions between the C-ter region of the small subunit and the helix L (aa 543–558) of the large subunit. The structure of a ternary complex of the DNA pol γ holoenzyme with DNA and the incoming nucleotide, showed significant conformational changes upon DNA binding, which increased the interactions between the p140 and the distal p55 monomer of the small subunits dimer. This interaction occurs within the accessory-interacting determinant (AID) subdomain, which is present in the spacer domain, between the polymerase and exonuclease portions of the enzyme. Overall, the ternary complex adopts a more compact structure. The biochemical properties of DNA pol γ are summarized in Table 3.2. Human DNA pol γ has a 3′→5′ proofreading exonuclease activity[100] and can utilize a variety of nucleic acid substrates, including homopolymeric RNA/DNA hybrids such as poly(rA)/oligo(dT), the typical substrate of viral RT's.[101–103] Dissociation studies showed a strong interaction between the catalytic and accessory subunits of *Drosophila DNA* pol γ. The accessory subunit was important to maintain the structural integrity and catalytic efficiency of the holoenzyme.[104] In fact, the *Drosophila* catalytic subunit alone, showed very low catalytic activity (2% of that of the holoenzyme), especially at near physiological salt concentrations. Intrinsic processivity of the catalytic subunit is low (< 30 nt), however, biochemical studies showed that the p55 subunit stimulates the processivity of DNA pol γ up to >1000 nt. Thanks to its 3′→5′ exonuclease activity, DNA pol γ is intrinsically a high-fidelity enzyme, as expected for an enzyme responsible for DNA replication. Again, the small subunit was shown to stimulate the proofreading activity of DNA pol γ. As noted above, DNA pol γ is responsible also for the repair of mtDNA damages. Since mitochondria are the sites of oxygen metabolism, mtDNA is subjected to high levels of endogeneous oxidative damage. The main repair pathway operating the removal of oxidized bases in the DNA is the base excision repair (BER). When acting on nuclear DNA, BER relies on the activity of two specialized family X DNA pols: DNA pol β and λ. These enzymes in fact possess intrinsic 5′-deoxyribose phosphate (dRP) lyase activity, which is essential for BER. Remarkably, DNA pol γ also has dRPlyase activity,[105,106] which is consistent with its role as the main mtDNA BER DNA pol.[107,108] This activity has been shown to be stimulated by the small subunit.

DNA polymerase θ. DNA pol θ is encoded by the human POLQ gene on chromosome 3, with an expressed mRNA of 8.5 kb encoding for a protein of 2592 amino acids. Human POLQ gene is homologous to the *D. melanogaster* mus308 protein product, a putative polymerase–helicase involved in repair of interstrand crosslinks. Flies carrying a mutation in this gene showed sensitivity to DNA crosslinking agents, elevated frequency of chromosomal aberration and altered DNA metabolism.[109] The C-terminal region of DNA pol θ contains the canonical polymerase motifs A, B, and C found in the family A type of DNA pols. The N-terminal region contains an ATPase-helicase domain.[34] Full-length human DNA pol θ was expressed in a baculovirus system and purified as a recombinant protein. Before recombinant DNA pol θ was available, a DNA pol was purified from HeLa cells, based on its ability to replicate a depurinated DNA template and showing all the characteristics of a family A enzyme. Initial classification, based on its cross-reactivity with polyclonal antibodies raised against the C-terminal half of DNA pol θ, identified it as human DNA pol θ[110] Contrary to the recombinant enzyme,[111] the natural enzyme possessed a fidelity comparable to replicative enzymes, thanks to a 3′→5′ exonuclease activity and showed efficient bypass of an AP site. Interestingly, SDS-PAGE and gel filtration analysis under native conditions suggested that the native enzyme was a heterotrimer composed of three subunits of 80, 90 and 100 kDa, respectively. The 100 kDa subunit was responsible for the DNA pol catalytic activity, as revealed by *in situ* activity gel analysis, and possibly represented a truncated form of the large subunit of DNA pol θ, containing the DNA pol domain alone. Whether any of the two other subunits were contributing the observed 3′→5′ exonuclease activity is presently unknown. The main biochemical properties of recombinant DNA pol θ are summarized in Table 3.2. The enzyme showed polymerase activity on nicked double-stranded DNA and on a singly primed DNA template. In addition, DNA pol θ exhibited a single-stranded DNA-dependent ATPase activity. DNA pol θ is very efficient in bypassing an AP site,[112] inserting A with 22% of the efficiency of a normal template, and continuing extension past the lesion as efficiently as with a normally paired base. Consistently with these findings, DNA pol θ was found to extend quite efficiently A:G, A:T and A:C mismatches.[113] DNA pol θ was unable to incorporate nucleotides opposite a cyclobutane pyrimidine dimer or a (6–4) photoproduct, but when combined with DNApol ι, an enzyme that can insert a base opposite UV-induced lesions, it could perform extension from the resulting primers allowing complete bypass of a (6–4) photoproduct.[113] According with its putative role as an extender in TLS, recombinant DNA pol θ showed error rates 10- to 100-fold higher than other family A members when copying undamaged DNA, depending on the sequence context.[111] Its average base substitution rate is comparable to the very inaccurate family Y DNA pol ι. While a role for DNA pol θ in TLS was inferred from *in vitro* data, experiments with knockout cells revealed a possible role for this enzyme in BER.[114] In fact, targeted disruption of the POLQ gene in chicken DT40 cells resulted in hypersensitivity to oxidative base damage, but not to UV or cisplatin treatments. This phenotype was synergistically increased by concomitant inactivation of the major BER enzyme, DNA pol β. The observation that DNA pol θ possesses a dRP-lyase activity, similarly to the major BER enzymes DNA pols β and λ, may be consistent with a role in BER. A role has been ascribed to DNA pol θ in somatic hypermutations of Ig genes.[115] Knockout mice have revealed that a deficiency in family Y DNA pol η caused an 80% reduction of mutations at A/T basepairs, suggesting that DNA pol η is the major enzyme generating A/T mutations during somatic hypermutation. The absence of DNA pol θ also resulted in measurable (20%) decreases of both A/T and C/G mutations. DNA Pol θ/Pol η double null mice, however, did not show a further decrease of A/T mutations as compared with DNA pol η knockout mice, suggesting that these two enzymes function in the same genetic pathway in the generation of these mutations. The current hypothesis is that the mismatch-extender DNA pol θ might be responsible for the extension of mispaired termini opposite an A or T generated by DNA pol η. The fact that inactivating DNA pol θ causes only a 20% reduction of mutations implies that in its absence another general mismatch extender can take up its role. Currently, at least three other DNA pols have been shown to efficiently extend mispaired termini: family B DNA pol ζ, family X DNA pol λ and family Y DNA pol κ, thus providing a likely backup for the inactivation of DNA pol θ.

The presence of a 5′dRP lyase activity might be more important in what appears to be the most important role of DNA pol θ: repair of DNA double strand breaks. In fact, it has been recently suggested that DNA pol θ has a prominent role in nonhomologous end joining (NHEJ).[116] This observation is supported by the crystallographic structures of ternary substrate complexes of DNA pol θ, showing how the enzyme grasps the primer to bypass DNA lesions or to extend poorly annealed DNA termini to mediate end joining.[36] This primer-grip strategy, coupled with pol θ ability to dimerize, could enable the enzyme to direct DNA synthesis from unmatched primer termini during repair of double strand breaks (see Ref. 117 for a review).

Evidence has been presented that pol θ plays a role in determining the timing of replication in human cells, suggesting that the enzyme functions during the earliest steps of DNA replication and influences the time of replication initiation.[118] It has been reported that DNA pol θ up-regulation perturbs DNA replication, promotes genetic instability and is associated with poor survival in breast cancer.[119] Therefore DNA pol θ overexpression may be a promising prognostic marker for breast cancer.

DNA polymerase υ. The gene for human DNA pol υ[120] comprises 24 exons within 160 kilobases of genomic DNA and encodes a protein of 900 amino acid residues. The C-terminal half of the protein shows 33% identity (48% similarity) with *E. coli* DNA pol I and 29% identity (43% similarity) with human DNA pol θ. Numerous alternatively spliced transcripts for DNA pol ν have been found in human cell lines and expression analysis by northern blotting and *in situ* hybridization showed highest expression of the full-length protein in human and mouse testis. The main biochemical properties of DNA pol ν are summarized in Table 3.2. DNA pol ν shows moderate processivity (100 nucleotides) and strand displacement activity. Recombinant DNA pol ν proved to be a low-fidelity enzyme incorporating T opposite template G with a frequency higher than most other DNA pols.[121] Kinetic studies showed that its misincorporation efficiency is also strongly influenced by the template sequence. In the context of highly error-prone sequences, DNA pol ν shows a strongly reduced catalytic rate for correct incorporation, with respect to sequence contexts where fidelity is higher, while its rate for mismatch formation is unaffected. Pre-steady state kinetic studies also confirmed that DNA pol ν can catalyze the formation of both correct base pairs or mismatches with similar efficiences. DNA pol ν is particularly efficient and accurate in translesion synthesis past a 5*S*-thymine glycol,[122] a common product of oxidative damage to DNA which blocks DNA synthesis by most DNA pols.

3.5.2 *Family B DNA polymerases*

Family B DNA pols are multisubunit enzymes (Table 3.1) which, with the exception of DNA pol ζ, are responsible for the replication of nuclear DNA in all eukaryotic organisms.

DNA polymerase α-primase. DNA pol α consists of four subunits (Table 3.1) whose primary structures are conserved in all eukaryotes, consistent with the primary role of this enzyme in the initiation of leading strand DNA replication and in the repeated priming of Okazaki fragments during lagging-strand DNA replication.[1] These functions rely on the unique properties of the p55 and p48 subunits, which constitute a dimeric DNA-dependent RNA polymerase called DNA primase.[123–125] Thus, the DNA pol α holoenzyme possesses two distinct yet functionally interacting active sites: one in the large subunit responsible for DNA synthesis and one in the dimeric primase, responsible for RNA synthesis. The biochemical properties of DNA pol α-primase are summarized in Table 3.2. DNA pol α has been among the first discovered eukaryotic DNA pols (See Chapter 1).[126,127] Even though no detailed pre-steady state enzymatic analysis is available for this enzyme, steady state kinetic studies combined with site directed mutagenesis, allowed the identification of important residues for catalysis and fidelity, as well as to the definition of a plausible model for the functional interaction between the polymerase and primase activities.[128–130] Site-directed mutagenesis helped to define a group of amino acid residues important for catalysis. Asp1002 and Asp1003 of Motif I are thought to coordinate the catalytic metal ion, with the participation of the Thr1003 side chain. In the -DFNSLYPII- Motif II sequence, Asp860 and Tyr865 are involved in direct interaction with the incoming dNTP, whereas Ser867 is involved in primer binding. Finally, Lys950 of Motif III has been implicated in binding the α-phosphate of the incoming dNTP. A low resolution structure of the p180-p70 complex has been obtained by electron microscopy. The three dimensional reconstruction shows the presence of two "lobes": one formed by the C-ter domain interacting with p70, the other formed by the catalytic core of the large subunit. These two lobes do not contact each other, but are connected by an arm, which keeps the C-ter-p70 lobe over the palm subdomain of p180. A certain flexibility has been noted for this arm, suggesting that its motion may be part of the catalytic cycle of DNA pol α. Kinetic studies with the primase subunits revealed that neither of the purified subunits displayed primase activity alone.[130] The p55 subunit stabilized the binding by p48 of the initiating ribonucleotide (usually adenine) to form the initiation complex. After initiation, the p48 subunit alone was able to extend the RNA primer and its Arg304 residue was inferred to be important for both nucleotide binding and primer elongation. The p55 subunit increased both the affinity for the incoming nucleotide and the catalytic rate of p48. Immunoprecipitation experiments indicated that only the p55 subunit directly contacted the large (p180) subunit of DNA pol α. Finally, it was shown that the DNA pol α-primase four subunits complex does not dissociate from the primer during the transition from RNA to DNA synthesis, suggesting that the switch from the primase to the polymerase active site occurs by an intramolecular mechanism.

The p70 subunit has no detectable enzymatic activity but it is essential for cell viability in *S. cerevisiae* and has been proposed to function at the initiation of S phase and to participate in intra-S phase checkpoint regulation.[131] Consistent with this hypothesis, p70 has been shown to interact with the origin recognition complex proteins Orc1 and Orc2. In addition, p180 and p70 were shown in yeast to bind the telomere capping proteins Stn1 and Cdc13, respectively. Similarly, human DNA Pol α p180 and p70 subunits were localized at telomeres and found to interact with telomere-binding proteins, suggesting a role of DNA pol α-primase in coupling telomere elongation with completion of DNA replication.[132]

DNA pol α is also subjected to a complex regulation during the cell cycle, both at the transcriptional and post-transcriptional level (see Section 4.6).

DNA polymerase δ. DNA pol δ is a heterotetrameric enzyme (Table 3.1), formed by one large, catalytic subunit (p125 in humans) and three smaller subunits (p66, p50 and p12) devoid of catalytic activities. DNA pol δ is the most conserved DNA pol of the B family.[133] Sequence identity within the large subunit is 90% among mammalian species and 49% between mammalian and yeast genes. Genetic experiments showed that the exchange of domains within the catalytic subunit between *S. cerevisiae* and *S. pombe* resulted in active chimeric enzymes, whereas the intact catalytic subunit of either *S. cerevisiae* (POL3) or mouse was unable to complement the disruption of the correspondent gene (Pol3) in *S. pombe*, suggesting, that species-specificity is mediated by different domains within the large subunit.[134] In human cells the mRNA for the large subunit has been shown to increase at the G1/S border and the protein itself has been shown to be phosphorylated in S-phase, suggesting a cell cycle-dependent regulation both at the transcriptional and post-transcriptional levels.[135] On the contrary, in *S. pombe* neither mRNA nor protein levels were found to change during the cell cycle for the large DNA pol δ subunit.[136] The second subunit of human DNA pol δ (p66) showed a 33% identity with the corresponding *S. pombe* gene (Cdc27).[137] Remarkably, the corresponding *S. cerevisiae* gene (POL32) showed very limited sequence similarity with either the *S. pombe* Cdc27 and the human p66 gene.[138] POL32 was not essential, but mutants in this gene showed hypersensitivity to DNA damaging agents and synthetic lethality when combined with conditional mutations in either the catalytic subunit (POL3) or the p50 subunit (POL31).[139] The third subunit of human DNA pol δ (p50) showed 25–28% sequence identity to the corresponding genes of *S. cerevisiae* (POL31) and *S. pombe* (Cdc1). The POL31 gene is allelic to HYS2, an essential gene whose conditional mutants were found to show hypersensitivity to the DNA replication inhibitor hydroxyurea.[139] The fourth subunit of human DNA pol δ (p12) shows 25% identity with the corresponding *S. pombe* gene (Cdm1).[140,141] So far, no homologous genes have been found in *S. cerevisiae*, whose DNA pol δ is still considered to be a three-subunit enzyme. The Cdm1 gene is not essential in *S. pombe*, but interfering-RNA mediated inhibition of the expression of the mammalian homolog p12 in murine cells, caused a significant decrease in proliferation upon growth factor stimulation, suggesting a role of p12 in DNA replication. The precise quaternary structure of DNA pol δ along with the exact functional roles of the different subunits is still not fully understood. In fact, different sub-complexes of DNA pol δ have been isolated and shown to be active.[141–143] An important characteristic of DNA pol δ is its functional stimulation by the auxiliary protein PCNA. Again, this stimulation appears not to be restricted to a unique form of DNA pol δ.[144–150] Reconstitution of human DNA pol δ subcomplexes have shown that the trimeric form lacking p12 exhibits less than 10% of the activity of the tetrameric holoenzyme. However, an alternative trimeric complex lacking p66 appeared to be fully active in the presence of PCNA. Similarly, a dimeric form composed of p125 and p50 has been shown to be active and fully responsive to PCNA. All three small subunits have been shown to interact with PCNA, whereas a direct interaction between the p125 catalytic subunit and PCNA is still controversial.[151–155] In addition, p50 binds tightly to p125, p12 binds both p50 and p125, whereas p66 interacts with p50. Interestingly, a damaged-induced degradation of p12 has been shown to lead to a trimeric form of DNA pol δ with altered catalytic properties,[156,157] suggesting that this small subunit might play a role in the functional regulation of the DNA pol δ holoenzyme. The biochemical properties of DNA pol δ are summarized in Table 3.2. The main distinguishing feature of DNA pol δ is its dependence on the auxiliary factor PCNA for processive DNA synthesis. The catalytic activity of the DNA pol δ holoenzyme in the absence of PCNA is almost undetectable on primer/template substrates containing long stretches of single strand DNA such as poly(dA)$_{200}$/oligo(dT)$_{20}$ or singly primed M13 DNA. This prevented a full characterization of this enzyme, until more sensitive assays were developed, employing 5'-radioactively labeled oligonucleotide primer/template substrates, enabling the detection of single nucleotide incorporation events. These assays, coupled to pre-steady state kinetic analysis techniques, allowed detailed studies of the biochemical properties of DNA pol δ.[158–166] The best characterized form is the dimeric p125–p50 subcomplex from calf thymus. This enzyme, similarly to most DNA pols, showed a steady state rate-limiting step following product formation determined by its dissociation rate from the DNA substrate, and a pre-steady state rate-limiting step (probably a conformational change) preceding the chemical step. The main effect of PCNA was to decrease the dissociation rate of DNA pol δ from the primer/template, consistent with its role as a processivity clamp (see Section 4.5 in this Chapter and also Chapter 2). Robust genetic and biochemical evidence supports a main role of DNA pol δ as the lagging strand replicase.[17,167] In addition, DNA pol δ has been shown to participate in several repair pathways including BER, nucleotide excision repair (NER), mismatch repair (MMR), homologous recombination (HR)[168–171] and cell cycle checkpoints.[172–175] Consistently, all these pathways also require the presence of PCNA.[176] Very recently, it has also been shown that DNA pol δ replicates both DNA strands after homologous recombination-dependent fork restart.[177]

DNA pol δ possesses the ability to perform strand displacement-coupled DNA synthesis, being able to invade a double-stranded DNA tract located downstream with respect to the direction of polymerization, such as in the case of gapped DNA structures, and to displace several nucleotides.[178] This activity is very important for the proposed role of DNA pol δ as the lagging strand replicase. According to the current model for Okazaki fragment synthesis, the action of DNA pol δ is not only critical for the extension of the newly synthesized Okazaki fragment, but also for the displacement of an RNA/DNA segment of about 30 nucleotides on the pre-existing downstream Okazaki fragment, in order to create an intermediate flap structure which is the target for the subsequent action of the Dna2 endonuclease and the flap endonuclease 1 (Fen-1).[179–182] This process has the advantage of removing the entire RNA/DNA hybrid fragment synthesized by the DNA pol α/primase, potentially containing nucleotide misincorporations due to the lack of a proofreading exonuclease activity of DNA pol α/primase. This results in a more accurate copy synthesized by DNA pol δ. The intrinsic strand displacement activity of DNA pol δ, in conjunction with Fen-1, PCNA and replication protein A (RP-A), has also been proposed to be essential for S-phase specific long-patch BER pathway.[183]

DNA polymerase ε. DNA pol ε is a multisubunit enzyme, which, similarly to DNA pol α-primase and DNA pol δ, is conserved in all eukaryotic organisms.[1] In all eukaryotic organisms studied so far, DNA pol ε is an heterotetramer composed of one large catalytic subunit and four small accessory subunits in a 1:1:1:1 stoichiometry.[184–189] The N-terminal half of the large subunit contains all the canonical polymerase and exonuclease boxes.

In addition, the DNA pol ε large subunit has an extended C-terminal domain, with very little homology to any other DNA pol, and which is likely involved in DNA pol ε specific protein–protein interactions. The second largest subunit (p59 in human cells) interacts with the C-terminus of the DNA pol ε large subunit and is devoid of enzymatic activity. However, it likely fulfills important roles in the determination of the correct (i.e. catalytically active) quaternary structure of the DNA pol ε holoenzyme, as suggested by the strong mutator phenotype of yeast strains harboring mutations in the corresponding gene. Similarly, the p17 and p12 subunits lack any detectable enzymatic activity and also interact with the C-terminal part of the large subunit. In addition, they interact with each other through a histone-fold motif, homologous to those found in histone H2A and H2B. As expected for a protein involved in many cellular processes, DNA pol ε has been shown to physically interact with several partners, such as the checkpoint protein TopBP1, the cell cycle regulator

Mdm2, the sister chromatid cohesion protein Trf4, the replication and repair proteins PCNA and DNA ligase I.[190–193] In addition, DNA pol ε was shown to interact with RNA polymerase II. The biochemical properties of DNA pol ε are summarized in Table 3.2.[194] DNA pol ε synthesizes DNA in a processive manner, even though its processivity can be stimulated by PCNA. It also possesses a 3′→5′ proofreading exonuclease activity, which is responsible for its high fidelity of DNA synthesis. Due to these properties, along with the compelling evidence for a role in DNA replication, DNA pol ε since its discovery has contended to DNA pol δ the role of the leading strand replicase. Genetic and biochemical evidence seems to have convincingly demonstrated that DNA pol ε is indeed acting at the leading strand, whereas DNA pol δ shares with DNA pol α-primase the task of replicating the discontinuous lagging strand.[23,195] However, a recent article[25] raises the entire question of strand assignments again and concludes that pol δ replicates both leading and lagging strands.

As is the case of DNA pol δ, also DNA pol ε is involved in many biochemical pathways besides DNA replication, such as BER, NER, HR (see Ref. 189 and references therein) and TLS.[196,197] A role for DNA pol ε has also been shown in cell cycle control, mainly based on genetic evidence in yeast.[198,199] For example, mutants in the C-terminal part of the catalytic subunit of *S. cerevisiae* (POL2), resulted in hydroxyurea sensitivity, due to a defect in S-phase checkpoint activation. It was shown that this effect was dependent on the interaction of DNA pol ε with the essential helicase SGS1. Also the second largest subunit (DBP2 in *S. cerevisiae*) has been linked to cell cycle regulation through a phosphorylation/dephosphorylation cycle, which was dependent on the main yeast S-phase cyclin dependent kinase CDC28 and on the CDC14 phosphatase.[200] It has been shown that the Dpb2 subunit of yeast DNA pol ε interacts with the Psf1 component of the GINS complex, which is part of the Cdc45-MCM-GINS (CMG) replicative helicase.[201] In a recent study, the single-particle EM technique was used to directly image a yeast replisome.[202] Contrary to the expectations, DNA pol ε appears to be positioned ahead and not behind of the CMG helicase, suggesting that an unforeseen function of DNA pol ε may underlie the unexpected replisome architecture.

Genetic evidence also points to an important role of *S. cerevisiae* DNA pol ε in gene silencing.[203] C-terminal POL2 mutants as well as mutations in the third subunit (DPB3), were shown to impact on gene silencing through alterations of the chromatin status. It is likely that proper holoenzyme assembly, rather than the catalytic activity, is essential for this function of DNA pol ε. Another link between DNA pol ε and chromatin structure comes from the discovery that its third subunit in human cells (p17) is also a component of the chromatin remodeling complex CHRAC.[187]

DNA polymerase ζ. Most information about DNA pol ζ comes from the budding yeast *S. cerevisiae*.[204,205] *S. cerevisiae* DNA pol ζ consists of two subunits: the catalytic polymerase subunit encoded by the REV3 gene and an accessory subunit encoded by the REV7 gene. Homologs of both REV3 and REV7 have been also identified in most eukaryotes (Table 3.1). Two transcripts have been identified for the human large subunit (Rev3L). One form encodes a 3130-amino-acid protein and the other a 3052-amino-acid protein, due to alternative splicing and translation from an internal initiation codon. The predicted size of mammalian Rev3L is 350 kDa, making it the largest eukaryotic DNA pol known so far, twice the size of *S. cerevisiae* REV3 (173 kDa). Deletion of mouse REV3L gene causes embryonic lethality, revealing the essential role of DNA pol ζ in mammalian development.[206] Yeast REV3 and mammalian Rev3L proteins share three regions of high sequence similarity: i) the N-terminal region, showing 36% identity between human and yeast, with several motifs also present in DNA pol δ; ii) the region involved in Rev7 binding, which is 29% identical between human and yeast; iii) the DNA pol domain, showing 39% identity and bearing the six conserved B-family DNA pol motifs and two zinc-finger motifs. Rev3L is expressed in many different tissues and cell lines. A 3D structural model of yeast DNA pol ζ based on electron microscopy reconstruction has been proposed.[207] With regard to its structural architecture, DNA pol ζ shows some similarities with high fidelity replicative DNA pols. The human Rev7 cDNA encodes a 211-amino-acid protein with a predicted molecular weight of 24 kDa. It shows 23% sequence identity to the corresponding *S. cerevisiae* REV7. Similar to its yeast counterpart, human Rev7 not only interacts with Rev3L, but also with several additional proteins such as Cdc20 and Mad2, suggesting that this protein has additional roles besides acting as a component of the DNA pol ζ holoenzyme. For example, human Rev7 was shown to bind the transcription factor ELK1 and a phosphorylated form of the MAP kinase JNK.

JNK phosphorylates ELK1, leading to the activation of several cell proliferation genes. Additional studies suggest that Rev7/Mad2 complex can inhibit the anaphase promoting complex through interactions with Cdc20. Genetic evidence in yeast and generation of knockdown mammalian cell lines by interfering RNA supported a role of DNA pol ζ in spontaneous mutagenesis and recombination-dependent DNA repair. Consistent with an essential role, DNA pol ζ –/– mice showed severe embryonic lethality. The biochemical properties of DNA pol ζ are summarized in Table 3.2. Contrary to DNA pol δ or ε, DNA pol ζ lacks a 3′→5′ exonuclease activity, and has relatively low fidelity and processivity (approximatively 2–3 nt). The best characterized *in vitro* activity of DNA pol ζ is its role as a general extender DNA pol during TLS.[57] DNA pol ζ has a remarkable ability in elongating aberrant primer/template termini arising from incorporation of nucleotides opposite several DNA lesions. Its efficiency in extending mismatched primer termini is >1000 fold higher than that of DNA pol α. Biochemical studies with *in vitro* reconstituted systems showed that a combination of DNA pol ζ with either a specialized TLS DNA pol such as DNA pol ι or even with the p60 and p55 subunits of the replicative DNA pol δ, can lead to efficient bypass of several blocking lesions such as cis–syn thymidine dimers or abasic sites.[208–210] These observations led to the two-polymerase model for TLS, whereby efficient bypass of certain DNA lesions can be achieved through the coordinated action of a specialized inserter DNA pol, responsible for the incorporation of a nucleotide opposite the lesion, and a general extender DNA pol, able to elongate the resulting aberrant primer terminus. DNA pol ζ, thus appears to be the principal extender enzyme in TLS. The *in vitro* lesion bypass activity of DNA pol ζ has been shown to be stimulated by PCNA,[211] even though no PCNA binding motifs exist in either Rev3L or Rev7 and no direct physical interaction between DNA pol ζ and PCNA has been shown. However, this stimulation may result from the interaction of PCNA with the DNA pol δ subunits associated to DNA pol ζ.[212]

Appropriate repair of DNA lesions and the inhibition of DNA repair activity at telomeres are crucial to prevent genomic instability. Very recently, Rev7 has been identified through functional genetic screening as a novel factor controlling DNA repair activities at mammalian telomeres.[213] Furthermore, Rev7 has been shown to counteract DNA-double strand break resection and to affect PARP inhibition.[214]

3.5.3 *Family X DNA polymerases*

Family X consists of specialized small DNA pols whose primary function is to fill gaps of one to a few nucleotides during DNA repair. However, while in lower eukaryotes as well as in plants only one X-family DNA pol is present, vertebrates have four members (TdT, DNA pols λ, μ and β) with different specific functions in a variety of processes, such as DNA repair, V(D)J recombination and translesion synthesis.[1]

DNA polymerase λ. DNA pol λ is the product of the gene POLL, which is localized on chromosome 10 in humans and chromosome 19 in mice and comprises nine exons spanning a genomic segment of 8 kb. The human DNA pol λ contains 575 amino acid residues, and the murine 573, accounting for a polypeptide of 67–70 kDa.[215,216] The structure of the C-terminal part of pol λ shows the typical folding with palm, finger, thumb and 8 kDa subdomains. The first 230 N-terminal amino acids of DNA pol λ contain a BRCT and a proline/serine rich domain.[217] The C-terminal part of human DNA pol λ shows 32–33% sequence identity to the corresponding region of human DNA pol β. Differences in the substrate specificity of mammalian family X DNA pols are proposed to partly depend on a loop (loop 1) upstream the polymerase active site. To investigate if this is the case for DNA pol λ, a variant of the human polymerase was constructed in which nine residues of loop 1 were replaced with four residues from the equivalent position in DNA pol β. Indeed, it was found that loop 1 modulates DNA pol λ fidelity by controlling dNTP-induced movements of the template strand and the primer terminal 3′-OH as the enzyme changes from an inactive to active conformation.[218] DNA pol λ is expressed at the highest level in testis, ovary, and fetal liver and in small amount in other tissues. The main biochemical properties of DNA pol λ are summarized in Table 3.2. DNA pol λ is endowed with multiple catalytic activities: in addition to template dependent DNA pol activity, it possesses terminal deoxynucleotidyl transferase, dRPlyase and polynucleotide synthetase activities. Moreover, it efficiently synthesizes DNA from an RNA primer.[219,220] The terminal deoxynucleotidyl transferase activity of DNA pol λ prefers single-stranded DNA and is not active on 3′-recessed DNA ends such as those in primer/template junctions. It is, at least in part, sequence specific and preferentially incorporates pyrimidine nucleotides.[221] DNA pol λ has been purified from calf thymus tissue and the natural enzyme was shown to efficiently bypass AP sites.[222] This is also true of the recombinant enzyme, which also introduces –1 deletions by a primer misalignment mechanism.[223] Thanks to its dRP-lyase activity, DNA pol λ can efficiently participate in BER of uracil-containing DNA in an *in vitro* reconstituted reaction. A remarkable feature of DNA pol λ is its preference for Mn^{2+}, over Mg^{2+}.[223] *In vitro* experiments suggested that DNA pol λ participates in double strand break DNA repair (DSBR) via NHEJ.[224,225] In addition, biochemical evidence supports a prominent role of DNA pol λ in the repair of oxidative DNA lesion such as 2-hydroxy-adenine and 8-oxo-guanine in cooperation with the auxiliary proteins RP-A and PCNA[226–228] (Figure 3.6).

Knock out mice for DNA pol $\lambda^{-/-}$ are fertile, and homozygous breeding has been performed up to the third generation without any noticeable problem. However, detailed analysis of immunoglobulin locus rearrangements showed that immunoglobulin heavy chain junctions from DNA pol λ-deficient animals have shorter length with normal N-additions, thus indicating a role of this enzyme during heavy chain rearrangement.[29,229] DNA pol $\lambda^{-/-}$ mouse embryonic fibroblasts were more sensitive to oxidative DNA damage and this phenotype was stronger when combined with inactivation of the related enzyme DNA pol β, suggesting backup functions for these two proteins in the repair of DNA oxidative lesions.[230,231] Interestingly, it has been shown that DNA pols λ, and β can bypass abasic site in the presence of the replicative DNA pol ε, likely by initiating the 3′OHs created at the lesion by the arrested DNA pol ε. However in the case of DNA pols λ and β, this bypass requires the presence of a DNA gap downstream from the product synthesized by the DNA pol ε, and the optimal gap for efficient bypass is different for the two DNA pols.[232]

DNA pol λ is unique in possessing all the enzymatic activities (DNA polymerization, terminal transferase, dRPlyase) which are individually present in the other X family members. Indeed, beside vertebrates all other eukaryotic organisms possess only one family X DNA pol, which, based on sequence homology, is the ortholog of DNA pol λ, which thus appears to be the most conserved DNA pol from this family throughout evolution.[233,234] Biochemical characterization of DNA pol λ from *Oriza sativa* and *Arabidopsis thaliana*, has revealed a remarkable functional similarity of the plant enzymes with their animal cell counterpart,

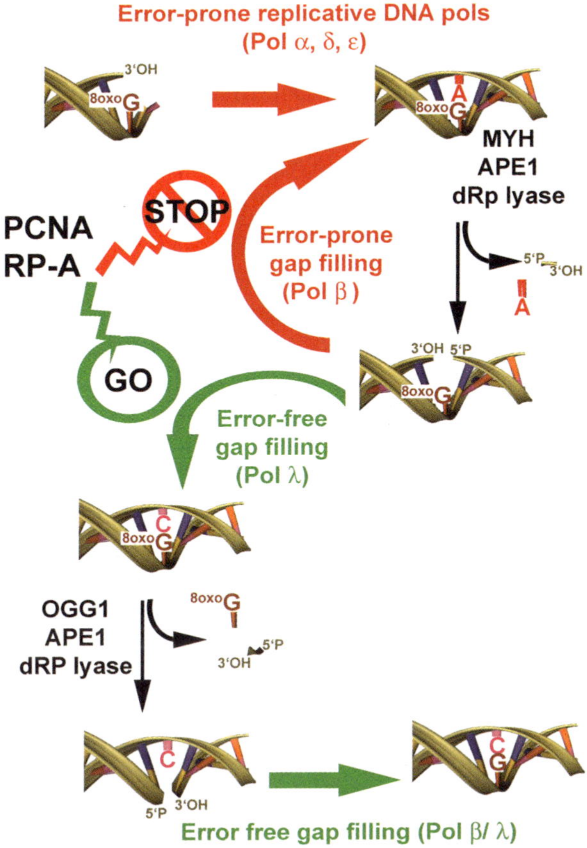

Figure 3.6 **A model for the coordinated action of DNA pol λ, PCNA and RP-A in the repair of A:8-oxo-G mismatches**. After the removal by the glycosylase MYH of the mismatched A incorporated opposite the 8-oxo-G lesion, the resulting gap has to be filled by a specialized DNA pol, able to incorporate the correct C opposite the lesion. PCNA and RP-A restrict the access to this gapped intermediate by DNA pol β, which would result in a futile cycle (red arrow) favoring the action of the more faithful DNA pol λ (green arrow). Once a "correct" C:8-oxo-G base pair is formed, this can be recognized by the OGG1-dependent BER pathway, leading to the replacement of the oxidized guanine with an undamaged one.

including the ability to participate in BER and to bypass abasic sites and 8-oxo-guanine.[235,236] Remarkably, plant DNA pol λ also shows terminal transferase activity. In animal cells, this enzymatic activity has always been linked to the V(D)J recombination process during immunoglobulin gene rearrangement, but its presence in plant cells argues for a more fundamental and evolutionarily conserved function. Human DNA pol λ is also regulated at the post-transcriptional level during the cell cycle (see Section 3.6).

DNA polymerase μ. Human DNA pol μ is encoded by the POLM gene located on chromosome 7, containing 11 exons over a 12-kb genomic fragment, encoding a protein of 494 amino acids (55 kDa).[237] Among the pol X family, it has the strongest homology to TdT with 42% amino acid identity. The human DNA pol μ possesses both DNA template dependent and independent (terminal transferase) activity and prefers Mn^{2+} over Mg^{2+} as a divalent cation. The main biochemical properties of DNA pol μ are summarized in Table 3.2. The DNA pol μ terminal transferase activity requires a template/primer junction for optimal efficiency, contrary to TdT and DNA pol λ, which are active on single-stranded DNA, and preferentially incorporates pyrimidines.[238] In the presence of Mn^{2+}, DNA pol μ behaves as a strong mutator, lacking base discrimination during nucleotide insertion on a DNA template-primer structure.[239,240] DNA pol μ is expressed predominantly in peripheral lymphoid tissues, although basal levels were detected in most other tissues analysed. DNA pol μ knock out mice have a specific alteration in the $IgM^{-/-}$ to $IgM^{+/+}$ transition in bone marrow.[241] Ig light chain gene rearrangement was impaired at the levels of Vκ-Jκ and Vλ-Jλ junctions. These alterations led to a profound defect in the peripheral B cell compartment, which caused 40% reduction in the splenic B cell fractions. DNA pol μ appears, therefore, as a key factor contributing to Ig gene rearrangement. DNA pol μ can efficiently extend a primer terminus located opposite a lesion containing a single N-2-acetyl-aminofluorene (AAF) adduct.[242] It can also efficiently extend mismatched bases by annealing the primer end with a microhomology region in the template 3 or 4 nt downstream resulting in frameshift DNA synthesis with a high frequency.[240] DNA pol μ efficiently bypasses also small DNA lesions such as 8-oxoguanine, AP-site and 1,N^6-ethenoadenine as well as bulky DNA lesions other than AAF, such as + and — trans-anti-benzo[α]pyrene-N^2-dG, cis-Pt and T-T dimer, through a deletion mechanism.[243,244] Deletion resulted from primer realignment before TLS. Bypass of T-T dimers is achieved mainly in a error-free manner with AA incorporation. A novel bypass mechanism over AP-lesions by DNA pol μ was recently described as a template-dependent sequence independent nucleotidyl transferase activity. By this mechanism DNA pol μ would use its terminal transferase activity to extend primers whose 3′terminal nucleotides are located opposite an AP site, in a manner that does not depend on the sequence of the template.[238] Biochemical data as well as the physical interaction between DNA pol μ, and the end-joining factors Ku and XRCC4-LigIV also support a role of DNA pol μ in NHEJ.[245] A major challenge concerning the NHEJ pathway is that chromosome breaks have dirty end structures, making them difficult to repair. A recent article shows that the two mammalian DNA pols μ and λ have a pivotal role in helping resolve such ends. Each is proficient in different contexts and has a different impact on repair fidelity.[246] Moreover, crystal structures of DNA pol μ with a 2-nt gapped DNA substrate explain how the unpaired base is accommodated in the active site, and yield insights into the behavior of this DNA pol within the context of NHEJ in double-strand break repair.[247]

DNA polymerase β. The human gene encoding DNA pol β spans 33 kb and contains 14 exons. DNA pol β is composed of a single 39-kDa polypeptide containing 335 amino acid residues consisting of two domains connected by a protease-sensitive hinge region. The N-terminal 8 kDa domain contains the dRPlyase activity and binds single-stranded DNA, whereas the C-terminal 31 kDa domain carries the polymerase active site. The main biochemical properties of DNA pol β[248] are summarized in Table 3.2. DNA pol β prefers gapped DNA substrates bearing a 5′-phosphate on the downstream strand in the gap.[249] In the absence of a downstream strand, DNA binding affinity of DNA pol β is strongly reduced. In contrast, high-affinity DNA binding does not require the primer strand. This suggests that DNA pol β binds short DNA gaps first through interaction with the 5′-phosphate on the downstream portion of the gap.[250] Only when such a first contact is established, an available 3′-OH end upstream of the gap will be bound and extended. This is consistent with the processive gap-filling DNA synthesis observed with short (<5 nucleotide) DNA gaps. A prominent role of DNA pol β in DNA repair has been known for more than 35 years.[251,252] In particular, DNA pol β is the main enzyme involved in the gap-filling step of short-patch BER.[253] This fundamental function of DNA pol β is very conserved, since the eukaryotic enzyme can substitute for the *E. coli* BER enzyme DNA pol I. It appears that DNA pol β also acts in meiosis, DSBR, neurogenesis and TLS.[254] Experimental evidence for these additional functions comes from different studies. First, DNA pol β has been implicated in meiotic events associated with synapsis and recombination.[255] Second, DNA pol β was found to bypass *in vitro* d(GpG)-cisplatin adduct placed on codon 13 of the human ras gene in a highly mutagenic way.[256,257] This correlates with the observation that upregulation of DNA pol β in certain cell types might lead to genomic instability and cancer.[258,259] Finally, mice carrying a targeted disruption of the DNA pol β gene had growth retardation and died of a respiratory failure immediately after birth.[260] Apoptotic cell death was observed in the developing central and peripheral nervous system in newly generated post-mitotic neurons of DNA pol β knock-out mice, indicating an essential role of this enzyme in neurogenesis. Indeed, DNA pol β is the only DNA pol expressed in post-mitotic neurons, likely due to its essential role in the repair of endogenous DNA lesions, such as oxidized bases and uracil.

Terminal deoxynucleotidyl transferase (TdT). TdT belongs to the pol family X, together with DNA pols β, λ and μ.[1] TdT and DNA pol μ are the closest derivatives with 41% identity at the amino acid level and are derived from a common ancestor, which presumably was a template-dependent polymerase.[30] TdT seems to be a recent acquisition through evolution. This is in agreement with the established role of TdT in the V(D)J recombination leading to the generation of the diversity of the antigen-binding region of immunoglobulins and T-cell receptors. This process has only evolved in higher organisms endowed with an immune system, such as vertebrates, and thus has appeared late in evolution. TdT is a nuclear enzyme found predominantly in developing B and T cells in the bone marrow and thymus. It is also found in fetal human B- and T-cell precursors. Human, bovine, mouse and rat TdT genes consist of 13 exons and their expression is restricted to B- and T-cell precursors through a complex regulatory mechanism.[261,262] Three alternative splice variants for TdT have been found in humans and other mammalian organisms, which are termed TdTL1, L2 and S, where L stand for Long (529 aa) and S for Short (509 aa) protein isoform.[263–265] In normal human pre-B and T cells, only TdTS or TdTL2 are expressed. However, this ordered pattern of expression is altered in transformed cells, so that TdTL1 can be detected in transformed B and T cell lines. This expression pattern is also different in evolution, so that in mice TdTL1 is expressed, whereas TdTL2 is not present. This evolutionary constraint to express only one of the two possible long isoforms, can be related to the fact that, whereas the short isoform can perform terminal nucleotide addition, the TdTL1 and 2 forms have a 3′→5′ exonuclease activity, which is also important for V(D)J recombination. Thus, it appears that only one exonuclease-containing TdT is required during B- and T-cell development, in combination with TdTS. Several conserved domains can be recognized in the TdT sequence.[266] From the N- to the C-terminus they are: a nuclear localisation signal (NLS), a BRCA1 C-terminal (BRCT) domain, exonuclease (Exo) motifs I and II (Exo I and II), a DNA-binding domain (HhH), the Pol X domain and the Exo III box. Three cAMP kinase consensus sites for threonine phosphorylation have been also found, but it is not known whether TdT is phosphorylated *in vivo*. The main biochemical properties of TdT are summarized in Table 3.2. TdT can add nucleotides in a non-templated manner to the 3′-OH end of almost any DNA, either single-stranded or double-stranded with both 3′-recessed or protruding ends. Also blunt end substrates can be utilized, but with lower efficiency. The nature of the nucleotide to be incorporated and the sequence of the 3′end of the DNA substrate can also influence the efficiency of the reaction. Finally, TdT activity is profoundly influenced by the metal activator: Co^{2+} seems to be preferred *in vitro*, but Mn^{2+} and Mg^{2+} are likely utilized *in vivo*. TdT has been shown to interact with several proteins: the DNA-dependent protein kinase (DNA-PK), the DNA repair and recombination protein Pso4, the replication and repair factor PCNA and the transcription factor p65 homologous protein TdIF1.[264] However, the significance of many of these interactions is still not clear. The best studied is the interaction with the heterotrimeric DNA-PK, because of its importance in the V(D)J recombination pathway. TdT binds to all three DNA-PK subunits (Ku70, Ku86 and DNA-PK catalytic subunit) and this interaction can regulate TdT activity by limiting both the length and composition of nucleotide additions.[267,268] The crystal structure of TdT with an annealed DNA synapsis containing one micro-homology (MH) base pair and one nascent base pair has been solved. This structure reveals how the N-terminal domain and loop 1 of TdT cooperate for bridging the two DNA ends, providing a templating base in *trans* and limiting the MH search region to only two base pairs. A network of ordered water molecules is proposed to assist the incorporation of any nucleotide independently of the *trans* templating base.[269]

3.5.4 *Family Y DNA polymerases*

As outlined in Section 3.3, replicative DNA pols are endowed with a high degree of accuracy during DNA synthesis, but at the cost of being unable to bypass lesions that significantly distort the DNA double helix. TLS is carried out by specialized DNA pols, particularly of the Y family, able to synthesize DNA past lesions that block replicative DNA pols, but at the cost of a much higher error rate. As mentioned in Section 3.2.2, Y family DNA pols share a common basic structural arrangement. Nevertheless, a large degree of functional divergence has occurred, rendering the different members of this family highly specialized for the bypass of specific lesions.

DNA polymerase η. DNA pol η in yeast is encoded by the RAD30 gene whose mutation causes hypersensitivity to UV-induced mutagenesis. Similarly, mutation in the human gene for DNA pol η is responsible for the so called variant form of the disease Xeroderma pigmentosum (XP-V), characterized by high-level of UV-induced mutagenesis and skin cancer incidence.[270–272] The strong phenotype caused by DNA pol η deficiency as well as biochemical studies on the purified enzyme, suggested that its main role *in vivo* is to replicate through *cis-syn* TT dimers, a strongly blocking DNA lesion commonly induced by UV light, in an error free manner, inserting an adenine opposite each "T" in the dimer.[273–275] Beside cys–syn TT dimers, DNA pol η seems to be involved in error free bypass of other major DNA lesions.[276] It bypasses the 6–4 TC or CC photoproducts, by preferentially inserting a guanine opposite the 3′C of the dimer. DNA pol η is also fairly proficient in error-free bypass of an 8-oxo-guanine lesion, inserting the correct "C" in about 75% of the cases. On the other hand, DNA pol η is strongly inhibited by O^6-methyl guanine and abasic sites and by bulky lesions such as benzo[a]pyrene 7,8-diol 9,10-epoxide, butadiene epoxide and the acrolein-derived adduct γ-hydroxy-1N2-propano-deoxyguanosine. The inhibition of DNA pol η by lesions disrupting the Watson-Crick base pairing geometry can be accounted for by the particular active site architecture of DNA pol η.[277] The ability of DNA pol η of bypassing cyclobutane pyrimidine dimers derives from its particularly wide active site, able to accommodate both bases of the dimer. As a consequence, most of the structural constraints present in the active site of replicative DNA pols and responsible for the induced-fit fidelity mechanism (see Section 3.3) are absent in DNA pol η. Thus, its ability to discriminate between correct and incorrect base pairs might strictly rely on the intrinsic stability of Watson-Crick hydrogen bonding. In agreement with this hypothesis, kinetic and thermodynamic analysis showed that DNA pol η binds the correct nucleotide only three- to five-fold better than the incorrect one,[278–280] corresponding to a free energy difference ($\Delta\Delta G$) of approx. 1.0 kcal/mol, close to the one provided by Watson-Crick hydrogen bonding of base pairs. The known biochemical properties of DNA pol η are summarized in Table 3.2. Immunolocalization studies in cells overexpressing DNA pol η at low levels, showed that this enzyme is strictly associated to the sites of ongoing DNA replication (the so called replication factories) in the nucleus during S-phase.[281] Distinct domains in the C-terminal part of the protein have been shown to be important for its subnuclear localization, including a bipartite nuclear localization signal. The recruitment of DNA pol η to replication factories and its ability to "switch" with the replicative enzymes in case the replication fork is stalled at a lesion, are thought to be mediated by its interaction with the essential replication and repair protein PCNA.[282] DNA pol η can interact with PCNA through the conserved PCNA-interacting protein (PIP) domain I/L-x-x-F-F, localized at the extreme C-terminus (aa 624–628) of the protein.[283] The addition of a single ubiquitin chain at Lys 164 of PCNA stabilizes this interaction.[284] A novel ubiquitin-binding motif has been characterized in DNA pol η and called UBZ.[285] The UBZ domain of DNA pol η lies close to the PIP-domain (aa 548–577) and is characterized by a mononuclear zinc-finger domain of the C2H2 type. This domain has been shown to bind ubiquitin, to increase DNA pol η affinity for the ubiquitinated form of PCNA and to be important for DNA pol η recruitment to replication factories.[286] DNA pol η itself is ubiquitinated, even though by enzymes different from those responsible for PCNA ubiquitination. Ubiquitinated DNA pol η cannot bind to ubiquitinated PCNA, suggesting a possible regulatory function for this modification. However, the physiological significance of DNA pol η ubiquitination is still under investigation. DNA pol η seems to be involved in homologous recombination.[287–289] Targeted disruption of the DNA pol η gene in chicken DT40 cells, a B-cell derived population, which can perform immunoglobulin gene rearrangements through a recombination-mediated gene conversion mechanism, resulted in a 2- to 6-fold reduction of these gene conversion events. Gain-of-function experiments proved that this defective phenotype could be rescued by reintroducing a catalytically active form of DNA pol η. However, immunoglobulin gene rearrangements in human cells do not require gene conversion and XP-V patients do not show immune system deficiencies. *In vitro* reconstituted systems with purified enzymes or XP-V cell extracts suggested that DNA pol η might be involved in the resynthesis step starting from D-loop recombination intermediates. Indeed breast cancer proteins PALB2 and BRCA2 stimulate DNA pol η in recombination-associated DNA synthesis at blocked replication forks.[290] Human DNA pol η also appears to be required for common fragile site stability during unperturbed DNA replication[291] by preventing the appearance of under-replicated DNA in mitosis.[291]

From a medical point of view, it has been observed that enhanced expression of DNA pol η contributes to cisplatin resistance of ovarian stem cells[292] suggesting that targeting this DNA pol might be exploited as strategy to increase the efficacy of cisplatin treatment.

DNA polymerase ι. Human DNA pol ι is encoded by the POLI gene located on chromosome 18 and accounting for a protein of 715 amino acids.[293] DNA pol ι has a low processivity and lacks an intrinsic 3′→5′ exonuclease. It can insert nucleotides opposite lesions present in DNA but is very inefficient in extending the resulting template/primers.[294,295] DNA pol ι physically interacts with and is stimulated by the auxiliary protein PCNA.[296] It also physically interacts with DNA pol η and co-localizes with DNA pol η into replication factories at sites of DNA damage.[281] The DNA pol η-ι interacting region has been mapped to their respective ubiquitin-domains (UBZ in DNA pol η and UBM1 and UBM2 in DNA pol ι) and, in the same work, it has been shown that ubiquitination of either DNA pols is a prerequisite for their physical and functional interaction.[297] Thus the ubiquitin status of DNA pol η and ι plays a key regulatory role in these enzymes functions and their posttranslational regulation.[298]

The fidelity of DNA pol ι is high opposite template A with error frequencies of 10^{-4} to 10^{-5}. Opposite templates G and C, DNA pol ι has a low fidelity, with error frequencies ranging from 10^{-1} to 10^{-2}. However, DNA pol ι has an unprecedented low fidelity opposite template T, with error frequencies $\geq 10^{-1}$, so that DNA pol ι preferentially inserts the incorrect G 10-fold more efficiently than the correct A. The known biochemical properties of DNA pol ι are summarized in Table 3.2. As reported in Section 3.3, DNA pol ι, in contrast to all known DNA pols, uses Hoogsteen base pairing instead of the regular Watson-Crick one.[57,82] This unusual property of DNA pol ι is related to its function in promoting replication through purine adducts. Only the template purines can form Hoogsteen-hydrogen bond with incoming pyrimidines, thus Hoogsteen base pairing can serve as a mechanism to replicate through adducts of purines, such as those covalently attached to the N^3 of adenine or the N^2 group of guanine, where the change from *anti* to *syn* conformation would place the adduct into the major groove, causing less steric hindrance. Recent data have shown that DNA pol ι levels of expression correlate significantly with tumor proliferation and invasion (Ref. 299 and references therein).

DNA polymerase κ. Human DNA pol κ is encoded by the POLK gene on chromosome 5 and contains 870 amino acid residues.[57] Like the other Y-family enzymes, DNA pol κ lacks an intrinsic 3′→5′ proofreading exonuclease. DNA pol κ has a higher processivity than DNA pol η or ι, and has higher fidelity.[300] In addition, DNA pol κ can efficiently extend mismatched primer termini, but often generating –1 frameshift deletions through a slippage mechanism.[301] This property is in agreement with the crystal structure of the catalytic core of the human DNA pol κ, suggesting that the enzyme's active site is constrained at the site of the templating base and incoming nucleotide, but that DNA pol κ is less constrained following translocation of the lesion.[302] The known biochemical properties of DNA pol κ are summarized in Table 3.2. Like pol η, pol κ can copy DNA containing lesions that substantially distort helix geometry, but with a different lesion bypass specificity.[303,304] For example, DNA pol κ does not bypass a cys-syn TT dimer, but it does bypass other lesions, including a benzo[a]pyrene diol epoxide (BPDE) adduct on the N2 of guanine, inserting a C. This has led to the suggestion that pol κ may participate in the bypass of lesions generated by polycyclic aromatic hydrocarbons in a manner that avoids PAH-induced mutations. Indeed, it was shown that the expression of the mouse DNA pol κ gene is under the control of the aryl-hydrocarbon receptor, which is responsible for the conversion of benzo[a] pyrene into BPDE in mammalian cells. DNA pol κ was shown to be able to efficiently elongate aberrant primer termini resulting from incorporation of a nucleotide opposite 8-oxo-guanine and O^6-methyl guanine by DNA pol δ, thus suggesting a role of this enzyme as an extender DNA pol in TLS.[305] *In vivo* evidence has been reported that DNA pol κ is responsible for error free bypass across DNA cross-links induced by mitomicin C.[306] Mouse DNA pol κ seems also to have a functional role in the repair of DNA breaks,[307] to contribute to replication checkpoint response and to be required for recovery after replication stress.[308]

REV1. The *S. cerevisiae* REV1 gene encodes a polypeptide of 985 amino acid residues.[309] The human Rev1 cDNA consists of 4255 bp and codes for a protein of 1251 amino acid residues with a calculated molecular weight of 138 kDa.[310] The human Rev1 gene is localized on chromosome 2 and is ubiquitously expressed in various human tissues. In *S. cerevisiae* REV1 depends on the functions of the REV3 and REV7 genes. The yeast and human Rev3 and Rev7 proteins are subunits of family B DNA pol ζ, that appears to carry out the extension step during TLS.[311,312,204] Yeast and human Rev1 proteins have been classified as cytidylate-transferases that insert dCMP opposite template G, A and AP sites. However, further biochemical analysis showed that *S. cerevisiae* Rev1 specifically inserts a C residue opposite template G, and it is approximately 25-, 40-, and 400-fold less efficient at inserting a C residue opposite an AP site, an O^6-methylguanine, and an 8-oxoguanine lesion. Also, Rev1 misincorporates G, A, and T residues opposite template G with a very low efficiency (10^{-3}–10^{-4}). Thus, yeast and human Rev1 can be considered as G template-specific DNA pols, since their preferred substrates are a templating G and an incoming dCTP nucleotide. The known biochemical properties of Rev1 are summarized in Table 3.2. Rev1 interacts with Rev7 and with the Y-family DNA pols, DNA pol ι , DNA pol η and DNA pol κ, through the same C-terminal region. Furthermore, Rev7 competes directly with DNA pol κ for binding to the Rev1. However, the polymerase activities of Rev1 and DNA pol κ are unaffected by their physical interaction when extending normal primer-termini or when bypassing a single thymine glycol lesion. Rev 1 is required for successful replication of G-quadruplex DNA (G4 DNA) in higher eukaryotes and human Rev1 disrupts G4 DNA structures and prevents refolding *in vitro*;[313] Recently Rev1 was also found to play several roles at stalled replication forks in response to replication stress in association with the FANCD2,[314] and the Fanconi anemia pathway also appears to control Rev1-mediated translesion synthesis.[315]

3.5.5 *PrimPol*

In 2013, a second DNA primase has been identified in human cells, based on *in silico* analysis.[18–20] The CCDC111 (FLJ33167) gene, located on chromosome 4q35.1 was predicted to encode a protein belonging to the archaea-eukaryotic primase superfamily. Subsequent molecular cloning and expression of the gene product, led to the identification of a peculiar enzyme, endowed with both DNA primase and polymerase activity, which has been subsequently named PrimPol.

PrimPol shows primase activity with either dNTPs or rNTPs. In addition, it can elongate a DNA primer/template. Both primase and DNA pol activities reside in the same active site. PrimPol is able to bypass an 8oxoG lesion, using either deoxy- or ribonucleotides, with the following preference: dA≈dC > rC > rA. In the presence of an abasic site, PrimPol does not incorporate a nucleotide opposite the lesion, but rather uses the downstream base as a template for synthesis, causing a –1 frameshift.[19] In addition, PrimPol has been shown to be unable to directly incorporate nucleotides opposite a *cys–syn* thymine dimer (CPD), but to efficiently elongate a primer terminating with two A's paired opposite the lesion. In contrast, PrimPol was able to fully bypass the 6–4 photoproduct, suggesting a role in TLS of UV-induced lesions.[20]

PrimPol contains a zinc finger motif, which has been shown to be essential for the primase activity. This domain binds to ssDNA and is also important for the fidelity and processivity of the pol activity. Interestingly, when the zinc finger is disrupted, PrimPol becomes able to bypass a CPD lesion. Kinetic studies revealed that nucleotide selectivity limits the chemical step, while the dissociation of the DNA products is the ultimate rate-limiting step of the reaction. In addition, PrimPol is very tolerant towards nucleotide modifications, being able to incorporate several nucleotide analogs.[316]

Intracellular localization studies detected PrimPol both in the mitochondria and in the nucleus. Indeed, PrimPol gene silencing impaired mitochondrial DNA replication, even if PrimPol knockout mice were viable. In the nucleus, PrimPol has been found to play an important role in DNA damage tolerance. It can bypass fork-blocking lesions, assisting replication fork restart. It is also required for the bypass of G quadruplex, allowing repriming downstream of these strong replication blocks.

3.6 Interaction with auxiliary factors

DNA pols always work within multiprotein complexes. Isolation of such complexes has shown that DNA pols associate with and are modulated by a particular class of proteins called "auxiliary proteins" (see also Chapter 2). The function of these auxiliary factors is either to enhance or regulate intrinsic properties of DNA pols, so that their action is best suited to a particular function, or to confer even novel properties to these enzymes.[317] The two main DNA pol auxiliary proteins are: proliferating cell nuclear antigen (PCNA) and replication protein A (RP-A).

Proliferating Cell Nuclear Antigen. PCNA was originally characterized as a processivity factor for the replicative DNA pol δ. Subsequent studies have revealed several additional multiple protein partners for PCNA, involved in several metabolic pathways including Okazaki fragment processing, DNA repair, TLS, DNA methylation, chromatin remodeling and cell cycle regulation.[176] The proteins interacting with PCNA contain a common binding motif (PCNA Interacting Protein- or PIP-box),[318] with the consensus amino-acid sequence: Q-xx-(h)-x-x-(a)-(a), (where "h" represents residues with moderately hydrophobic side chains, e.g., L, I, M; "a" represents residues with highly hydrophobic, aromatic side chains, e.g., F, Y; and "x" is any residue). An additional PIP-box, termed KA-box, has been identified, which is also present in several PCNA

Table 3.3 Effects of PCNA on the catalytic properties of eukaryotic DNA polymerases

DNA polymerase	Effect of PCNA
DNA pol δ	Increases processivity, decreases fidelity of TLS
DNA pol ε	Increases primer binding
DNA pol β	Stimulates TLS
DNA pol ζ	Stimulates TLS[a]
DNA pol η	Increases the affinity for the nucleotide substrate, stimulates TLS
DNA pol ι	Increases the affinity for the nucleotide substrate, stimulates TLS
DNA pol κ	Increases the affinity for the nucleotide substrate, stimulates TLS
DNA pol λ	Increases processivity, increases bypass fidelity

a. Physical interaction is mediated by the Pol31 and Pol32 small subunits of DNA pol δ[212]

interacting proteins and whose consensus is: K-A-(A/L/I)-(A/L/Q)-x-x-(L/V).[319] More than one PIP-box can be present in the same protein and selective interaction with these multiple domains has been suggested to differentially regulate the activity of the interacting partners.[320] Among the PCNA interacting proteins are several DNA pols, and PCNA proved to affect their catalytic activity in different ways (Table 3.3). The first DNA pol shown to interact with PCNA was the major replicative enzyme DNA pol δ.[112,144,321,322] A structural model for this interaction was built using the solved three-dimensional structures of the B family RB69 bacteriophage DNA pol and its PCNA homolog gp45[323] (Figure 3.7).

PCNA was found to stimulate the replicative DNA pol, ε[324] and the TLS enzymes DNA pols η, ι, κ and λ (see Sections 3.5.2–4). In the case of DNA pols δ and λ, PCNA acts as a processivity factor, allowing synthesis of long stretches of DNA by the polymerase preventing its dissociation from the template strand. In the case of DNA pol ε, PCNA acts by increasing the rate of association of the enzyme to the primer-template, thus increasing the affinity for the DNA template, rather than the processivity. Finally, PCNA was shown to stimulate the TLS enzymes DNA pol η, ι and κ by increasing their affinities for the incoming nucleotide. Its ubiquitination is essential for its role in translesion synthesis.[325] Furthermore, posttranslational modification of lysine residues in key proteins other than PCNA is required for bypass replication of DNA damage (for

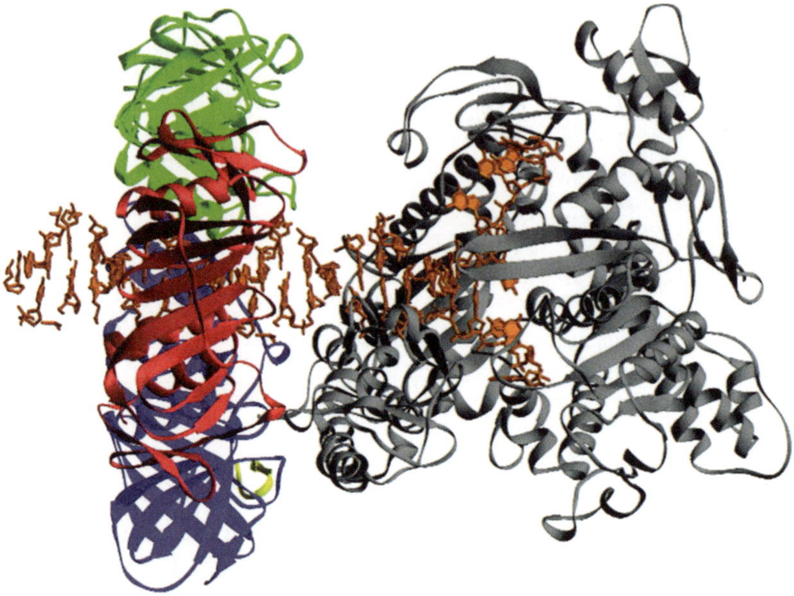

Figure 3.7 **Interaction between RB69 DNA pol and the gp45 processivity factor**. The DNA double helix is orange. The RB69 DNA pol is grey, with its C-terminal side facing the gp45 protein. Each monomer of the trimeric gp45 is rendered in blue, red and green, respectively. (Reproduced from Ref. 323 with permission.)

review see Ref. 326). PCNA was found to promote TLS by DNA pol δ over thymidine dimers, abasic sites and 8-oxo-G lesions.[162,165,327] Interestingly, PCNA was found to specifically increase the rate of dCTP incorporation opposite the 8-oxo-G lesion and TLS activity over an abasic site by DNA pol λ.[227,328]

Replication Protein A. RP-A is a heterotrimer consisting of three subunits (p70, p32 and p14) and is the main eukaryotic single-stranded DNA binding protein. RP-A fulfills multiple roles in the cell, during DNA replication initiation and elongation, DNA repair and DNA recombination.[329] RP-A is also able to influence the catalytic activities of DNA pols. It has been shown that RP-A modulates the DNA synthetic activity of DNA pol α[330] by increasing the stability of DNA pol α binding to the primer and by reducing its overall misincorporation rate. RP-A was found to exert similar effects on DNA pol λ.[331] In addition, RP-A was able, together with PCNA, to significantly increase the fidelity of 8-oxo-G and 2-OH-A bypass by DNA pol λ.[226,227] One study[332] also showed that PCNA and RP-A cooperate to efficiently suppress the tendency by DNA pol δ to generate large deletions when replicating templates containing direct repeats. The activity of the recently discovered error-prone DNA polymerase-primase Primpol is also regulated by RPA.[333]

DNA polymerase δ interacting protein 2 (PolDIP2). Very recently the protein PolDIP2 has been identified as a novel auxiliary factor for DNA pols λ, η and PrimPol. PolDIP2 is a 368 amino acids protein, identified in a yeast two-hybrid screening as an interacting partner of the p50 subunit of DNA pol δ.[334] PolDIP2 was also found to interact with the NADPH oxidase Nox4, one of the major sources of endogenous reactive oxygen species, and to increase its activity.[335] PolDIP2 has been shown to physically interact with Pols η and ζ, and Rev1.[336] Depletion of PolDIP2 caused a decrease in cell survival after UV irradiation. *In vitro*, PolDIp2 stimulates the processivity of DNA pols η and λ during normal synthesis or bypass of oxidative lesions and PolDIP2 knockdown sensitizes cells to oxidative stress.[337] PolDIP2 was found also to stimulate the activity of PrimPol as well as its error-free bypass of 8-oxo-G lesions.[338] In this work it has been established that PolDIP2 binds to PrimPol catalytic domain and that depletion of PolDIP2 in human cells causes a decrease in replication fork rates, similar to what observed in PrimPol –/– cells. PolDIP2 knockout mice showed perinatal lethality with reduced size of the embryos, underlying the essential role of this protein during development.

The Cdc13-Stn1-Ten1 (CST) complex. The CST complex is a heterotrimeric single-stranded DNA binding complex with strong structural similarities with RP-A. In particular, the Stn1 and Ten1 subunits bear close resemblance to the Rpa2 and Rpa3 subunits of RP-A.[339,340] The CST complex has been originally identified as a specific telomeric G-strand binding comple[341] and it is conserved from yeast to humans.[342] CST is required for telomere maintenance, since mutations in its subunits are often lethal and cause chromosomal aberrations consistent with compromised telomere processing.[343] The Cdc13 and Stn1 subunits have been originally identified as DNA pol α interacting proteins.[344] Indeed, it has been shown that CST associates with DNA pol α and regulates its catalytic activity at telomeres. During telomeric ends elongation, Telomerase elongates the G-rich strands, which are subsequently copied by DNA pol α. The N-ter OB-fold and the C-ter winged-helix domains CST interact with the p70 subunit of DNA pol α, stimulating the polymerase activity and inhibiting the primase.[345]

3.7 Incorporation of ribonucleotides in DNA

In their initial description of DNA polymerization Kornberg and colleagues pointed out that DNA synthesis does not begin de novo, but rather requires a "primer" (see Chapter 1). In eukaryotes this primer is a short RNA chain synthesized by a RNA primase associated to the DNA pol α or a short RNA/DNA chain catalyzed by a polymerase-primase. These "primers" are used to initiate DNA synthesis of Okazaki fragments and are, undoubtedly, a major source of ribonucleotides introduced in DNA during DNA replication.

For DNA pols, the probability of ribonucleotide insertion and stable incorporation depends on the cellular concentration of rNTPs and dNTPs,[346] and measurements in yeast and mammalian cells have shown that rNTPs concentrations are much higher than dNTPs concentrations (see Ref. 347 and references therein). Indeed, increasing the rNTPs:dNTPs ratio to that estimated in mammalian cells, resulted in a several fold increase in ribonucleotide incorporation *in vitro* by the human X family DNA polymerase μ.[348] This capacity is not restricted to DNA pol μ, since the yeast replicative DNA pols α, δ and ε stably incorporate one rNMP for every 625, 5000 or 1250 dNMPs, respectively.[349] Notably, human DNA pols δ and ε have ribonucleotide incorporation properties similar to those of yeast replicases.[350,351] Furthermore, *in vivo* studies indicate that ribonucleotides are incorporated into the genome mainly by DNA pol ε and removed in an RNase H2-dependent manner.[352] Human DNA pol λ was also reported to be able to incorporate ribonucleotides during DNA repair[353] and ribonucleotides have also been detected in mitochondrial DNA, consistent with their incorporation during replication by DNA pol γ.[354]

Besides DNA replication, also DNA repair can be a source of ribonucleotide incorporation. Recent studies point to a significant role of DNA pol β in mediating ribonucleotide incorporation in mammalian cells, following oxidative DNA damage.[355,356] Several comprehensive reviews on origins, repair and consequences of ribonucleotides in DNA have recently been published.[22,357–360]

3.8 Eukaryotic DNA polymerases are tightly regulated in the cell cycle

As it is evident from the properties described so far, the multitude of DNA pols present in eukaryotic cells often participate in different, yet interconnected, metabolic pathways.

Moreover, frequently their action must be restricted both spatially and temporally within the cell cycle (Figure 3.8), in order to avoid unscheduled activities, that would otherwise lead to deleterious effects.

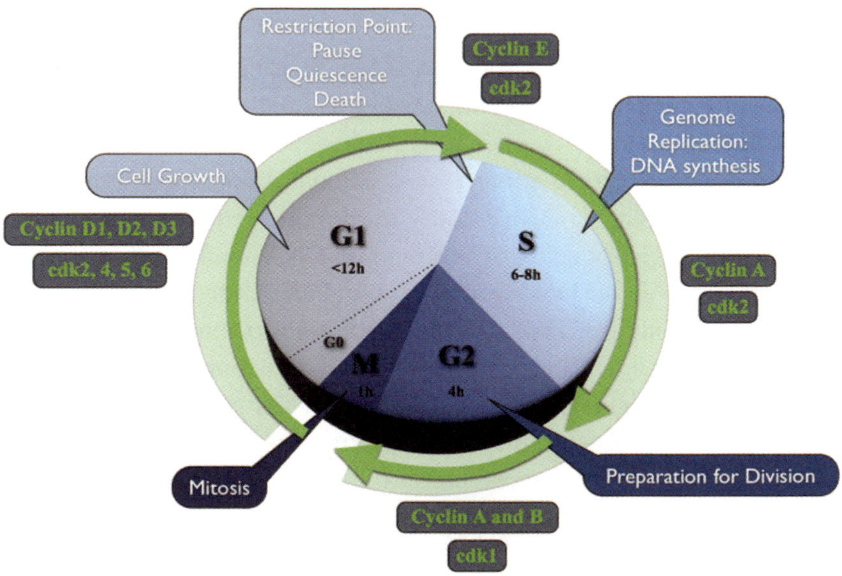

Figure 3.8 **The eukaryotic cell cycle and its main regulatory proteins the cyclin/CDKs.** G1, G2, gap phases; S, DNA synthesis (replication) phase; M, mitosis. Green arrows indicate the order of events from G1 to M phase. Boxes indicate the main events taking place at each phase, as well as the main cyclin/CDK complexes which trigger that specific phase transition.

Finally, a relevant degree of functional redundancy exists among the different eukaryotic DNA pols. All these facts imply the existence of finely tuned mechanisms of regulation to allow the selection of the right enzyme for any given situation. Beside transcriptional regulation through the alternative use of specific transcription factors, a major regulatory mechanism is through phosphorylation. Phosphorylation of DNA pols can regulate their stability, intracellular localization and functional interactions with other protein partners, as illustrated below for representative DNA pols belonging to the different families.

DNA poymerase γ. DNA pol γ transcriptional regulation has been studied for both the catalytic and the small subunit.[87,94,361] Both subunits of DNA pol γ are encoded by nuclear DNA. In *D. melanogaster* the mRNA levels of the catalytic and small subunit are differentially regulated during development. The mRNA level for the large subunit decreases as development proceeds and it is relatively low when mtDNA replication starts to take place in embryos. On the contrary, the mRNA levels of the small subunit show an opposite trend, reaching their peak when mtDNA replication is activated. The genes for both subunits are located within a gene cluster, which contains three promoter regions (P1–P3). The P1 promoter directs the expression of the small subunit gene, and contains a genetic element called DNA replication-related element (DRE), bound by its cognate transcription factor DREF and known to regulate the expression of genes encoding components of the nuclear DNA replication machinery. The expression of the DNA pol γ large subunit is regulated by the P3 region, which contains a weaker promoter and does not have any DRE sequence. Thus, the different strength and structure of the promoter regions for the small (P1) and the large (P3) subunits of DNA pol γ can explain the differences in mRNA levels during *Drosophila* embryonic development. Moreover, the presence of a DRE element in P1, suggests a common regulatory mechanism for nuclear and mtDNA replication at the transcriptional level.

DNA polymerase α. Given the central role of DNA pol α in the replication process, its regulation has been studied in detail both at the transcriptional and post-transcriptional levels (see Refs. 131 and 362 and references therein). The promoter region of the DNA pol α large subunit gene from human cells was shown to contain a 248-bp region containing the binding sites for the transcription factors Ap1, AP2 and E2F and to be inducible upon serum stimulation of cell proliferation. However, when cells are actively cycling, the levels of DNA pol α transcripts do not fluctuate during the different phases of the cell cycle. The catalytic and the B (the second largest subunit) subunits of the DNA pol α four subunit complex, have been shown to be phosphorylated in a cell cycle-dependent manner. Specifically, in *S. cerevisiae* the B subunit is phosphorylated early in S-phase and dephosphorylated at the exit from mitosis. The phosphorylation sites of the B subunit of human DNA pol α were mapped to Ser141, Ser147, S152 and Thr156 and shown to be targeted by different cyclin/cdk complexes. Interestingly, the functional consequences of these phosphorylation events were not equivalent. Phosphorylation by cylinA/cdk2 and cyclinA/cdk1 complexes resulted in a DNA pol α-primase four subunits complex incapable of promoting intiation of DNA replication in the SV40 system. On the opposite, phosphorylation by cyclinE/cdk2 resulted in higher competence in the initiation of DNA replication. Studies of the patterns of phosphorylation during the cell cycle revealed that the human B subunit is phosphorylated by cyclinE/cdk2 at the G1/S border, consistent with its requirement for the initiation of DNA replication, whereas it is phosphorylated by cyclinA/cdk2 in G2, when replication is completed. These phosphorylation events did not change the basic enzymatic properties of the DNA pol α-primase four subunits complex, nor did they promote its degradation. Thus, their functional significance is likely to modulate protein–protein interactions that take place at the replication fork. A multiprotein complex competent for DNA replication and containing DNA pol α-primase, DNA pol δ and DNA pol ε, together with other replication fork components, has been purified from HeLa cells at different stages of the cell cycle.[174] Intriguingly, its association with the chromatin (and hence its competence to function as the replication fork) was shown to be regulated depending on which particular cyclin/cdk complex was bound. When cyclinA/cdk2 was present, the complex was loaded onto chromatin, whereas when it was replaced by cyclinA/cdk1 or cyclinB/cdk2, the complex was found in the soluble nuclear fraction. Thus, cyclin/cdk phosphorylation events might help to regulate the chromatin association state of DNA pol α. The protein levels of the DNA pol α p180 subunit are also regulated by proteosomal degradation. In yeast, the p180 subunit (Cdc17) is modified by the Ubc4/Not4 E2-E3 ubiquitination complex and thus targeted to degradation. Ubiquitination of Cdc17 is counteracted by with the replisome complex protein Mcm10.[363,364] *Drosophila* DNA pol α has also been found to promote methylation of histone H4 Lys 20, through physical interaction with the SETD7 methylase. This suggests that replicative DNA pols may also be involved in the regulation of the epigenetic state of the chromatin, affecting the transcriptional program.[365]

DNA polymerase λ. DNA pol λ expression is regulated during embryonic development in mice, but the transcript levels do not appear to change during the cell cycle of non-embryonic proliferating cells.[215] However, DNA pol λ protein levels are regulated in a cell cycle-dependent manner through phosphorylation.[366] DNA pol λ was shown to physically interact with cdk2 *in vivo* and to be phosphorylated *in vitro* by cycinE/cdk2, cyclinA/cdk2 and cyclinA/cdk1 complexes. The main phosphorylation sites for cdk were mapped to Ser167, Ser177 and Ser230 of the N-terminal proline-rich domain and Thr553 in the C-terminal catalytic domain. The phosphorylation status of DNA pol λ was shown to fluctuate during the cell cycle, with a hyperphosphorylated form present in the G2 phase. Experiments with phosphorylation-defective mutants[367] suggested that DNA pol λ is targeted to the proteasomal degradation pathway through ubiquitination unless the Thr553 residue is phosphorylated. In particular, DNA pol λ is targeted for proteosomal degradation through ubiquitination by the E3 ligase Mule. Phosphorylation of DNA pol λ by Cdk/cyclins promotes its recruitment to chromatin, thus avoiding degradation and resulting in its stabilization during cell cycle progression in the late S and G2 phases.[368] This most likely allows pol λ to correctly conduct repair of damaged DNA during and after S phase.

DNA polymerase η. As mentioned in Section 3.4.4, DNA pol η plays a major role in the TLS-dependent tolerance to UV-induced DNA damages. Upon UV irradiation, cellular mRNA or protein levels of DNA pol η do not change, but the protein was shown to be stabilized upon UV-treatment[369] and to relocalize to stalled DNA replication forks in irradiated cells.[281] DNA pol η translocation to stalled replication forks was shown also in response to non-damaging treatments such as nucleotide deprivation of hydroxyurea treatment, suggesting that such a dynamical behavior was part of a general response to the block of DNA replication. The intracellular translocation of DNA pol η is controlled through phosphorylation by protein kinase C (PKC) and the S-phase checkpoint kinase ATR.[370] DNA pol η phosphorylation was specifically increased by UV-treatment and inhibition of ATR or PKC intracellular activity reduced DNA pol η accumulation at stalled replication forks. In addition, mutations at two PKC phosphorylation sites (Ser587 and Thr617) in DNA pol η prevented the recruitment of the mutated protein to blocked DNA replication forks. XP-V cells are naturally sensitive to UV, since they are deprived of DNA pol η. Transient expression of the Ser587Ala and Thr617Ala mutants in such a DNA pol η null genetic background, failed to complement the UV-sensitivity, contrary to wild type DNA pol η. Thus, phosphorylation is an important mechanism regulating the subcellular localization of DNA pol η in response to DNA damage. Localization of DNA pol η to sites of replication fork stalling is also indirectly regulated through phosphorylation of the E3 ubiquitin ligase Rad18 by the Cdc7 kinase. Phosphorylation of Rad18 is required for binding to DNA pol η. The DNA pol η-Rad18 complex is then recruited to stalled replication fork, where Rad18 monoubiquitinates PCNA, which then recruits DNA pol η for TLS.[371] Monoubiquitination of DNA pol η itself, in close proximity of its ubiquitin binding domain, prevents interaction with PCNA, but this modification is inhibited as part of the DNA damage response. This provides a further mechanism for regulating the levels of DNA pol η available for TLS.[372]

3.9 Chapter summary

The diversification, throughout evolution, from single-cell to multicellular organisms, required the acquisition of more complex levels of regulation for the duplication of the genetic information of the cells, given the different proliferation rates during embryonic development as well as in the various tissues of an adult organism. In eukaryotic cells, DNA replication is both spatially and temporally regulated, being restricted to a specific phase of the cell cycle (the S phase). The exact duplication of the genetic information before cellular duplication is accomplished through a complex network of enzymes and proteins, which constitute the replisome.

The core enzymes of the replisome are the DNA pols, the only enzymes capable of using a DNA strand to generate its complementary copy. The polymerization reaction is quite complex, involving different kinetic steps, whose ordered sequence is ensured by a combination of thermodynamic barriers along the reaction pathway. The different structural features among DNA pols can have important effects on these energetic barriers, thus resulting in different catalytic properties. Moreover, the existence of a series of distinct kinetic steps during the reaction is essential for the intrinsic fidelity of DNA pols, which are able to "sense" a distorted base-pair and to stop the polymerization process. Additional factors such as proofreading exonuclease activities and physical interactions with auxiliary proteins, increase the ability of replicative DNA pols to produce a faithful copy of the genetic information of the mother cell. Beside an essential role in DNA replication, DNA pols are also key players in the DNA repair mechanisms, where a certain amount of damaged information has to be replaced with an intact copy. Different DNA pols act in different types of DNA repair (see Table 4.3, Chapter 4). The extreme case is the one of translesion DNA synthesis, where different specialized DNA pols have evolved unconventional catalytic properties to selectively bypass particular DNA lesions. DNA pols are also tightly linked to the cell cycle regulatory pathways and to the cellular response mechanisms to DNA damage. This regulation operates both at the transcriptional and post-transcriptional level. Among the latter, both ubiquitination and phosphorylation play important roles either alone or in combination with each other. In spite of their different properties and functions, all eukaryotic DNA pols share a common structural "core" module, with distinct subdomains (palm, fingers, thumb) which can be recognized also in prokaryotic and viral DNA polymerases, testifying their evolutionary diversification from a common ancestral enzyme.

References

1. Hubscher, U., Maga, G. & Spadari, S. Eukaryotic DNA polymerases. *Annu Rev Biochem* **71**, 133–163 (2002).
2. Koonin, E. V. Temporal order of evolution of DNA replication systems inferred by comparison of cellular and viral DNA polymerases. *Biol Direct* **1**, 39 (2006).

3. O'Keefe, R. T., Henderson, S. C. & Spector, D. L. Dynamic organization of DNA replication in mammalian cell nuclei: spatially and temporally defined replication of chromosome-specific alpha-satellite DNA sequences. *J. Cell Biol* **116**, 1095–1110 (1992).
4. Stillman, B. Cell cycle control of DNA replication. *Science* **274**, 1659–1664 (1996).
5. Toone, W. M., Aerne, B. L., Morgan, B. A. & Johnston, L. H. Getting started: regulating the initiation of DNA replication in yeast. *Annual Review of Microbiology* **51**, 125–149 (1997).
6. Lindahl, T. & Wood, R. D. Quality control by DNA repair. *Science* **286**, 1897–1905 (1999).
7. Kelly, T. J. & Brown, G. W. Regulation of chromosome replication. *Annu Rev Biochem* **69**, 829–880 (2000).
8. Lisby, M., Barlow, J. H., Burgess, R. C. & Rothstein, R. Choreography of the DNA damage response: spatiotemporal relationships among checkpoint and repair proteins. *Cell* **118**, 699–713 (2004).
9. Gabaldon, T. & Huynen, M. A. From endosymbiont to host-controlled organelle: the hijacking of mitochondrial protein synthesis and metabolism. *PLoS Comput Biol* **3**, e219 (2007).
10. Falkenberg, M., Larsson, N. G. & Gustafsson, C. M. DNA replication and transcription in mammalian mitochondria. *Annu Rev Biochem* **76**, 679–699 (2007).
11. Scarpulla, R. C. Transcriptional paradigms in mammalian mitochondrial biogenesis and function. *Physiol Rev* **88**, 611–638 (2008).
12. Tsurimoto, T. & Stillman, B. Functions of replication factor C and proliferating-cell nuclear antigen: functional similarity of DNA polymerase accessory proteins from human cells and bacteriophage T4. *Proc Natl Acad Sci USA* **87**, 1023–1027 (1990).
13. Nethanel, T. & Kaufmann, G. Two DNA polymerases may be required for synthesis of the lagging DNA strand of simian virus 40. *J Virol* **64**, 5912–5918 (1990).
14. Syvaoja, J. *et al.* DNA polymerases alpha, delta, and epsilon: three distinct enzymes from HeLa cells. *Proc Natl Acad Sci USA* **87**, 6664–6668 (1990).
15. Bambara, R. A., Murante, R. S. & Henricksen, L. A. Enzymes and reactions at the eukaryotic DNA replication fork. *J Biol Chem* **272**, 4647–4650 (1997).
16. Hübscher, U. & Sogo, J. The eukaryotic DNA replication fork. *News Physiol Sci* **12**, 125–131 (1997).

17. Nick McElhinny, S. A., Gordenin, D. A., Stith, C. M., Burgers, P. M. & Kunkel, T. A. Division of labor at the eukaryotic replication fork. *Mol Cell* **30**, 137–144 (2008).
18. Wan, L. *et al.* hPrimpol1/CCDC111 is a human DNA primase-polymerase required for the maintenance of genome integrity. *Embo Reports* **14**, 1104–1112 (2013).
19. Garcia-Gomez, S. *et al.* PrimPol, an Archaic Primase/Polymerase Operating in Human Cells. *Molecular Cell* **52**, 541–553 (2013).
20. Bianchi, J. *et al.* PrimPol Bypasses UV Photoproducts during Eukaryotic Chromosomal DNA Replication. *Molecular Cell* **52**, 566–573 (2013).
21. Johansson, E. & Dixon, N. Replicative DNA Polymerases. *Csh Perspect Biol* **5** (2013).
22. Williams, J. S. & Kunkel, T. A. Ribonucleotides in DNA: Origins, repair and consequences. *DNA Repair* **19**, 27–37 (2014).
23. Georgescu, R. E. *et al.* Mechanism of asymmetric polymerase assembly at the eukaryotic replication fork. *Nat Struct Mol Biol* **21**, 664–670 (2014).
24. Yu, C. H. *et al.* Strand-Specific Analysis Shows Protein Binding at Replication Forks and PCNA Unloading from Lagging Strands when Forks Stall. *Molecular Cell* **56**, 551–563 (2014).
25. Johnson, R. E., Klassen, R., Prakash, L. & Prakash, S. A Major Role of DNA Polymerase delta in Replication of Both the Leading and Lagging DNA Strands. *Molecular Cell* **59**, 163–175 (2015).
26. Stillman, B. Reconsidering DNA Polymerases at the Replication Fork in Eukaryotes. *Molecular Cell* **59**, 139–141 (2015).
27. Yeeles, J. T., Janska, A., early, A. & Diffley, J. F. How the Eukaryotic Replisome Achieves Rapid and Efficient DNA Replication. *Molecular Cell* **65**, 105–116 (2017).
28. Benedict, C. L., Gilfillan, S., Thai, T. H. & Kearney, J. F. Terminal deoxynucleotidyl transferase and repertoire development. *Immunol Rev* **175**, 150–157 (2000).
29. Bertocci, B., De Smet, A., Weill, J. C. & Reynaud, C. A. Nonoverlapping functions of DNA polymerases mu, lambda, and terminal deoxynucleotidyl-transferase during immunoglobulin V(D)J recombination in vivo. *Immunity* **25**, 31–41 (2006).
30. Di Santo, R. & Maga, G. Human terminal deoxynucleotidyl transferases as novel targets for anticancer chemotherapy. *Curr Med Chem* **13**, 2353–2368 (2006).
31. Steitz, T. A. DNA polymerases: structural diversity and common mechanisms. *J Biol Chem* **274**, 17395–17398 (1999).

32. Franklin, M. C., Wang, J. & Steitz, T. A. Structure of the replicating complex of a pol alpha family DNA polymerase. *Cell* **105**, 657–667 (2001).
33. Steitz, T. A. A mechanism for all polymerases. *Nature* **391**, 231–232 (1998).
34. Seki, M., Marini, F. & Wood, R. D. POLQ (Pol theta), a DNA polymerase and DNA-dependent ATPase in human cells. *Nucleic Acids Res* **31**, 6117–6126 (2003).
35. Newman, J. A., Cooper, C. D. O., Aitkenhead, H. & Gileadi, O. Structure of the Helicase Domain of DNA Polymerase Theta Reveals a Possible Role in the Microhomology-Mediated End-Joining Pathway. *Structure* **23**, 2319–2330 (2015).
36. Zahn, K. E., Averill, A. M., Aller, P., Wood, R. D. & Doublie, S. Human DNA polymerase theta grasps the primer terminus to mediate DNA repair. *Nat Struct Mol Biol* **22**, 304–U351 (2015).
37. Beagan, K. & Mcvey, M. Linking DNA polymerase theta structure and function in health and disease. *Cellular and Molecular Life Sciences* **73**, 603–615 (2016).
38. Lee, Y. S., Gao, Y. & Yang, W. How a homolog of high-fidelity replicases conducts mutagenic DNA synthesis. *Nat Struct Mol Biol* **22**, 298–U242 (2015).
39. Coloma, J., Johnson, R. E., Prakash, L., Prakash, S. & Aggarwal, A. K. Human DNA polymerase alpha in binary complex with a DNA:DNA template-primer. *Sci Rep-Uk* **6** (2016).
40. Swan, M. K., Johnson, R. E., Prakash, L., Prakash, S. & Aggarwal, A. K. Structural basis of high-fidelity DNA synthesis by yeast DNA polymerase delta. *Nat Struct Mol Biol* **16**, 979–986 (2009).
41. Brautigam, C. A. & Steitz, T. A. Structural and functional insights provided by crystal structures of DNA polymerases and their substrate complexes. *Curr Opin Struct Biol* **8**, 54–63 (1998).
42. Zhao, Y. *et al.* Crystal structure of an archaebacterial DNA polymerase. *Struct Fold Des* **7**, 1189–1199 (1999).
43. Rodriguez, A. C., Park, H. W., Mao, C. & Beese, L. S. Crystal structure of a pol α family DNA polymerase from the hyperthermophilic archaeon Thermococcus sp. 9 degrees N-7. *J Mol Biol* **299**, 447–462 (2000).
44. Hashimoto, H. *et al.* Crystal structure of DNA polymerase from hyperthermophilic archaeon Pyrococcus kodakaraensis KOD1. *J Mol Biol* **306**, 469–477 (2001).
45. Hogg, M. *et al.* Structural basis for processive DNA synthesis by yeast DNA polymerase epsilon. *Nat Struct Mol Biol* **21**, 49–56 (2014).
46. Jain, R. *et al.* Crystal Structure of Yeast DNA Polymerase e Catalytic Domain. *Plos One* **9** (2014).

47. Asturias, F. J. *et al.* Structure of Saccharomyces cerevisiae DNA polymerase epsilon by cryo-electron microscopy. *Nat Struct Mol Biol* **13**, 35–43 (2006).
48. Ramadan, K., Shevelev, I. & Hubscher, U. The DNA-polymerase-X family: controllers of DNA quality? *Nat Rev Mol Cell Biol* **5**, 1038–1043 (2004).
49. Moon, A. F. *et al.* Structural insight into the substrate specificity of DNA Polymerase mu. *Nat Struct Mol Biol* **14**, 45–53 (2007).
50. Beard, W. A. & Wilson, S. H. Structure and mechanism of DNA polymerase Beta. *Chem Rev* **106**, 361–382 (2006).
51. Garcia-Diaz, M. *et al.* A structural solution for the DNA polymerase lambda-dependent repair of DNA gaps with minimal homology. *Mol Cell* **13**, 561–572 (2004).
52. Garcia-Diaz, M., Bebenek, K., Krahn, J. M., Kunkel, T. A. & Pedersen, L. C. A closed conformation for the Pol lambda catalytic cycle. *Nat Struct Mol Biol* **12**, 97–98 (2005).
53. Picher, A. J. *et al.* Promiscuous mismatch extension by human DNA polymerase lambda. *Nucleic Acids Res* **34**, 3259–3266 (2006).
54. Garcia-Diaz, M., Bebenek, K., Krahn, J. M., Pedersen, L. C. & Kunkel, T. A. Structural analysis of strand misalignment during DNA synthesis by a human DNA polymerase. *Cell* **124**, 331–342 (2006).
55. Garcia-Diaz, M. *et al.* Template strand scrunching during DNA gap repair synthesis by human polymerase lambda. *Nat Struct Mol Biol* **16**, 967–972 (2009).
56. Moon, A. F. *et al.* The X family portrait: Structural insights into biological functions of X family polymerases. *DNA Repair (Amst)* (2007).
57. Yang, W. Damage repair DNA polymerases Y. *Curr Opin Struct Biol* **13**, 23–30 (2003).
58. Beard, W. A. & Wilson, S. H. DNA lesion bypass polymerases open up. *Structure* **9**, 759–764 (2001).
59. Friedberg, E. C., Fischhaber, P. L. & Kisker, C. Error-prone DNA polymerases: novel structures and the benefits of infidelity. *Cell* **107**, 9–12 (2001).
60. Lipps, G., Weinzierl, A. O., von Scheven, G., Buchen, C. & Cramer, P. Structure of a bifunctional DNA primase-polymerase. *Nat Struct Mol Biol* **11**, 157–162 (2004).
61. Kilkenny, M. L., Longo, M. A., Perera, R. L. & Pellegrini, L. Structures of human primase reveal design of nucleotide elongation site and mode of Pol alpha tethering. *P Natl Acad Sci USA* **110**, 15961–15966 (2013).

62. Baranovskiy, A. G. *et al.* Insight into the Human DNA Primase Interaction with Template-Primer. *Journal of Biological Chemistry* **291**, 4793–4802 (2016).
63. Iyer, L. M., Koonin, E. V., Leipe, D. D. & Aravind, L. Origin and evolution of the archaeo-eukaryotic primase superfamily and related palm-domain proteins: structural insights and new members. *Nucleic Acids Res* **33**, 3875–3896 (2005).
64. Weiner, B. E. *et al.* An iron-sulfur cluster in the C-terminal domain of the p58 subunit of human DNA primase. *J Biol Chem* **282**, 33444–33451 (2007).
65. Agarkar, V. B., Babayeva, N. D., Pavlov, Y. I. & Tahirov, T. H. Crystal structure of the C-terminal domain of human DNA primase large subunit Implications for the mechanism of the primase-polymerase alpha switch. *Cell Cycle* **10**, 926–931 (2011).
66. Baranovskiy, A. G. *et al.* Mechanism of Concerted RNA-DNA Primer Synthesis by the Human Primosome. *Journal of Biological Chemistry* **291**, 10006–10020 (2016).
67. Fan, L., Sanschagrin, P. C., Kaguni, L. S. & Kuhn, L. A. The accessory subunit of mtDNA polymerase shares structural homology with aminoacyl-tRNA synthetases: implications for a dual role as a primer recognition factor and processivity clamp. *Proc Natl Acad Sci USA* **96**, 9527–9532 (1999).
68. Carrodeguas, J. A., Theis, K., Bogenhagen, D. F. & Kisker, C. Crystal structure and deletion analysis show that the accessory subunit of mammalian DNA polymerase gamma, Pol gamma B, functions as a homodimer. *Mol Cell* **7**, 43–54 (2001).
69. Lee, Y. S., Kennedy, W. D. & Yin, Y. W. Structural insight into processive human mitochondrial DNA synthesis and disease-related polymerase mutations. *Cell* **139**, 312–324 (2009).
70. Mäkiniemi, M. *et al.* A novel family of DNA-polymerase B subunits. *Trends Biochem Sci* **24**, 14–16 (1999).
71. Nuutinen, T. *et al.* The solution structure of the amino-terminal domain of human DNA polymerase epsilon subunit B is homologous to C-domains of AAA+ proteins. *Nucleic Acids Res* **36**, 5102–5110 (2008).
72. Suwa, Y. *et al.* Crystal Structure of the Human Pol alpha B Subunit in Complex with the C-terminal Domain of the Catalytic Subunit. *Journal of Biological Chemistry* **290**, 14328–14337 (2015).
73. Loeb, L. A. & Kunkel, T. A. Fidelity of DNA synthesis. *Annu Rev Biochem* **51**, 429–457 (1982).

74. Johnson, K. A. Conformational coupling in DNA polymerase fidelity. *Annu. Rev Biochem* **62**, 685–713 (1993).
75. Kool, E. T. Active site tightness and substrate fit in DNA replication. *Annu Rev Biochem* **71**, 191–219 (2002).
76. Joyce, C. M. & Benkovic, S. J. DNA polymerase fidelity: kinetics, structure, and checkpoints. *Biochemistry* **43**, 14317–14324 (2004).
77. Florian, J., Goodman, M. F. & Warshel, A. Computer simulation studies of the fidelity of DNA polymerases. *Biopolymers* **68**, 286–299 (2003).
78. Shevelev, I. V. & Hubscher, U. The 3′ 5′ exonucleases. *Nat Rev Mol Cell Biol* **3**, 364–376 (2002).
79. Beard, W. A. & Wilson, S. H. Structural insights into the origins of DNA polymerase fidelity. *Structure* **11**, 489–496 (2003).
80. Freisinger, E., Grollman, A. P., Miller, H. & Kisker, C. Lesion (in)tolerance reveals insights into DNA replication fidelity. *Embo J* **23**, 1494–1505 (2004).
81. Zhang, Y., Yuan, F., Wu, X. & Wang, Z. Preferential incorporation of G opposite template T by the low-fidelity human DNA polymerase iota. *Mol Cell Biol* **20**, 7099–7108 (2000).
82. Washington, M. T., Johnson, R. E., Prakash, L. & Prakash, S. Human DNA polymerase iota utilizes different nucleotide incorporation mechanisms dependent upon the template base. *Mol Cell Biol* **24**, 936–943 (2004).
83. Swan, M. K., Johnson, R. E., Prakash, L., Prakash, S. & Aggarwal, A. K. Structure of the human Rev1-DNA-dNTP ternary complex. *J Mol Biol* **390**, 699–709 (2009).
84. Guo, D., Xie, Z., Shen, H., Zhao, B. & Wang, Z. Translesion synthesis of acetylaminofluorene-dG adducts by DNA polymerase zeta is stimulated by yeast Rev1 protein. *Nucleic Acids Res* **32**, 1122–1130 (2004).
85. Washington, M. T. *et al.* Efficient and error-free replication past a minor-groove N2-guanine adduct by the sequential action of yeast Rev1 and DNA polymerase zeta. *Mol Cell Biol* **24**, 6900–6906 (2004).
86. Ye, F., Carrodeguas, J. A. & Bogenhagen, D. F. The γ subfamily of DNA polymerases: cloning of a developmentally regulated cDNA encoding Xenopus laevis mitochondrial DNA polymerase γ. *Nucl Acids Res* **24**, 1481–1488 (1996).
87. Kaguni, L. S. DNA polymerase gamma, the mitochondrial replicase. *Annu Rev Biochem* **73**, 293–320 (2004).
88. Takeuchi, R., Kimura, S., Saotome, A. & Sakaguchi, K. Biochemical properties of a plastidial DNA polymerase of rice. *Plant Mol Biol* **64**, 601–611 (2007).

89. Seow, F. *et al.* The plastidic DNA replication enzyme complex of Plasmodium falciparum. *Mol Biochem Parasitol* **141**, 145–153 (2005).
90. Carrillo, N. & Bogorad, L. Chloroplast DNA replication in vitro: site-specific initiation from preferred templates. *Nucleic Acids Res* **16**, 5603–5620 (1988).
91. Sala, F. *et al.* Specific and reversible inhibition by aphidicolin in the alpha-like DNA polymerase of plant cells. *FEBS Lett* **117**, 93–98 (1980).
92. Simpson, L. The mitochondrial genome of kinetoplastid protozoa: genomic organization, transcription, replication, and evolution. *Annu Rev Microbiol* **41**, 363–382 (1987).
93. Liu, B., Liu, Y., Motyka, S. A., Agbo, E. E. & Englund, P. T. Fellowship of the rings: the replication of kinetoplast DNA. *Trends Parasitol* **21**, 363–369 (2005).
94. Graziewicz, M. A., Longley, M. J. & Copeland, W. C. DNA polymerase gamma in mitochondrial DNA replication and repair. *Chem Rev* **106**, 383–405 (2006).
95. Klingbeil, M. M., Motyka, S. A. & Englund, P. T. Multiple mitochondrial DNA polymerases in Trypanosoma brucei. *Mol Cell* **10**, 175–186 (2002).
96. Wernette, C. M. & Kaguni, L. S. A mitochondrial DNA polymerase from embryos of Drosophila melanogaster. Purification, subunit structure, and partial characterization. *J Biol Chem* **261**, 14764–14770 (1986).
97. Wang, Y. & Kaguni, L. S. Baculovirus expression reconstitutes Drosophila mitochondrial DNA polymerase. *J Biol Chem* **274**, 28972–28977 (1999).
98. Ropp, P. A. & Copeland, W. C. Cloning and characterization of the human mitochondrial DNA polymerase, DNA polymerase γ. *Genomics* **36**, 449–458 (1996).
99. Yakubovskaya, E. *et al.* The EM structure of human DNA polymerase gamma reveals a localized contact between the catalytic and accessory subunits. *EMBO J* **26**, 4283–4291 (2007).
100. Kaguni, L. S. & Olson, M. W. Mismatch-specific 3′----5′ exonuclease associated with the mitochondrial DNA polymerase from Drosophila embryos. *Proc Natl Acad Sci USA* **86**, 6469–6473 (1989).
101. Wernette, C. M., Conway, M. C. & Kaguni, L. S. Mitochondrial DNA polymerase from Drosophila melanogaster embryos: kinetics, processivity, and fidelity of DNA polymerization. *Biochemistry* **27**, 6046–6054 (1988).
102. Graves, S. W., Johnson, A. A. & Johnson, K. A. Expression, purification, and initial kinetic characterization of the large subunit of the human mitochondrial DNA polymerase γ. *Biochemistry* **37**, 6050–6058 (1998).

103. Longley, M. J., Ropp, P. A., Lim, S. E. & Copeland, W. C. Characterization ot the native and recombinant catalytic subunit of human DNA polymerase gamma: identification of residues critical for exonuclease activity and dideoxynucleotide sensitivity. *Biochemistry* **37**, 10529–10539 (1998).
104. Lim, S. E., Longley, M. J. & Copeland, W. C. The mitochondrial p55 accessory subunit of human DNA polymerase γ enhances DNA binding, promotes processive DNA synthesis, and confers N-ethylmaleimide resistance. *J Biol Chem* **274**, 38197–38203 (1999).
105. Longley, M. J., Prasad, R., Srivastava, D. K., Wilson, S. H. & Copeland, W. C. Identification of 5′-deoxyribose phosphate lyase activity in human DNA polymerase γ and its role in mitochondrial base excision repair *in vitro*. *Proc Natl Acad Sci USA* **95**, 12244–12248 (1998).
106. Pinz, K. G. & Bodenhagen, D. F. Characterization of a catalytically slow AP lyase activity in DNA polymerase gamma and other family A DNA polymerases. *J Biol Chem* **275**, 12509–12514 (2000).
107. Pinz, K. G., Shibutani, S. & Bogenhagen, D. F. Action of mitochondrial DNA polymerase gamma at sites of base loss or oxidative damage. . *J Biol Chem* **270**, 9202–9206 (1995).
108. Pinz, K. G. & Bogenhagen, D. F. Efficient repair of abasic sites in DNA by mitochondrial enzymes. *Mol Cell Biol* **18**, 1257–1265 (1998).
109. Boyd, J., Sakaguchi, K. & Harris, P. mus308 mutants of Drosophila exhibit hypersensitivity to DNA crosslinking agents and are defective in a deoxyribonuclease. *Genetics* **125**, 813–819 (1990).
110. Maga, G., Shevelev, I., Ramadan, K., Spadari, S. & Hubscher, U. DNA polymerase theta purified from human cells is a high-fidelity enzyme. *J Mol Biol* **319**, 359–369 (2002).
111. Arana, M. E., Seki, M., Wood, R. D., Rogozin, I. B. & Kunkel, T. A. Low-fidelity DNA synthesis by human DNA polymerase theta. *Nucleic Acids Res* **36**, 3847–3856 (2008).
112. Seki, M. *et al.* High-efficiency bypass of DNA damage by human DNA polymerase Q. *EMBO J* **23**, 4484–4494. Epub 2004 Oct 4421 (2004).
113. Seki, M. & Wood, R. D. DNA polymerase theta (POLQ) can extend from mismatches and from bases opposite a (6–4) photoproduct. *DNA Repair (Amst)* **7**, 119–127 (2008).
114. Yoshimura, M. *et al.* Vertebrate POLQ and POLbeta cooperate in base excision repair of oxidative DNA damage. *Mol Cell* **24**, 115–125 (2006).
115. Masuda, K. *et al.* DNA polymerases eta and theta function in the same genetic pathway to generate mutations at A/T during somatic hypermutation of Ig genes. *J Biol Chem* **282**, 17387–17394 (2007).

116. Mateos-Gomez, P. A. *et al.* Mammalian polymerase theta promotes alternative NHEJ and suppresses recombination. *Nature* **518**, 254-U285 (2015).
117. Beard, W. A. & Wilson, S. H. Structures of human DNA polymerases nu and theta expose their end game. *Nat Struct Mol Biol* **22**, 273–275 (2015).
118. Fernandez-Vidal, A. *et al.* A role for DNA polymerase theta in the timing of DNA replication. *Nat Commun* **5** (2014).
119. Lemee, F. *et al.* DNA polymerase theta up-regulation is associated with poor survival in breast cancer, perturbs DNA replication, and promotes genetic instability. *P Natl Acad Sci USA* **107**, 13390–13395 (2010).
120. Marini, F., Kim, N., Schuffert, A. & Wood, R. D. POLN, a nuclear PolA family DNA polymerase homologous to the DNA cross-link sensitivity protein Mus308. *J Biol Chem* **278**, 32014–32019 (2003).
121. Arana, M. E., Takata, K., Garcia-Diaz, M., Wood, R. D. & Kunkel, T. A. A unique error signature for human DNA polymerase nu. *DNA Repair (Amst)* **6**, 213–223 (2007).
122. Takata, K., Shimizu, T., Iwai, S. & Wood, R. D. Human DNA polymerase N (POLN) is a low fidelity enzyme capable of error-free bypass of 5S-thymine glycol. *J Biol Chem* **281**, 23445–23455 (2006).
123. Mizuno, T., Yamagishi, K., Miyazawa, H. & Hanaoka, F. Molecular architecture of the mouse DNA polymerase alpha-primase complex. *Mol Cell Biol* **19**, 7886–7896 (1999).
124. Arezi, B. & Kuchta, R. D. Eukaryotic DNA primase. *Trends Biochem Sci* **25**, 572–576 (2000).
125. Frick, D. N. & Richardson, C. C. Dna Primases. *Annu Rev Biochem* **70**, 39–80 (2001).
126. Wong, S. W., Paborsky, L. R., Fisher, P. A., Wang, T. S. & Korn, D. Structural and enzymological characterization of immunoaffinity-purified DNA polymerase alpha.DNA primase complex from KB cells. *J Biol Chem* **261**, 7958–7968 (1986).
127. Wang, T. S., Wong, S. W. & Korn, D. Human DNA polymerase alpha: predicted functional domains and relationships with viral DNA polymerases. *Faseb J* **3**, 14–21 (1989).
128. Copeland, W. C. & Wang, T. S. F. Catalytic subunit of human DNA polymerase α overproduced in baculo-virus-infected insect cells. Sturctural and enzymological characterization. *J Biol Chem* **266**, 22739–22748 (1991).
129. Copeland, W. C. & Wang, T. S. F. Mutational analysis of the human DNA polymerase α. The most conserved region in the α-like DNA polymerases

is involved in metal-specific catalysis. *J Biol Chem* **268**, 11028–11040 (1993).

130. Copeland, B. C. & Wang, T. S. F. Enzymatic chraracterization of the individual mammalian primase subunits reveal a biphasic mechanism for initiation of DNA replication. *J Biol Chem* **268**, 26179–26189 (1993).
131. Foiani, M., Lucchini, G. & Plevani, P. The DNA polymerase alpha-primase complex couples DNA replication, cell-cycle progression and DNA-damage response. *Trends Biochem Sci* **22**, 424–427 (1997).
132. Grossi, S., Puglisi, A., Dmitriev, P. V., Lopes, M. & Shore, D. Pol12, the B subunit of DNA polymerase alpha, functions in both telomere capping and length regulation. *Genes Dev* **18**, 992–1006 (2004).
133. Cullmann, G., Hindges, R., Berchtold, M. W. & Hubscher, U. Cloning of a mouse cDNA encoding DNA polymerase delta: refinement of the homology boxes. *Gene* **134**, 191–200 (1993).
134. Moussy, G., De Recondo, A. M. & Baldacci, G. Inter-species DNA polymerase delta chimeras are functional in Saccharomyces cerevisiae. *Eur J Biochem* **231**, 45–49 (1995).
135. Zeng, X. R., Hao, H., Jiang, Y. & Lee, M. Y. Regulation of human DNA polymerase delta during the cell cycle. *J Biol Chem* **269**, 24027–24033 (1994).
136. Park, H., Francesconi, S. & Wang, T. S. Cell cycle expression of two replicative DNA polymerases α and δ from *Schizosaccharomyces pombe*. *Mol Biol Cell* **4**, 145–157 (1993).
137. Shikata, K. *et al.* The human homologue of fission Yeast cdc27, p66, is a component of active human DNA polymerase delta. *J Biochem* **129**, 699–708 (2001).
138. Hughes, P., Tratner, I., Ducoux, M., Piard, K. & Baldacci, G. Isolation and identification of the third subunit of mammalian DNA polymerase delta by PCNA-affinity chromatography of mouse FM3A cell extracts. *Nucleic Acids Res* **27**, 2108–2114 (1999).
139. Gerik, K. J., Li, X., Pautz, A. & Burgers, P. M. Characterization of the two small subunits of Saccharomyces cerevisiae DNA polymerase delta. *J Biol Chem* **273**, 19747–19755 (1998).
140. Reynolds, N., Watt, A., Fantes, P. A. & MacNeill, S. A. Cdm1, the smallest subunit of DNA polymerase d in the fission yeast Schizosaccharomyces pombe, is non-essential for growth and division. *Curr Genet* **34**, 250–258 (1998).
141. Zuo, S., Bermudez, V., Zhang, G., Kelman, Z. & Hurwitz, J. Structure and activity associated with multiple forms of Schizosaccharomyces pombe DNA polymerase delta. *J Biol Chem* **275**, 5153–5162 (2000).

142. Goulian, M., Herrmann, S. M., Sackett, J. W. & Grimm, S. L. Two forms of DNA polymerase delta from mouse cells. Purification and properties. *J Biol Chem* **265**, 16402–16411 (1990).
143. Xie, B. *et al.* Reconstitution and characterization of the human DNA polymerase delta four-subunit holoenzyme. *Biochemistry* **41**, 13133–13142 (2002).
144. Brown, W. C. & Campbell, J. L. Interaction of proliferating cell nuclear antigen with yeast DNA polymerase delta. *J Biol Chem* **268**, 21706–21710 (1993).
145. Eissenberg, J. C., Ayyagari, R., Gomes, X. V. & Burgers, P. M. Mutations in yeast proliferating cell nuclear antigen define distinct sites for interaction with DNA polymerase δ and DNA polymerase ε. *Mol Cell Biol* **17**, 6367–6378 (1997).
146. Sun, Y. *et al.* Expression and characterization of the small subunit of human DNA polymerase delta. *J Biol Chem* **272**, 13013–13018 (1997).
147. Burgers, P. M. & Gerik, K. J. Structure and processivity of two forms of Saccharomyces cerevisiae DNA polymerase delta. *J Biol Chem* **273**, 19756–19762 (1998).
148. Wu, S. M. *et al.* Characterization of the p125 subunit of human DNA polymerase δ and its deletion mutants. Interaction with cyclin-dependent kinase-cyclins [published erratum appeared in J Biol Chem 1998 Jul 10;273:17978]. *J Biol Chem* **273**, 9561–9569 (1998).
149. Perderiset, M. *et al.* Mutant DNA polymerase d from thermosensitive S. pombe strains display reduced stimulation by profilerating cell nuclear antigen. *Biochem J* **335**, 581–588 (1998).
150. Mozzherin, D. J., Tan, C.-K., Downey, K. M. & Fisher, P. A. Architecture of the active DNA polymerase delta-proliferationg cell nuclear antigen template-primer complex. *J Biol Chem* **274**, 19862–19867 (1999).
151. Zhang, S. J. *et al.* A conserved region in the amino terminus of DNA polymerase delta is involved in proliferating cell nuclear antigen binding. *J Biol Chem* **270**, 7988–7992 (1995).
152. Tratner, I., Piard, K., Grenon, M., Perderiset, M. & Baldacci, G. PCNA and DNA polymerase δ catalytic subunit from *Schizosaccharomyces pombe* do not interact directly. *Biochem Biophys Research Comm* **231**, 321–328 (1997).
153. Zhou, J. Q., He, H., Tan, C. K., Downey, K. M. & So, A. G. The small subunit is required for functional interaction of DNA polymerase delta with the proliferating cell nuclear antigen. *Nucleic Acids Res* **25**, 1094–1099 (1997).
154. Podust, V. N., Chang, L. S., Ott, R., Dianov, G. L. & Fanning, E. Reconstitution of human DNA polymerase delta using recombinant

baculoviruses: the p12 subunit potentiates DNA polymerizing activity of the four-subunit enzyme. *J Biol Chem* **277**, 3894–3901 (2002).
155. Li, H. *et al.* Functional roles of p12, the fourth subunit of human DNA polymerase delta. *J Biol Chem* **281**, 14748–14755 (2006).
156. Meng, X. *et al.* DNA damage alters DNA polymerase delta to a form that exhibits increased discrimination against modified template bases and mismatched primers. *Nucleic Acids Research* **37**, 647–657 (2009).
157. Zhang, S. *et al.* A novel DNA damage response: rapid degradation of the p12 subunit of dna polymerase delta. *The Journal of Biological Chemistry* **282**, 15330–15340 (2007).
158. Podust, V., Mikhailov, V., Georgaki, A. & Hubscher, U. DNA polymerase delta and epsilon holoenzymes from calf thymus. *Chromosoma* **102**, S133–141 (1992).
159. Podust, L. M., Podust, V. N., Floth, C. & Hubscher, U. Assembly of DNA polymerase delta and epsilon holoenzymes depends on the geometry of the DNA template. *Nucleic Acids Res* **22**, 2970–2975 (1994).
160. Zhou, J. Q., Tan, C. K., So, A. G. & Downey, K. M. Purification and characterization of the catalytic subunit of human DNA polymerase delta expressed in baculovirus-infected insect cells. *J Biol Chem* **271**, 29740–29745 (1996).
161. Mozzherin, D. J. *et al.* Proliferating cell nuclear antigen promotes misincorporation catalyzed by calf thymus DNA polymerase δ. *J Biol Chem* **271**, 31711–31717 (1996).
162. Mozzherin, D. J., Shibutani, S., Tan, C. K., Downey, K. M. & Fisher, P. A. Proliferating cell nuclear antigen promotes DNA synthesis past template lesions by mammalian DNA polymerase δ. *Proc Natl Acad Sci USA* **94**, 6126–6131 (1997).
163. Chen, X., Zao, S., Kelman, Z., Hurwitz, J. & Goodman, M. F. Fidelity of eukaryotic DNA polymerase δ holoenzyme from *Schizosaccharomyces pombe*. *J Biol Chem* **275**, 17677–17685 (2000).
164. Einolf, H. J. & Guengerich, F. P. Kinetic analysis of nucleotide incorporation by mammalian DNA polymerase delta. *J Biol Chem* **275**, 16316–16322 (2000).
165. Einolf, H. J. & Guengerich, F. P. Fidelity of nucleotide insertion at 8-oxo-7,8-dihydroguanine by mammalian DNA polymerase delta. Steady-state and pre-steady-state kinetic analysis. *J Biol Chem* **276**, 3764–3771 (2001).
166. Hashimoto, K., Shimizu, K., Nakashima, N. & Sugino, A. Fidelity of DNA polymerase delta holoenzyme from Saccharomyces cerevisiae: the sliding

clamp proliferating cell nuclear antigen decreases its fidelity. *Biochemistry* **42**, 14207–14213 (2003).

167. Daigaku, Y. *et al.* A global profile of replicative polymerase usage. *Nat Struct Mol Biol* **22**, 192–198 (2015).
168. Hindges, R. & Hübscher, U. DNA polymerase δ, an essential enzyme for DNA transactions. *Biol Chem* **378**, 345–362 (1997).
169. Wood, R. D. & Shivji, M. K. K. Which DNA polymeases are used for DNA-repair in eukaryotes? *Carcinogenesis* **18**, 605–610 (1997).
170. Prindle, M. J. & Loeb, L. A. DNA polymerase delta in dna replication and genome maintenance. *Environ Mol Mutagen* **53**, 666–682 (2012).
171. Zhang, Q., Chen, H. Q., Chen, Y. & Zhou, Y. J. Multiple Forms of Human DNA Polymerase Delta Sub-Assembling in Cellular DNA Transactions. *Curr Protein Pept Sc* **17**, 746–755 (2016).
172. Francesconi, S., De Recondo, A. M. & Baldacci, G. DNA polymerase delta is required for the replication feedback control of cell cycle progression in Schizosaccharomyces pombe. *Mol Gen Genet* **246**, 561–569 (1995).
173. Francesconi, S., Park, H. & Wang, T. S. Fission yeast with DNA polymerase delta temperature-sensitive alleles exhibits cell division cycle phenotype. *Nucleic Acids Res* **21**, 3821–3828 (1993).
174. Frouin, I. *et al.* Cell cycle-dependent dynamic association of cyclin/Cdk complexes with human DNA replication proteins. *Embo J* **21**, 2485–2495 (2002).
175. Frouin, I., Montecucco, A., Spadari, S. & Maga, G. DNA replication: a complex matter. *EMBO Rep* **4**, 666–670 (2003).
176. Maga, G. & Hubscher, U. Proliferating cell nuclear antigen (PCNA): a dancer with many partners. *J Cell Sci* **116**, 3051–3060 (2003).
177. Miyabe, I. *et al.* Polymerase delta replicates both strands after homologous recombination-dependent fork restart. *Nat Struct Mol Biol* **22**, 932–938 (2015).
178. Podust, V. N. & Hubscher, U. Lagging strand DNA synthesis by calf thymus DNA polymerases alpha, beta, delta and epsilon in the presence of auxiliary proteins. *Nucleic Acids Res* **21**, 841–846 (1993).
179. Bae, S. H. & Seo, Y. S. Characterization of the enzymatic properties of the yeast Dna2 helicase/endonuclease suggests a new model for Okazaki fragment processing. *J Biol Chem* **275**, 38022–38031 (2000).
180. Maga, G. *et al.* Okazaki fragment processing: modulation of the strand displacement activity of DNA polymerase δ by the concerted action of replication protein A, proliferating cell nuclear antigen and flap endonuclease 1. *Proc Natl Acad Sci USA* **98**, 14298–14303 (2001).

181. Ayyagari, R., Gomes, X. V., Gordenin, D. A. & Burgers, P. M. Okazaki Fragment Maturation in Yeast. I. DISTRIBUTION OF FUNCTIONS BETWEEN FEN1 AND DNA2. *J Biol Chem* **278**, 1618–1625 (2003).
182. Jin, Y. H., Ayyagari, R., Resnick, M. A., Gordenin, D. A. & Burgers, P. M. Okazaki fragment maturation in yeast. II. Cooperation between the polymerase and 3′-5′-exonuclease activities of Pol delta in the creation of a ligatable nick. *J Biol Chem* **278**, 1626–1633 (2003).
183. Matsumoto, Y. Molecular mechanism of PCNA-dependent base excision repair. *Prog Nucleic Acid Res Mol Biol* **68**, 129–138 (2001).
184. Li, Y., Asahara, H., Patel, V. S., Zhou, S. & Linn, S. Purification, cDNA cloning, and gene mapping of the small subunit of human DNA polymerase ε. *J Biol Chem* **272**, 32337–32344 (1997).
185. Jokela, M. *et al.* The small subunits of human and mouse DNA polymerase ε are homologous to the second largest subunit of the yeast *Saccharomyces cerevisiae* DNA polymerase ε. *Nucl Acids Res* **26**, 730–734 (1998).
186. Dua, R., Edwards, S., Levy, D. L. & Campbell, J. L. Subunit interactions within the Saccharomyces cerevisiae DNA polymerase epsilon (pol epsilon) complex. Demonstration of a dimeric pol epsilon. *J Biol Chem* **275**, 28816–28825 (2000).
187. Li, Y., Pursell, Z. F. & Linn, S. Identification and cloning of two histone fold motif-containing subunits of HeLa DNA polymerase epsilon. *J Biol Chem* **275**, 31554 (2000).
188. Ohya, T., Maki, S., Kawasaki, Y. & Sugino, A. Structure and function of the fourth subunit (Dpb4p) of DNA polymerase epsilon in Saccharomyces cerevisiae. *Nucleic Acids Res* **28**, 3846–3852 (2000).
189. Pursell, Z. F. & Kunkel, T. A. DNA polymerase epsilon: a polymerase of unusual size (and complexity). *Prog Nucleic Acid Res Mol Biol* **82**, 101–145 (2008).
190. Maga, G., Jonsson, Z. O., Stucki, M., Spadari, S. & Hubscher, U. Dual mode of interaction of DNA polymerase epsilon with proliferating cell nuclear antigen in primer binding and DNA synthesis. *J Mol Biol* **285**, 259–267 (1999).
191. Vlatkovic, N. *et al.* MDM2 interacts with the C-terminus of the catalytic subunit of DNA polymerase ε. *Nucl. Acids Res.* **28**, 3581–3586 (2000).
192. Makiniemi, M. *et al.* BRCT domain-containing protein TopBP1 functions in DNA replication and damage response. *J Biol Chem* **6**, 6 (2001).
193. Fuss, J. & Linn, S. Human DNA polymerase epsilon colocalizes with proliferating cell nuclear antigen and DNA replication late, but not early, in S phase. *J Biol Chem* **277**, 8658–8666 (2002).

194. Lee, S. H., Pan, Z. Q., Kwong, A. D., Burgers, P. M. & Hurwitz, J. Synthesis of DNA by DNA polymerase epsilon in vitro. *J Biol Chem* **266**, 22707–22717 (1991).
195. Pursell, Z. F., Isoz, I., Lundstrom, E. B., Johansson, E. & Kunkel, T. A. Yeast DNA polymerase epsilon participates in leading-strand DNA replication. *Science* **317**, 127–130 (2007).
196. Sabouri, N. & Johansson, E. Translesion synthesis of abasic sites by yeast DNA polymerase epsilon. *J Biol Chem* **284**, 31555–31563 (2009).
197. Sabouri, N., Viberg, J., Goyal, D. K., Johansson, E. & Chabes, A. Evidence for lesion bypass by yeast replicative DNA polymerases during DNA damage. *Nucleic Acids Res* **36**, 5660–5667 (2008).
198. Navas, T. A., Sanchez, Y. & Elledge, S. J. RAD9 and DNA polymerase ε form parallel sensory branches for transducing the DNA damage checkpoint signal in *Saccharomyces cerevisiae*. *Genes Dev.* **10**, 2632–2643 (1996).
199. Navas, T. A., Zhou, Z. & Elledge, S. J. DNA polymerase epsilon links the DNA replication machinery to the S phase checkpoint. *Cell* **80**, 29–39 (1995).
200. Kesti, T., McDonald, W. H., Yates, J. R., 3rd & Wittenberg, C. Cell cycle-dependent phosphorylation of the DNA polymerase epsilon subunit, Dpb2, by the Cdc28 cyclin-dependent protein kinase. *J Biol Chem* **279**, 14245–14255 (2004).
201. Sengupta, S., van Deursen, F., de Piccoli, G. & Labib, K. Dpb2 Integrates the Leading-Strand DNA Polymerase into the Eukaryotic Replisome. *Current Biology* **23**, 543–552 (2013).
202. Sun, J. C. *et al.* The architecture of a eukaryotic replisome. *Nat Struct Mol Biol* **22**, 976–982 (2015).
203. Smith, J. S., Caputo, E. & Boeke, J. D. A genetic screen for ribosomal DNA silencing defects identifies multiple DNA replication and chromatin-modulating factors. *Mol Cell Biol* **19**, 3184–3197 (1999).
204. Gan, G. N., Wittschieben, J. P., Wittschieben, B. O. & Wood, R. D. DNA polymerase zeta (pol zeta) in higher eukaryotes. *Cell Res* **18**, 174–183 (2008).
205. Makarova, A. V. & Burgers, P. M. Eukaryotic DNA polymerase zeta. *DNA Repair* **29**, 47–55 (2015).
206. Wittschieben, J. *et al.* Disruption of the developmentally regulated Rev3I gene causes embryonic lethality. *Curr Biol* **10**, 1217–1220 (2000).
207. Gomez-Llorente, Y. *et al.* The Architecture of Yeast DNA Polymerase zeta. *Cell Rep* **5**, 79–86 (2013).

208. Johnson, R. E., Washington, M. T., Haracska, L., Prakash, S. & Prakash, L. Eukaryotic polymerases iota and zeta act sequentially to bypass DNA lesions. *Nature* **406**, 1015–1019 (2000).
209. Johnson, R. E., Haracska, L., Prakash, S. & Prakash, L. Role of DNA polymerase zeta in the bypass of a (6-4) TT photoproduct. *Mol Cell Biol* **21**, 3558–3563 (2001).
210. Johnson, R. E., Prakash, L. & Prakash, S. Pol31 and Pol32 subunits of yeast DNA polymerase delta are also essential subunits of DNA polymerase zeta. *P Natl Acad Sci USA* **109**, 12455–12460 (2012).
211. Garg, P., Stith, C. M., Majka, J. & Burgers, P. M. Proliferating cell nuclear antigen promotes translesion synthesis by DNA polymerase zeta. *J Biol Chem* **280**, 23446–23450 (2005).
212. Makarova, A. V., Stodola, J. L. & Burgers, P. M. A four-subunit DNA polymerase zeta complex containing Pol delta accessory subunits is essential for PCNA-mediated mutagenesis. *Nucleic Acids Research* **40**, 11618–11626 (2012).
213. Boersma, V. *et al.* MAD2L2 controls DNA repair at telomeres and DNA breaks by inhibiting 5′ end resection. *Nature* **521**, 537-U291 (2015).
214. Xu, G. T. *et al.* REV7 counteracts DNA double-strand break resection and affects PARP inhibition. *Nature* **521**, 541-U308 (2015).
215. Garcia-Diaz, M. *et al.* DNA polymerase lambda (Pol lambda), a novel eukaryotic DNA polymerase with a potential role in meiosis. *J Mol Biol* **301**, 851–867 (2000).
216. Garcia-Diaz, M. *et al.* DNA polymerase lambda, a novel DNA repair enzyme in human cells. *J Biol Chem* **277**, 13184–13191 (2002).
217. Garcia-Diaz, M. e. a. Structure-function studies of DNA polymerase lambda. *DNA Repair (Amst)* **4**, 1358–1367 (2005).
218. Bebenek, K., Garcia-Diaz, M., Zhou, R. Z., Povirk, L. F. & Kunkel, T. A. Loop 1 modulates the fidelity of DNA polymerase lambda. *Nucleic Acids Research* **38**, 5419–5431 (2010).
219. Garcia-Diaz, M., Bebenek, K., Kunkel, T. A. & Blanco, L. Identification of an intrinsic 5′-deoxyribose-5-phosphate lyase activity in human DNA polymerase lambda: a possible role in base excision repair. *J Biol Chem* **276**, 34659–34663 (2001).
220. Ramadan, K., Shevelev, I. V., Maga, G. & Hubscher, U. De novo DNA synthesis by human DNA polymerase lambda, DNA polymerase mu and terminal deoxyribonucleotidyl transferase. *J Mol Biol* **339**, 395–404 (2004).

221. Maga, G. *et al.* DNA elongation by the human DNA polymerase lambda polymerase and terminal transferase activities are differentially coordinated by proliferating cell nuclear antigen and replication protein A. *J Biol Chem* **280**, 1971–1981 (2005).
222. Ramadan, K., Shevelev, I. V., Maga, G. & Hubscher, U. DNA polymerase lambda from calf thymus preferentially replicates damaged DNA. *J Biol Chem* **277**, 18454–18458 (2002).
223. Blanca, G. *et al.* Human DNA polymerases lambda and beta show different efficiencies of translesion DNA synthesis past abasic sites and alternative mechanisms for frameshift generation. *Biochemistry* **43**, 11605–11615 (2004).
224. Lee, J. W. *et al.* Implication of DNA polymerase lambda in alignment-based gap filling for nonhomologous DNA end joining in human nuclear extracts. *J Biol Chem* **279**, 805–811 (2004).
225. Nick McElhinny, S. A. *et al.* A gradient of template dependence defines distinct biological roles for family X polymerases in nonhomologous end joining. *Mol Cell* **19**, 357–366 (2005).
226. Crespan, E., Hubscher, U. & Maga, G. Error-free bypass of 2-hydroxyadenine by human DNA polymerase lambda with Proliferating Cell Nuclear Antigen and Replication Protein A in different sequence contexts. *Nucleic Acids Res* **35**, 5173–5181 (2007).
227. Maga, G. *et al.* 8-oxo-guanine bypass by human DNA polymerases in the presence of auxiliary proteins. *Nature* **447**, 606–608 (2007).
228. Maga, G. *et al.* Replication protein A and proliferating cell nuclear antigen coordinate DNA polymerase selection in 8-oxo-guanine repair. *Proc Natl Acad Sci USA* **105**, 20689–20694 (2008).
229. Bertocci, B. *et al.* Cutting edge: DNA polymerases mu and lambda are dispensable for Ig gene hypermutation. *J Immunol* **168**, 3702–3706 (2002).
230. Braithwaite, E. K. *et al.* DNA polymerase lambda protects mouse fibroblasts against oxidative DNA damage and is recruited to sites of DNA damage/repair. *J Biol Chem* **280**, 31641–31647 (2005).
231. Braithwaite, E. K. *et al.* DNA polymerase lambda mediates a back-up base excision repair activity in extracts of mouse embryonic fibroblasts. *J Biol Chem* **280**, 18469–18475 (2005).
232. Villani, G. *et al.* In Vitro Gap-directed Translesion DNA Synthesis of an Abasic Site Involving Human DNA Polymerases epsilon, lambda, and beta. *Journal of Biological Chemistry* **286**, 32094–32104 (2011).

233. Uchiyama, Y., Kimura, S., T., Y., Ishibashi, T. & Sakaguchi, K. Plant DNA polymerase lambda, a DNA repair enzyme that functions in plant meristematic and meiotic tissues. *Eur J Biochem* **1**, 2799–2807 (2004).
234. Uchiyama, Y., Takeuchi, R., Kodera, H. & Sakaguchi, K. Distribution and roles of X-family DNA polymerases in eukaryotes. *Biochimie* **91**, 165–170 (2009).
235. Uchiyama, Y., Kimura, S., Yamamoto, T., Ishibashi, T. & Sakaguchi, K. Plant DNA polymerase lambda, a DNA repair enzyme that functions in plant meristematic and meiotic tissues. *European Journal of Biochemistry* **271**, 2799–2807 (2004).
236. Amoroso, A. *et al.* Oxidative DNA Damage Bypass in Arabidopsis thaliana Requires DNA Polymerase lambda and Proliferating Cell Nuclear Antigen 2. *Plant Cell* **23**, 806–822 (2011).
237. Dominguez, O. *et al.* DNA polymerase mu (Pol mu), homologous to TdT, could act as a DNA mutator in eukaryotic cells. *Embo J* **19**, 1731–1742 (2000).
238. Covo, S., Blanco, L. & Livneh, Z. Lesion bypass by human DNA polymerase mu reveals a template-dependent, sequence-independent nucleotidyl transferase activity. *J Biol Chem* **279**, 859–865 (2004).
239. Ruiz, J. F. *et al.* DNA polymerase mu, a candidate hypermutase? *Philos Trans R Soc Lond B Biol Sci* **356**, 99–109 (2001).
240. Zhang, Y., Wu, X., Yuan, F., Xie, Z. & Wang, Z. Highly frequent frameshift DNA synthesis by human DNA polymerase mu. *Mol Cell Biol* **21**, 7995–8006 (2001).
241. Bertocci, B., De Smet, A., Berek, C., Weill, J. C. & Reynaud, C. A. Immunoglobulin kappa light chain gene rearrangement is impaired in mice deficient for DNA polymerase mu. *Immunity* **19**, 203–211 (2003).
242. Duvauchelle, J. B., Blanco, L., Fuchs, R. P. & Cordonnier, A. M. Human DNA polymerase mu (Pol mu) exhibits an unusual replication slippage ability at AAF lesion. *Nucleic Acids Res* **30**, 2061–2067 (2002).
243. Zhang, Y. *et al.* Lesion bypass activities of human DNA polymerase mu. *J Biol Chem* **277**, 44582–44587 (2002).
244. Havener, J. M. *et al.* Translesion synthesis past platinum DNA adducts by human DNA polymerase mu. *Biochemistry* **42**, 1777–1788 (2003).
245. Mahajan, K. N., Nick McElhinny, S. A., Mitchell, B. S. & Ramsden, D. A. Association of DNA polymerase mu (pol mu) with Ku and ligase IV: role for pol mu in end-joining double-strand break repair. *Mol Cell Biol* **22**, 5194–5202 (2002).

246. Pryor, J. M. *et al.* Essential role for polymerase specialization in cellular nonhomologous end joining. *P Natl Acad Sci USA* **112**, E4537–E4545 (2015).
247. Moon, A. F., Gosavi, R. A., Kunkel, T. A., Pedersen, L. C. & Bebenek, K. Creative template-dependent synthesis by human polymerase mu. *P Natl Acad Sci USA* **112**, E4530–E4536 (2015).
248. Beard, W. A., Prasad, R. & Wilson, S. H. Activities and mechanism of DNA polymerase beta. *Methods Enzymol* **408**, 91–107 (2006).
249. Ahn, J., Kraynov, V. S., Zhong, X., Werneburg, B. G. & Tsai, M. D. DNA polymerase beta: effects of gapped DNA substrates on dNTP specificity, fidelity, processivity and conformational changes. *Biochem J* **331**, 79–87 (1998).
250. Jezewska, M. J., Rajendran, S. & Bujalowski, W. Interactions of the 8-kDa domain of rat DNA polymerase beta with DNA. *Biochemistry* **40**, 3295–3307 (2001).
251. Waser, J., Hubscher, U., Kuenzle, C. C. & Spadari, S. DNA polymerase beta from brain neurons is a repair enzyme. *Eur J Biochem* **97**, 361–368 (1979).
252. Hubscher, U., Kuenzle, C. C., Limacher, W., Scherrer, P. & Spadari, S. Functions of DNA polymerases alpha, beta, and gamma in neurons during development. *Cold Spring Harb Symp Quant Biol* **43 Pt 1**, 625–629 (1979).
253. Fortini, P. *et al.* Different DNA polymerases are involved in the short- and long-patch base excision repair in mammalian cells. *Biochemistry* **37**, 3575–3580 (1998).
254. Canitrot, Y., Frechet, M., Servant, L., Cazaux, C. & Hoffmann, J. S. Overexpression of DNA polymerase beta: a genomic instability enhancer process. *FASEB J.* **13**, 1107–1111 (1999).
255. Plug, A. W., Clairmont, C. A., Sapi, E., Ashley, T. & Sweasy, J. B. Evidence for a role for DNA polymerase beta in mammalian meiosis. *Proc Natl Acad Sci USA* **94**, 1327–1331 (1997).
256. Hoffmann, J. S. *et al.* DNA polymerase beta bypasses in vitro a single d(GpG)-cisplatin adduct placed on codon 13 of the HRAS gene. *Proc Natl Acad Sci USA* **92**, 5356–5360 (1995).
257. Hoffmann, J. S., Pillaire, M. J., Garcia-Estefania, D., Lapalu, S. & Villani, G. In vitro bypass replication of the cisplatin-d(GpG) lesion by calf thymus DNA polymerase beta and human immunodeficiency virus type I reverse transcriptase is highly mutagenic. *J Biol Chem* **271**, 15386–15392 (1996).

258. Frechet, M. *et al.* Deregulated DNA polymerase beta strengthens ionizing radiation-induced nucleotidic and chromosomal instabilities. *Oncogene* **21**, 2320–2327 (2002).
259. Frechet, M., Canitrot, Y., Cazaux, C. & Hoffmann, J. S. DNA polymerase beta imbalance increases apoptosis and mutagenesis induced by oxidative stress. *FEBS Lett* **505**, 229–232 (2001).
260. Sugo, N., Aratani, Y., Nagashima, Y., Kubota, Y. & Koyama, H. Neonatal lethality with abnormal neurogenesis in mice deficient in DNA polymerase beta. *EMBO J* **19**, 1397–1404 (2000).
261. Ernst, P. *et al.* Mechanisms of transcriptional regulation in lymphocyte progenitors: insight from an analysis of the terminal transferase promoter. *Cold Spring Harb Symp Quant Biol* **64**, 87–97 (1999).
262. Ernst, P. *et al.* A potential role for Elf-1 in terminal transferase gene regulation. *Mol Cell Biol* **16**, 6121–6131 (1996).
263. Thai, T. H. & Kearney, J. F. Distinct and opposite activities of human terminal deoxynucleotidyltransferase splice variants. *J Immunol* **173**, 4009–4019 (2004).
264. Thai, T. H. & Kearney, J. F. Isoforms of terminal deoxynucleotidyltransferase: developmental aspects and function. *Adv Immunol* **86**, 113–136 (2005).
265. Thai, T. H., Purugganan, M. M., Roth, D. B. & Kearney, J. F. Distinct and opposite diversifying activities of terminal transferase splice variants. *Nat Immunol* **3**, 457–462 (2002).
266. Delarue, M. *et al.* Crystal structures of a template-independent DNA polymerase: murine terminal deoxynucleotidyltransferase. *Embo J* **21**, 427–439 (2002).
267. Mahajan, K. N. *et al.* Association of terminal deoxynucleotidyl transferase with Ku. *Proc Natl Acad Sci USA* **96**, 13926–13931 (1999).
268. Mickelsen, S. *et al.* Modulation of terminal deoxynucleotidyltransferase activity by the DNA-dependent protein kinase. *J Immunol* **163**, 834–843 (1999).
269. Gouge, J. *et al.* Structural basis for a novel mechanism of DNA bridging and alignment in eukaryotic DSB DNA repair. *Embo Journal* **34**, 1126–1142 (2015).
270. McDonald, J. P. *et al.* Novel human and mouse homologs of Saccharomyces cerevisiae DNA polymerase eta. *Genomics* **60**, 20–30 (1999).
271. Masutani, C. *et al.* Xeroderma pigmentosum variant (XP-V) correcting protein from Hela cells has a thymine dimer bypass DNA polymerase activity. *EMBO J* **18**, 3491–3501 (1999).

272. Kannouche, P. *et al.* Domain structure, localization, and function of DNA polymerase eta, defective in xeroderma pigmentosum variant cells. *Genes Dev* **15**, 158–172 (2001).
273. Masutani, C., Kusumoto, R., Iwai, S. & Hanaoka, F. Mechanism of accurate translesion synthesis by human DNA polymerase eta. *EMBO J* **19**, 3100–3109 (2000).
274. Prakash, S. *et al.* Role of yeast and human DNA polymerase eta in error-free replication of damaged DNA. *Cold Spring Harb Symp Quant Biol* **65**, 51–59 (2000).
275. Washington, M. T., Johnson, R. E., Prakash, S. & Prakash, L. Accuracy of thymine-thymine dimer bypass by Saccharomyces cerevisiae DNA polymerase eta. *Proc Natl Acad Sci USA* **97**, 3094–3099 (2000).
276. Prakash, S., Johnson, R. E. & Prakash, L. Eukaryotic translesion synthesis DNA polymerases: specificity of structure and function. *Annu Rev Biochem* **74**, 317–353 (2005).
277. Yang, W. An Overview of Y-Family DNA Polymerases and a Case Study of Human DNA Polymerase eta. *Biochemistry* **53**, 2793–2803 (2014).
278. Washington, M. T., Johnson, R. E., Prakash, S. & Prakash, L. Fidelity and processivity of Saccharomyces cerevisiae DNA polymerase eta. *J Biol Chem* **274**, 36835–36838 (1999).
279. Johnson, R. E., Washington, M. T., Prakash, S. & Prakash, L. Fidelity of human DNA polymerase eta. *J Biol Chem* **275**, 7447–7450 (2000).
280. Matsuda, T., Bebenek, K., Masutani, C., Hanoaka, F. & Kunkel, T. A. Low fidelity DNA synthesis by human DNA polymerase eta. *Nature* **404**, 1011–1013 (2000).
281. Kannouche, P. *et al.* Localization of DNA polymerases eta and iota to the replication machinery is tightly co-ordinated in human cells. *EMBO J* **21**, 6246–6256 (2002).
282. Haracska, L., Kondratick, C. M., Unk, I., Prakash, S. & Prakash, L. Interaction with PCNA is essential for yeast DNA polymerase eta function. *Mol Cell* **8**, 407–415 (2001).
283. Haracska, L. *et al.* Physical and functional interactions of human DNA polymerase eta with PCNA. *Mol Cell Biol* **21**, 7199–7206 (2001).
284. Kannouche, P. L., Wing, J. & Lehmann, A. R. Interaction of human DNA polymerase eta with monoubiquitinated PCNA: a possible mechanism for the polymerase switch in response to DNA damage. *Mol Cell* **14**, 491–500 (2004).
285. Acharya, N. *et al.* Roles of PCNA-binding and ubiquitin-binding domains in human DNA polymerase eta in translesion DNA synthesis. *Proc Natl Acad Sci USA* **105**, 17724–17729 (2008).

286. Sabbioneda, S. *et al.* Ubiquitin-binding motif of human DNA polymerase eta is required for correct localization. *Proc Natl Acad Sci USA* **106**, E20; author reply E21 (2009).
287. Kawamoto, T. *et al.* Dual roles for DNA polymerase eta in homologous DNA recombination and translesion DNA synthesis. *Mol Cell* **20**, 793–799 (2005).
288. McIlwraith, M. J. *et al.* Human DNA polymerase eta promotes DNA synthesis from strand invasion intermediates of homologous recombination. *Mol Cell* **20**, 783–792 (2005).
289. Rattray, A. J. & Strathern, J. N. Homologous recombination is promoted by translesion polymerase poleta. *Mol Cell* **20**, 658–659 (2005).
290. Buisson, R. *et al.* Breast Cancer Proteins PALB2 and BRCA2 Stimulate Polymerase h in Recombination-Associated DNA Synthesis at Blocked Replication Forks. *Cell Rep* **6**, 553–564 (2014).
291. Rey, L. *et al.* Human DNA Polymerase eta Is Required for Common Fragile Site Stability during Unperturbed DNA Replication. *Molecular and Cellular Biology* **29**, 3344–3354 (2009).
292. Srivastava, A. K. *et al.* Enhanced expression of DNA polymerase eta contributes to cisplatin resistance of ovarian cancer stem cells. *P Natl Acad Sci USA* **112**, 4411–4416 (2015).
293. McDonald, J. P. *et al.* DNA polymerase iota and related rad30-like enzymes. *Philos Trans R Soc Lond B Biol Sci* **356**, 53–60 (2001).
294. Vaisman, A., Tissier, A., Frank, E. G., Goodman, M. F. & Woodgate, R. Human DNA polymerase iota promiscuous mismatch extension. *J Biol Chem* **276**, 30615–30622 (2001).
295. Zhang, Y., Yuan, F., Wu, X., Taylor, J. S. & Wang, Z. Response of human DNA polymerase iota to DNA lesions. *Nucleic Acids Res* **29**, 928–935 (2001).
296. Haracska, L. *et al.* Targeting of human DNA polymerase iota to the replication machinery via interaction with PCNA. *Proc Natl Acad Sci USA* **98**, 14256–14261 (2001).
297. McIntyre, J. *et al.* Ubiquitin mediates the physical and functional interaction between human DNA polymerases eta and iota. *Nucleic Acids Research* **41**, 1649–1660 (2013).
298. McIntyre, J. *et al.* Posttranslational Regulation of Human DNA Polymerase i. *Journal of Biological Chemistry* **290**, 27332–27344 (2015).
299. Sun, H. Y. *et al.* Elevated DNA polymerase iota (Poli) is involved in the acquisition of aggressive phenotypes of human esophageal squamous cell cancer. *Int J Clin Exp Patho* **8**, 3591–3601 (2015).

300. Ohashi, E. *et al.* Fidelity and processivity of DNA synthesis by DNA polymerase kappa, the product of the human DINB1 gene. *J Biol Chem* **275**, 39678–39684 (2000).
301. Levine, R. L. *et al.* Translesion DNA synthesis catalyzed by human pol eta and pol kappa across 1,N6-ethenodeoxyadenosine. *J Biol Chem* **276**, 18717–18721 (2001).
302. Uljon, S. N. *et al.* Crystal structure of the catalytic core of human DNA polymerase kappa. *Structure (Camb)* **12**, 1395–1404 (2004).
303. Zhang, Y. *et al.* Error-free and error-prone lesion bypass by human DNA polymerase kappa in vitro. *Nucleic Acids Res* **28**, 4138–4146 (2000).
304. Suzuki, N. *et al.* Translesion synthesis by human DNA polymerase kappa on a DNA template containing a single stereoisomer of dG-(+)- or dG-(-)-anti-N(2)-BPDE (7,8-dihydroxy-anti-9,10-epoxy-7,8,9,10-tetrahydrobenzo[a] pyrene). *Biochemistry* **41**, 6100–6106 (2002).
305. Haracska, L., Prakash, L. & Prakash, S. Role of human DNA polymerase kappa as an extender in translesion synthesis. *Proc Natl Acad Sci USA* **99**, 16000–16005 (2002).
306. Takeiri, A. *et al.* In vivo evidence that DNA polymerase kappa is responsible for error-free bypass across DNA cross-links induced by mitomycin C. *DNA Repair* **24**, 113–121 (2014).
307. Zhang, X. L. *et al.* Mouse DNA polymerase kappa has a functional role in the repair of DNA strand breaks. *DNA Repair* **12**, 377–388 (2013).
308. Betous, R. *et al.* DNA polymerase kappa-dependent DNA synthesis at stalled replication forks is important for CHK1 activation. *Embo Journal* **32**, 2172–2185 (2013).
309. Nelson, J. R., Lawrence, C. W. & Hinkle, D. C. Deoxycytidyl transferase activity of yeast REV1 protein. *Nature* **382**, 729–731 (1996).
310. Lin, W., Xin, H., Zhang, Y., Yuan, F. & Wang, Z. The human Rev1 gene codes for a DNA template dependent dCMP transferase. *Nucl Acids Res* **27**, 4468–4475 (1999).
311. Lawrence, C. W. *et al.* Roles of DNA polymerase zeta and Rev1 protein in eukaryotic mutagenesis and translesion replication. *Cold Spring Harb Symp Quant Biol* **65**, 61–69 (2000).
312. Zhu, F. & Zhang, M. DNA polymerase zeta: new insight into eukaryotic mutagenesis and mammalian embryonic development. *World J Gastroenterol* **9**, 1165–1169 (2003).
313. Eddy, S. *et al.* Human Rev1 polymerase disrupts G-quadruplex DNA. *Nucleic Acids Research* **42**, 3272–3285 (2014).

314. Yang, Y. R. *et al.* FANCD2 and REV1 cooperate in the protection of nascent DNA strands in response to replication stress. *Nucleic Acids Research* **43**, 8325–8339 (2015).
315. Kim, H., Yang, K. L., Dejsuphong, D. & D'Andrea, A. D. Regulation of Rev1 by the Fanconi anemia core complex. *Nat Struct Mol Biol* **19**, 164–170 (2012).
316. Mislak, A. C. & Anderson, K. S. Insights into the Molecular Mechanism of Polymerization and Nucleoside Reverse Transcriptase Inhibitor Incorporation by Human PrimPol. *Antimicrob Agents Ch* **60**, 561–569 (2016).
317. Hübscher, U., Maga, G. & Podust, V. N. in *DNA Replication in Eukaryotic Cells* (ed M. L. De Pamphilis) 525–543 (Cold Spring Harbor Laborytory Press, 1996).
318. Jonsson, Z. O. & Hubscher, U. Proliferating cell nuclear antigen: more than a clamp for DNA polymerases. *Bioessays* **19**, 967–975 (1997).
319. Xu, H., Zhang, P., Liu, L. & Lee, M. Y. W. T. A novel PCNA-binding motif identified by the panning of a random peptide display library. *Biochemistry* **40**, 4512–4520 (2001).
320. Masuda, Y. *et al.* Different types of interaction between PCNA and PIP boxes contribute to distinct cellular functions of Y-family DNA polymerases. *Nucleic Acids Research* **43**, 7898–7910 (2015).
321. Prelich, G. *et al.* Functional Identity of Proliferating Cell Nuclear Antigen and a DNA Polymerase-Delta Auxiliary Protein. *Nature* **326**, 517–520 (1987).
322. Bravo, R., Frank, R., Blundell, P. A. & Macdonaldbravo, H. Cyclin Pcna Is the Auxiliary Protein of DNA Polymerase-Delta. *Nature* **326**, 515–517 (1987).
323. Shamoo, Y. & Steitz, T. A. Builiding a replisome from interacting pieces: sliding clamp complexed to a peptide from DNA polymerase and a polymerase editing complex. *Cell* **99**, 155–166 (1999).
324. Maga, G. & Hubscher, U. DNA polymerase epsilon interacts with proliferating cell nuclear antigen in primer recognition and elongation. *Biochemistry* **34**, 891–901 (1995).
325. Chen, J. J., Bozza, W. & Zhuang, Z. H. Ubiquitination of PCNA and Its Essential Role in Eukaryotic Translesion Synthesis. *Cell Biochem Biophys* **60**, 47–60 (2011).
326. McIntyre, J. & Woodgate, R. Regulation of translesion DNA synthesis: Posttranslational modification of lysine residues in key proteins. *DNA Repair* **29**, 166–179 (2015).

327. O'Day, C. L., Burgers, P. M. & Taylor, J. S. PCNA-induced DNA synthesis past cis-syn and trans-syn-I thymine dimers by calf thymus DNA polymerase delta in vitro. *Nucleic Acids Res* **20**, 5403–5406 (1992).
328. Maga, G. *et al.* Human DNA polymerase lambda functionally and physically interacts with proliferating cell nuclear antigen in normal and translesion DNA synthesis. *J Biol Chem* **277**, 48434–48440 (2002).
329. Wold, M. S. Replication protein A: a heterotrimeric, single-stranded DNA-binding protein required for eukaryotic DNA metabolism. *Annu Rev Biochem* **66**, 61–92 (1997).
330. Maga, G., Frouin, I., Spadari, S. & Hubscher, U. Replication protein A as a "fidelity clamp" for DNA polymerase alpha. *J Biol Chem* **276**, 18235–18242 (2001).
331. Maga, G., Shevelev, I., Villani, G., Spadari, S. & Hubscher, U. Human replication protein A can suppress the intrinsic in vitro mutator phenotype of human DNA polymerase lambda. *Nucleic Acids Res* **34**, 1405–1415 (2006).
332. Fortune, J. M., Stith, C. M., Kissling, G. E., Burgers, P. M. & Kunkel, T. A. RPA and PCNA suppress formation of large deletion errors by yeast DNA polymerase delta. *Nucleic Acids Res* **34**, 4335–4341 (2006).
333. Guilliam, T. A. *et al.* Human PrimPol is a highly error-prone polymerase regulated by single-stranded DNA binding proteins. *Nucleic Acids Research* **43**, 1056–1068 (2015).
334. Liu, L., Rodriguez-Belmonte, E. M., Mazloum, N., Xie, B. & Lee, M. Y. Identification of a novel protein, PDIP38, that interacts with the p50 subunit of DNA polymerase delta and proliferating cell nuclear antigen. *J Biol Chem* **278**, 10041–10047 (2003).
335. Lyle, A. N. *et al.* Poldip2, a Novel Regulator of Nox4 and Cytoskeletal Integrity in Vascular Smooth Muscle Cells. *Circ Res* **105**, 249-U116 (2009).
336. Tissier, A. *et al.* Crosstalk between replicative and translesional DNA polymerases: PDIP38 interacts directly with Poleta. *DNA Repair (Amst)* **9**, 922–928 (2010).
337. Maga, G. *et al.* DNA polymerase delta-interacting protein 2 is a processivity factor for DNA polymerase lambda during 8-oxo-7,8-dihydroguanine bypass. *P Natl Acad Sci USA* **110**, 18850–18855 (2013).
338. Guilliam, T. A., Bailey, L. J., Brissett, N. C. & Doherty, A. J. PolDIP2 interacts with human PrimPol and enhances its DNA polymerase activities. *Nucleic Acids Research* **44**, 3317–3329 (2016).

339. Gelinas, A. D. *et al.* Telomere capping proteins are structurally related to RPA with an additional telomere-specific domain. *P Natl Acad Sci USA* **106**, 19298–19303 (2009).
340. Sun, J. *et al.* Stn1-Ten1 is an Rpa2-Rpa3-like complex at telomeres. *Gene Dev* **23**, 2900–2914 (2009).
341. Giraud-Panis, M. J., Teixeira, M. T., Geli, V. & Gilson, E. CST Meets Shelterin to Keep Telomeres in Check. *Molecular Cell* **39**, 665–676 (2010).
342. Price, C. M. *et al.* Evolution of CST function in telomere maintenance. *Cell Cycle* **9**, 3157–3165 (2010).
343. Miyake, Y. *et al.* RPA-like Mammalian Ctc1-Stn1-Ten1 Complex Binds to Single-Stranded DNA and Protects Telomeres Independently of the Pot1 Pathway. *Molecular Cell* **36**, 193–206 (2009).
344. Goulian, M., Heard, C. J. & Grimm, S. L. Purification and properties of an accessory protein for DNA polymerase alpha/primase. *J Biol Chem* **265**, 13221–13230 (1990).
345. Liu, B. *et al.* A primase subunit essential for efficient primer synthesis by an archaeal eukaryotic-type primase. *Nat Commun* **6** (2015).
346. Kumar, D., Viberg, J., Nilsson, A. K. & Chabes, A. Highly mutagenic and severely imbalanced dNTP pools can escape detection by the S-phase checkpoint. *Nucleic Acids Research* **38**, 3975–3983 (2010).
347. Ferraro, P., Franzolin, E., Pontarin, G., Reichard, P. & Bianchi, V. Quantitation of cellular deoxynucleoside triphosphates. *Nucleic Acids Research* **38** (2010).
348. Nick McElhinny, S. A. & Ramsden, D. A. Polymerase mu is a DNA-directed DNA/RNA polymerase. *Mol Cell Biol* **23**, 2309–2315 (2003).
349. Nick McElhinny, S. A. *et al.* Abundant ribonucleotide incorporation into DNA by yeast replicative polymerases. *Proc Natl Acad Sci USA* (2010).
350. Clausen, A. R., Zhang, S., Burgers, P. M., Lee, M. Y. & Kunkel, T. A. Ribonucleotide incorporation, proofreading and bypass by human DNA polymerase delta. *DNA Repair* **12**, 121–127 (2013).
351. Goksenin, A. Y. *et al.* Human DNA Polymerase ε Is Able to Efficiently Extend from Multiple Consecutive Ribonucleotides. *The Journal of Biological Chemistry* **287**, 42675–42684 (2012).
352. McElhinny, S. A. N. *et al.* Genome instability due to ribonucleotide incorporation into DNA. *Nat Chem Biol* **6**, 774–781 (2010).
353. Gosavi, R. A., Moon, A. F., Kunkel, T. A., Pedersen, L. C. & Bebenek, K. The catalytic cycle for ribonucleotide incorporation by human DNA Pol lambda. *Nucleic Acids Research* (2012).

354. Kasiviswanathan, R. & Copeland, W. C. Ribonucleotide discrimination and reverse transcription by the human mitochondrial DNA polymerase. *The Journal of Biological Chemistry* **286**, 31490–31500 (2011).
355. Cilli, P., Minoprio, A., Bossa, C., Bignami, M. & Mazzei, F. Formation and Repair of Mismatches Containing Ribonucleotides and Oxidized Bases at Repeated DNA Sequences. *Journal of Biological Chemistry* **290**, 26259–26269 (2015).
356. Crespan, E. *et al.* Impact of ribonucleotide incorporation by DNA polymerases beta and lambda on oxidative base excision repair. *Nat Commun* **7** (2016).
357. Potenski, C. J. & Klein, H. L. How the misincorporation of ribonucleotides into genomic DNA can be both harmful and helpful to cells. *Nucleic Acids Research* **42**, 10226-U10798 (2014).
358. Jinks-Robertson, S. & Klein, H. L. Ribonucleotides in DNA: hidden in plain sight. *Nat Struct Mol Biol* **22**, 176–178 (2015).
359. Vaisman, A. & Woodgate, R. Redundancy in ribonucleotide excision repair: Competition, compensation, and cooperation. *DNA Repair* **29**, 74–82 (2015).
360. Cerritelli, S. M. & Crouch, R. J. The Balancing Act of Ribonucleotides in DNA. *Trends in Biochemical Sciences* **41**, 434–445 (2016).
361. Iyengar, B., Luo, N., Farr, C. L., Kaguni, L. S. & Campos, A. R. The accessory subunit of DNA polymerase gamma is essential for mitochondrial DNA maintenance and development in Drosophila melanogaster. *Proc Natl Acad Sci USA* **99**, 4483–4488 (2002).
362. Muzi-Falconi, M., Giannattasio, M., Foiani, M. & Plevani, P. The DNA polymerase alpha-primase complex: multiple functions and interactions. *Scientific World Journal* **3**, 21–33 (2003).
363. Haworth, J., Alver, R. C., Anderson, M. & Bielinsky, A. K. Ubc4 and Not4 Regulate Steady- State Levels of DNA Polymerase-alpha to Promote Efficient and Accurate DNA Replication. *Molecular Biology of the Cell* **21**, 3205–3219 (2010).
364. Ricke, R. M. & Bielinsky, A. K. Mcm10 regulates the stability and chromatin association of DNA polymerase-alpha. *Mol Cell* **16**, 173–185 (2004).
365. Sahashi, R. *et al.* DNA polymerase alpha interacts with PrSet7 and mediates H4K20 monomethylation in Drosophila. *Journal of Cell Science* **127**, 3066–3078 (2014).
366. Frouin, I., Toueille, M., Ferrari, E., Shevelev, I. & Hubscher, U. Phosphorylation of human DNA polymerase lambda by the cyclin-dependent kinase Cdk2/cyclin A complex is modulated by its association with proliferating cell nuclear antigen. *Nucleic Acids Res* **33**, 5354–5361 (2005).

367. Wimmer, U., Ferrari, E., Hunziker, P. & Hubscher, U. Control of DNA polymerase lambda stability by phosphorylation and ubiquitination during the cell cycle. *EMBO Rep* **9**, 1027–1033 (2008).
368. Markkanen, E. *et al.* Regulation of oxidative DNA damage repair by DNA polymerase lambda and MutYH by cross-talk of phosphorylation and ubiquitination. *P Natl Acad Sci USA* **109**, 437–442 (2012).
369. Skoneczna, A., McIntyre, J., Skoneczny, M., Policinska, Z. & Sledziewska-Gojska, E. Polymerase eta is a short-lived, proteasomally degraded protein that is temporarily stabilized following UV irradiation in Saccharomyces cerevisiae. *J Mol Biol* **366**, 1074–1086 (2007).
370. Chen, Y. W. *et al.* Human DNA polymerase eta activity and translocation is regulated by phosphorylation. *Proc Natl Acad Sci USA* **105**, 16578–16583 (2008).
371. Day, T. A. *et al.* Phosphorylated Rad18 directs DNA Polymerase eta to sites of stalled replication. *Journal of Cell Biology* **191**, 953–966 (2010).
372. Bienko, M. *et al.* Regulation of translesion synthesis DNA polymerase eta by monoubiquitination. *Molecular Cell* **37**, 396–407 (2010).

Chapter 4

Human DNA Polymerases in Different DNA Transactions

4.1 Sixteen DNA polymerases and telomerase: share of workload and redundancies

Complex cellular functions are carried out by networks of protein machineries. Therefore the DNA pols involved in a particular DNA transaction (see Box 4.1) have to act in a concerted way with many other proteins.[1,2] The dynamic action of DNA pols in DNA transaction machineries such as the replisome, the repairosomes, the recombination complexes and the mismatch repair is essential. Moreover, these machineries are not static since they will act upon request at the right time (e.g. during the cell cycle, after a genotoxic insult, during development), at the right place (e.g. nucleus, nucleolus, open chromatin and mitochondria), at the right concentration (e.g. up- or down regulation in response to a DNA damage) and, finally, they should interact with the right partners. These specific transactions are in many cases guaranteed by posttranslational modifications either within the DNA pols and/or their interacting partners. PCNA, the sliding clamp interacting with several DNA pols (see Chapter 2, Table 2.7 for details) is a well-studied partner of several DNA pols.[3,4] Depending on the ubiquitination and SUMOylation status, PCNA can help to favor error-prone translesion synthesis (monoubiquitination), error-free translesion synthesis (polyubiquitination), SUMOylation (inhibition of recombination if modified at lysine 164 or inhibition of cohesion establishment if modified at lysine 127) (reviewed in Ref. 5). DNA pols themselves are the targets of many posttranslational modifications such as phosphorylation, acetylation, methylation, ubiquitination and SUMOylation. Even though our knowledge of these modifications and their consequences in the interactions of DNA pols with partners is still very limited, we will try to dissect a few aspects of these transactions in more details in this chapter.

In addition, we try to focus on the regulation of DNA pols (Section 4.5), their interaction at the replication forks (Section 4.2), and their engagements during translesion synthesis (Section 4.4). The different

Box 4.1 DNA transactions requiring DNA polymerases

DNA replication:

This is the process in which the DNA is duplicated in the S-phase of the cell cycle before a cell goes into mitosis. One strand, called the leading strand, is synthesized in a continuous way and the other, called the lagging strand, in a discontinuous way. This is due to the unidirectional direction in which all DNA pols synthesize DNA, namely in the 5′→3′ direction. The short stretches on the lagging strand are called Okazaki fragments and they have to be primed every 200 bases leading to over 10^7 initiation events.

Base excision repair (BER):

This is an essential repair pathway that deals with DNA alteration that frequently occur on one strand of the DNA. They include depurinations, depyrimidations, oxidations and deaminations. A group of different DNA glycosylases recognizes these alterations and remove the wrong base and renders the DNA a substrate for an efficient nuclease called apurinic endonuclease 1. In case of depurinations and depyrimidations leading to abasic sites, they are directly cut by apurinic endonuclease 1. Two types of DNA synthesis then occur: if only one nucleotide incorporation takes place the process is

(*Continued*)

Box 4.1 (*Continued*)

called short patch BER while if strand displacement of 2–12 nucleotides with subsequent DNA synthesis by DNA pols occurs, the process is called long patch BER.

Nucleotide excision repair (NER):

When lesions cause serious helix distortions on one strand of the DNA the very complex NER has to act. These distortions are first recognized by specific proteins. Next the DNA is unwound followed by the excision of about 30 bases by specific endonucleases on each side. The resulting gap is filled by DNA pols and finally ligated.

Non-homologous end joining (NHEJ):

This is a repair pathway that can restore double-stranded breaks without requiring the information of a homologous DNA sequence. This pathway is therefore error-prone. The ends are recessed by specific exonucleases and the broken ends bound by specific proteins. Finally, after DNA synthesis by DNA pols the ends are rejoined. This repair pathway is mainly active in the G1 phase of the cell cycle and is non-replication associated.

Homologous recombination (HR):

In this repair pathway the sister chromatid can be employed for the repair of the broken strand. First a resection by an exonuclease occurs and the 3′ overhang can invade a homologous DNA stretch on the sister chromatid. DNA synthesis by DNA pols restores the missing information. Finally, the complex is resolved by cutting and ligation to restore the repaired DNA. This repair pathway is error free and mainly active in the late S- and the G2 phase of the cell cycle.

Mismatch repair (MMR):

This repair pathway can restore mismatches and gaps that can occur after DNA replication. The mismatches are recognized by specific proteins that guarantee the removal of the wrongly incorporated nucleotide by an exonuclease. The DNA pol δ or ε holoenzyme (DNA pol δ or ε, RF-C and PCNA) gets therefore a second chance to correctly incorporate the base.

(*Continued*)

Box 4.1 (*Continued*)

Translesion DNA synthesis (TLS):

Lesions on the DNA that are in front of the replication DNA pols are resolved by specific translesion DNA pols that can first insert a nucleotide opposite the lesion. Even though these translesion DNA pols are per se not accurate their relatively correct synthesis over a lesion might be good enough to minimize mutations. Next special DNA pols are required to extend from the lesion and to hand over the DNA to the replication machinery again.

Replication of G4 structures

G4s are secondary structures formed by the stacking of quartets of Hoogsteen-bonded guanine bases. These structures contribute to cellular function, including regulation of DNA transcription and replication. However, they can also pose significant impediments to DNA replication and there is evidence that several specialized DNA helicases and DNA pols are involved in replicating G4 structures.

DNA crosslink repair:

In situations where the two DNA strands are covalently linked a complex set of pathways have to act, in which the many Fanconi's anemia proteins come into action. They work together with other DNA repair and damage avoidance pathways such as HR, NER and TLS.

V(D)J recombination and somatic hypermutation:

V(D)J recombination is a physiological event during development of pre B cells to guarantee the enormous variability of the antibodies. This double-stranded break repair is as already observed in NHEJ inaccurate, but this is of advantage to create an enormous amount of different antibodies. In addition somatic hypermutation of the immunoglobulin genes requires the activity of DNA pols to finally introduce mutations at both A/T and C/G base pairs. In summary nature has borrowed the low fidelity of DNA pols to enlarge the variability of antibodies.

Chromosome end replication:

Due to the unidirectional DNA synthesis by all DNA pols a problem arises at the end of linear chromosomes. This end replication problem is solved by specific enzymes called telomerases.

Table 4.1 Many of the known DNA transactions need more than one DNA polymerase[a]

DNA transaction[b]	**DNA polymerases involved**
DNA replication	DNA pols α, δ, ε and PrimPol,
Base excision repair (short patch)	DNA pols β, λ and θ
Base excision repair (long patch)	DNA pols β, δ, ε and λ
Nucleotide excision repair	DNA pols δ, ε, η and κ
Non-homologous end joining	DNA pols λ and μ
Homologous recombination	DNA pol η
Mismatch repair	DNA pol δ
Translesion synthesis:	
Insertion[c]	DNA pols η, ι, κ, δ and Rev1, PrimPol
Extension[c]	DNA pols ζ and κ
Replication of G4 structures	DNA pol η, κ and Rev 1
Interstrand crosslink repair	DNA pols θ and ν
V(D)J recombination and somatic hypermutation	DNA pols μ, λ, η, θ and TdT
Chromosome end replication	telomerase, DNA pol α, DNA pol η, DNA pol λ[d]

a: For details see subsequent Sections and the references therein.
b: The DNA transactions are explained in Box 4.1.
c: Depends on the lesion type. For details see Sections below.
d: G. Maga; U. Hübscher, personal communication.

DNA repair events all need DNA synthesis and therefore DNA pols (Section 4.3). Moreover, the physiological double-stranded break repair of V(D)J recombination requires DNA pols (Section 4.3) Since DNA replication very often encounters DNA damages during the synthesis event, next the interplay between DNA pols at the replication fork and translesion DNA pols will be summarized (Section 4.6). Finally, the DNA pols themselves are targets for checkpoint control mechanisms (Section 4.7). Table 4.1 overviews different DNA transactions and the DNA pols required in higher eukaryotes, such as mammals. More details will be given in the Sections below.

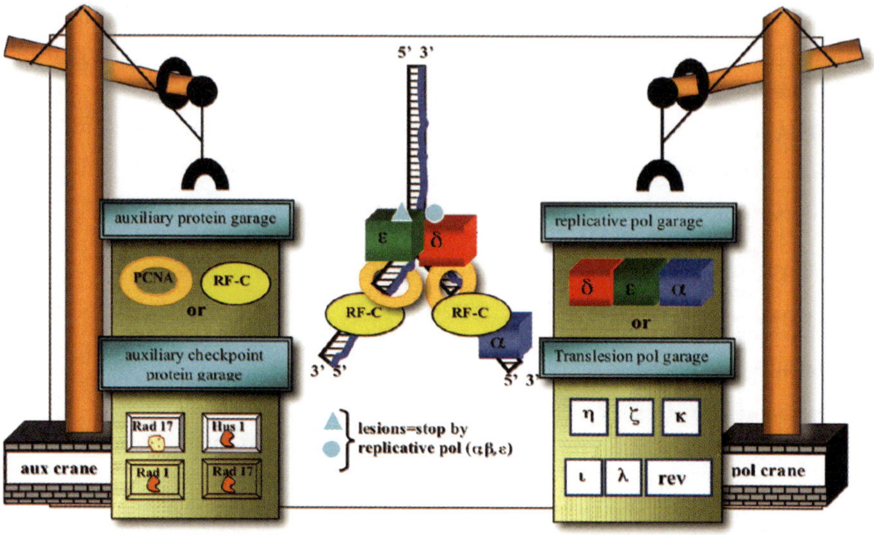

Figure 4.1 **A model for the interplay between DNA polymerases involved in replication and translesion DNA synthesis**. During DNA synthesis, the DNA replication machinery (schematically drawn as the DNA pol e/d holoenzymes (DNA pol ε/δ, RF-C, and PCNA)) eventually meets lesions on the DNA (represented with blue symbols). DNA pols ε and δ are unable to traverse DNA lesions and their arrest causes the block of the replication fork. For details of the DNA pol switch see text.

Figure 4.1 shows a model that was proposed by us in 2003,[3,6] where we postulated replication foci that can store the DNA pols and the respective auxiliary proteins (see Chapter 2.6 and 2.7 for details) to be exchanged when a replication fork encounters a lesion on the DNA (see also Section 4.6 for more details about the DNA pol switches in prokaryotes and eukaryotes). The current model predicts the existence of subnuclear compartments or foci (the "garages" in the drawing), where replicative or translesion (TLS) pols and their cognate auxiliary proteins are stored. When a replication fork stalls, checkpoints are activated leading to the recruitment of specific factors at the lesion through a yet unidentified machinery (drawn as an "auxiliary crane"). As a consequence, replicative pols are lifted off and replaced by specialized TLS pols by a "polymerase crane". After damage bypass, the normal replication machinery is reconstituted through an inverse mechanism. PCNA constitutes the common element to all these pathways, providing essential interactions with both replicative and TLS pols, as well as acting in the checkpoint process (see also posttranslational modification of PCNA in Section 4.6.2).

4.2 DNA replication in living organisms requires three DNA polymerase molecules at the replication fork

A database of information and resources for the eukaryotic replication community has been established.[7] The database (http://www.dnareplication.net) summarizes organism-sorted data on replication proteins in the categories of nomenclature, biochemical properties, motifs, interactions, modifications, structure, cell localizations and expression, as well as general comments.

4.2.1 *Prokaryotes*

We will briefly illustrate here the prokaryotic replisomes, before analyzing the structure of the eukaryotic replication fork. Comparison with eukaryotes, in fact, highlights those functions that have been conserved throughout evolution and thus represent the essential components of all replisomes. The replisomes in prokaryotes are the best studied so far.[8,9] Details are given for the bacterium *Escherichia coli (E. coli)*. The priority will be on the actions of DNA pols.

4.2.1.1 The *Escherichia coli* replisome

As indicated in Chapter 2.5 the *E. coli* replicase DNA pol III acts as a holoenzyme. DNA pol III holoenzyme contains 10 different polypeptides (Table 2.6.). These proteins are grouped into the three major functional units: (i) the DNA polymerase III core consisting of three subunits, (ii) the sliding clamp consisting of a homodimeric ring and (iii) the clamp loader/polymerase connector complex consisting of six different subunits (the τ subunit is a dimer). It was shown that the *E. coli* replisome contains two

DNA pols and it was postulated, according to the "trombone model".[10] that they have asymmetric functions in replicating the two DNA strands.[11] The *E. coli* helicase, the hexameric DNA B protein, which encircles the lagging strand, plays an important role for the asymmetry of the two DNA pol III molecules. Moreover, the two DNA pol III assemblies act in a symmetrical way, again supporting the "trombone model". The Dna B helicase contacts the t-subunit of the DNA pol III holenzyme and thus, together with the β-clamp, holds the leading strand DNA pol III on the DNA,[12] allowing processive DNA synthesis. The Dna B helicase/t-subunit interaction can speed up the helicase movement[13] and also generate the asymmetry in the dimeric DNA pol III acting on double-stranded DNA. In subsequent years the dynamics at the replication fork were refined (reviewed in Ref. 14) and are schematically drawn in Figure 4.2. While leading strand synthesis is continuous and needs only a few priming events, the lagging strand has to be initiated every 1000 bases and is consequently discontinuous. First, the primase synthesizes an RNA piece on the lagging strand and dissociates after synthesis [Figure 4.2.(a)]. Second the γ-complex, which is the clamp loader, uses the energy of ATP to open the ring like β-clamp and loads it onto the primer placed at the lagging strand. The γ-complex leaves the DNA and lets the β-clamp to wait for the next DNA pol III [Figure 4.2.(b)]. Third, the DNA pol III that has completed the previous Okazaki fragment leaves the DNA and cycles to the β-clamp that has been loaded by the γ-complex [Figure 4.2.(c)]. Fourth, the lagging strand DNA pol III synthesizes the next Okazaki fragment [Figure 4.2(d)]. Finally, the next round of Okazaki fragment synthesis can start again.

Published data, however, suggested that the DNA pol III holoenzyme assembles into a replisome that contains three DNA pol III (Figure 4.3). They all seem to have the capacity to act simultaneously.[15] In addition, the trimeric DNA pol III functionally interacts with other replisome components such as the Dna B helicase, the Dna G primase and the β-subunit sliding clamp. The data suggested that two DNA pol III act as leading and lagging strand replicases, respectively. The third DNA pol III, even though not bound to DNA but to the replisome, may be a reserve enzyme for purposes such as when the lagging strand has problems to follow the speed of the leading strand or when the leading and/or lagging strand are

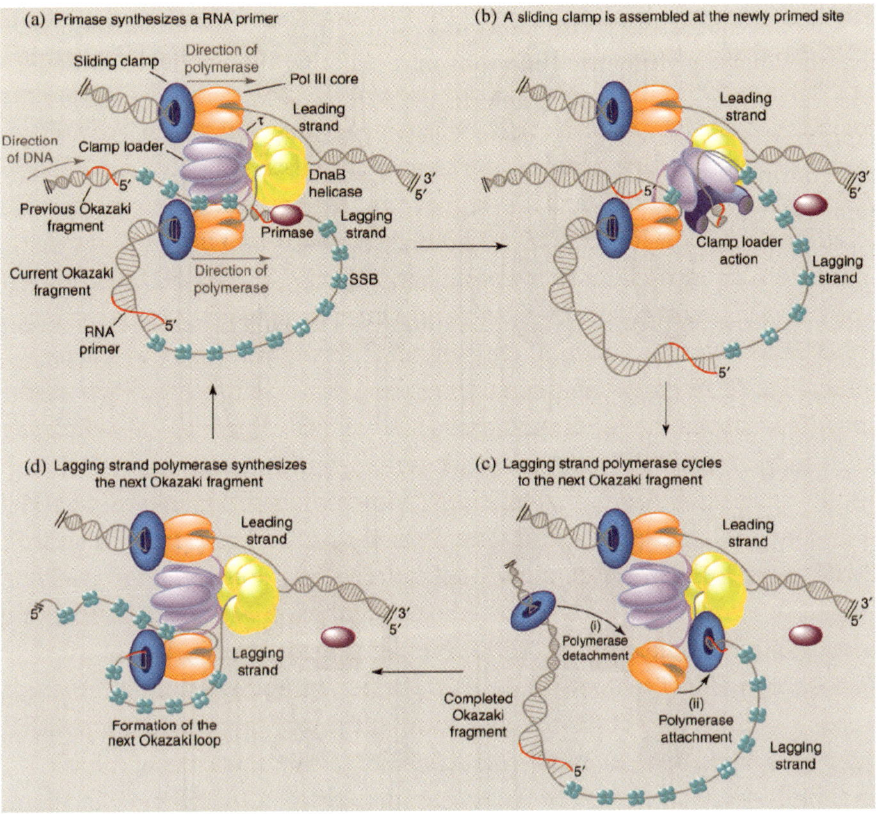

Figure 4.2 **Protein dynamics at the *E. coli* replication fork**. The leading strand synthesis requires a few priming events, while the lagging strand is initiated about every 1000 bases leading to discontinuous DNA pieces called Okazaki fragments. (a) Primase synthesizes an RNA primer and it dissociates after primer synthesis is completed. (b) The γ-complex clamp loader uses the energy of ATP to transiently open the β-clamp ring thus loading the β-clamp on the DNA. The γ-complex dissociates from the DNA thus leaving the β-clamp on the primer-template junction. (c) The lagging strand DNA pol III leaves, dissociates from the β-clamp as soon as an Okazaki fragment is completed and cycles to the next Okazaki fragment synthesized by the primase. (d) Finally the lagging strand DNA pol III synthesizes the next Okazaki fragment. (Reproduced with permission from Ref. 14.)

stalled due to DNA damage (see also Section 4.6). Finally, single molecule studies of fork dynamic in *E. coli* DNA replication indicated[16] that when DNA pol III is coupled to the Dna B helicase it has a processivity

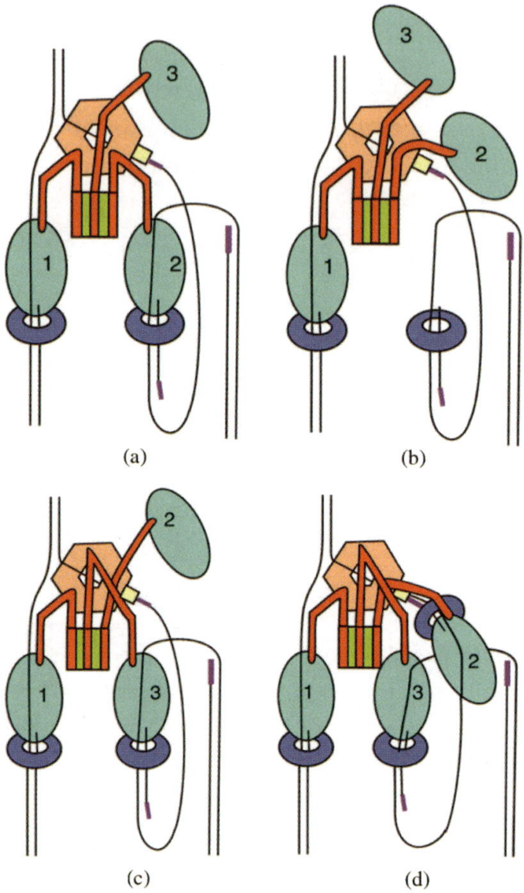

Figure 4.3 **The three DNA polymerase III molecules and dynamics at the *E. coli* replication fork**. (a) The three DNA pol III molecules are held together and to the replicative Dna B helicase with three τ-subunits (orange) of the clamp loader complex (green). Primase (yellow) synthesizes RNA (purple). The DNA pol III molecules engaged in DNA synthesis are complexed to the β-clamp (blue). (b) A transient dissociation of one DNA pol III leaves a β-clamp and 3′ end free to engage translesion DNA pols. (c) After translesion DNA synthesis the third DNA pol III binds again to the β-clamp thus (d) allowing three DNA pol III to act, namely one on the leading strand and two on the lagging strands of the replication fork. (Reproduced with permission from Ref. 9.)

of 10,500 bases. Upon addition of the Dna G primase the processivity is reduced three-fold. This reduction is caused by the protein–protein interaction of Dna B and Dna G and is not due to the primase activity of Dna G. In summary it was proposed that the modulation of the Dna B helicase activity by its interaction with Dna G primase prevents leading strand synthesis from outpacing the lagging stand due to the slower synthesis on the lagging strand.[16]

4.2.2 *Eukaryotes*

In eukaryotes three different DNA pols are engaged in DNA replication, being DNA pol α, DNA pol δ and DNA pol ε.[1] In a historical context (for details see Chapter 1) DNA pol α was the first DNA pol to have a role in eukaryotic DNA replication.[17] A few years later the primase activity was identified in one of the four subunit DNA pol α complex.[18] Recently, however, an additional DNA pol-primase, called PrimPol, has been discovered,[19–21] but its role in unperturbed DNA replication is still under debate. In the yeast *S. cerevisiae* DNA pol δ was found to be essential[22] and soon thereafter the essentiality of DNA pol ε was again documented in *S. cerevisiae.*[23] These results suggested that at least three different DNA pols share the workload of DNA replication in the entire eukaryotic kingdom. While DNA pol α was assigned to be the initiating DNA pol, due to is associated primase, there was a debate over more than 15 years about the *in vivo* functions of DNA pols δ and ε. Biochemical studies suggested a function of DNA pol δ at the lagging strand since it was able to idle at the lagging strand to provide a ligatable nick during lagging strand DNA replication.[24] Moreover, DNA pols δ and ε have both high processivity, but the effect of PCNA and their binding affinity to DNA was different, suggesting that DNA pol ε with its high affinity to DNA was more suited as the leading strand DNA pol, while DNA pol δ with a lower affinity to DNA might be a candidate for the frequent relocation at the lagging strand of the replication fork.[25]

The eukaryotic DNA replication mechanistic details were elaborated first in the simian virus 40 (SV 40) *in vitro* reconstituted system (reviewed in Ref. 26). This viral model suggested that DNA pol α is the initiating

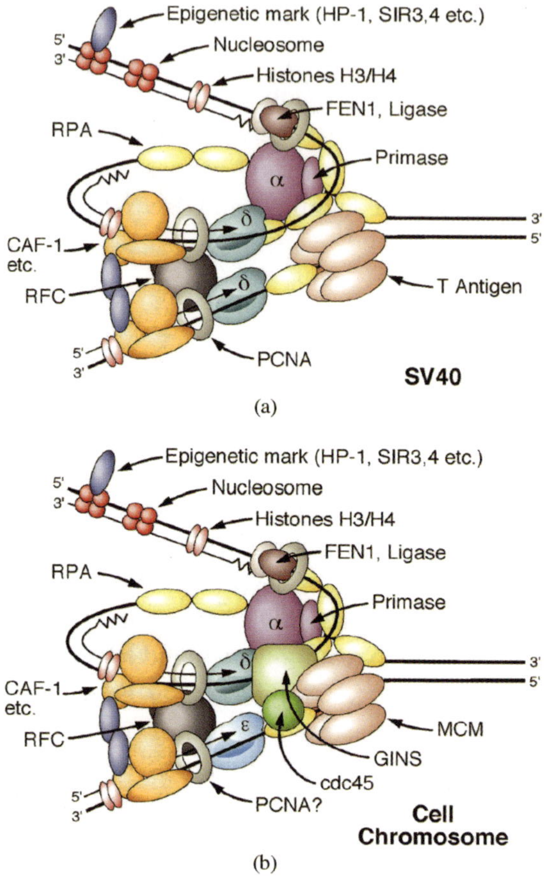

Figure 4.4 **SV40 replication needs DNA polymerases α and δ, while the eukaryotic cellular replication fork relies on the three DNA polymerases α, δ and ε.** Note that in the case of SV40 DNA pol d replicates both strands (a), while in the cellular context DNA pol ε is the leading and DNA pol δ the lagging strand replicase (b), respectively. For details see text. (Reproduced with permission from Ref. 31).

pol at both strands and that a DNA pol switch from DNA pol α to DNA pol δ allows the latter to replicate both strands. Moreover, no DNA pol ε was required for SV 40 replication *in vitro* [Figure 4.4 (a)]. In *S. cerevisiae*, thanks to the power of genetics, convincing data were reported that DNA pol ε is the leading strand DNA pol[27] and that DNA pol δ the lagging strand DNA pol,[28] respectively. Special amino-acids were mutated separately in DNA pol ε and δ that did not affect the DNA synthesis rate, but reduced their fidelity. By using these mutants, it was possible to analyze the frequency of mutation in a reporter gene close to the origin of replication and to qualify DNA pol ε as the leading strand and DNA pol δ as the lagging strand enzyme at the DNA replication fork in eukaryotic cells. Likely these data can be transferred to higher eukaryotic cells including human [Figure 4.4 (b)]. Nevertheless, a recent article[29] raises the entire question of strand assignments again and concludes that DNA pol δ replicates both the leading and lagging strands. The authors attribute to different mismatch repair mechanisms on the leading and lagging strand the apparent lack of contribution of pol δ to leading strand replication (see also Ref. 30 for a review). Clearly more investigation is required to completely unravel the roles of different DNA pols at the eukaryotic replication fork.

Published work addressed which DNA pols are recruited for chromosome end replication.[32] Data suggested that the leading and the lagging strand DNA pols arrived differently at the telomeres. DNA pol ε, the leading strand enzyme, arrived earlier at the telomeres than the lagging strand DNA pols α and δ. Furthermore, the S-phase specific recruitment of Trt1, Pot1 and Sts1, components of the telomeres replication machinery, matched the arrival of DNA pol α. These data suggested that the recruitment of the lagging strand replication machinery is delayed until the telomerase enzyme itself is recruited.

4.2.3 *Proofreader versus non-proofreader DNA polymerases*

DNA pols are not absolutely accurate *per se* (see also Chapter 2). An important activity that enables the replicative enzymes to reduce the misincorporation to a low level is the associated 3′→5′ exonuclease, also called proofreading activity (reviewed in Ref. 33). The replication fidelity largely depends on this activity. DNA pol δ and ε each contains a proofreading activity but not the initiating DNA pol α.[34] Proofreading *in trans* is an attractive hypothesis in which a non-proofreading DNA pol will cooperate with a proofreading DNA pol.[35] A mutant of DNA pol α with normal activity and processivity, but reduced fidelity, has a strong mutator phenotype. This mutator phenotype was increased when the proofreading exonuclease of DNA pol δ, but not of DNA pol ε′ was inactivated. This leads to the hypothesis that the 3′→5′ exonuclease of DNA pol δ could proofread for the error-prone DNA pol α after initiation of an Okazaki fragment.[36] Such intramolecular proofreading might also be of relevance in DNA repair and translesion DNA synthesis.

4.3 Different DNA repair pathways have their own DNA polymerases, but can also borrow them from the replication machinery

Forty years ago, when only DNA pols α, β and γ were known, data were presented indicating that DNA pol β was a repair enzyme.[37] DNA pol β was isolated from rat adult neuronal nuclei and was shown to perform ultraviolet-induced repair DNA synthesis *in vitro*. Since DNA pol β was virtually the exclusive DNA pol in these post-mitotic nuclei, it was concluded that DNA pol β was responsible for the observed DNA repair. This was further substantiated by demonstrating a virtually complete suppression of DNA repair in irradiated nuclei by 2’,3’-dideoxyribosylthymine 5′-triphosphate (d_2TTP), a potent DNA pol β inhibitor. It was therefore speculated that the ability to perform repair DNA synthesis is not unique to the neuronal DNA pol β but is a general function of DNA pol β in all tissues.[38] This statement is still true, but, as summarized in Table 4.2, has to be refined for different DNA repair pathways and in view of the discoveries of many new DNA pols since then.

The DNA repair pathways mentioned below are complex events that cannot be described in details here. We focus only on the re-synthesis step in all these pathways. For more detailed information on the mechanisms of DNA repair, the reader is referred to the excellent book *DNA Repair and Mutagenesis* (Friedberg *et al.*, ASM Press Washington, 2006).

Base excision repair (BER). The mammalian genome is always damaged and suffers from many endogenous and exogenous insults that modify DNA leading to base loss or base alterations.[39] It has been estimated that approximately 10,000–100,000 modifications occur daily in

Table 4.2 DNA polymerases involved in the different DNA repair pathways[a]

DNA repair pathway	DNA pols engaged
Base excision repair (short one nucleotide patch)	DNA pols β, λ and θ
Base excision repair (long 2–12 nucleotide patch)	DNA pols β, ε, δ and λ
Nucleotide excision repair	DNA pols ε, δ, η and κ
Mismatch repair	DNA pol δ
Non-homologous end joining	DNA pols λ, μ and θ
Recombination of immunoglobulin genes	DNA pols λ, μ, and TdT
Hypermutation of immunoglobulin genes	DNA pols θ and η
Homologous recombination	DNA pol η
Interstrand cross-link repair	DNA pols θ and ν

a: see also Table 4.1.

the DNA of a single cell.[40] In order to remove damage and to maintain the integrity of the genome, different DNA repair pathways have evolved. The major repair pathway protecting cells against a single-base DNA damage is base BER. It is often initiated by a damage type specific DNA glycosylase that cleaves the *N*-glycosydic bond of the damaged base, leaving an apurinic/apyrimidinic site, also referred to as an abasic site or AP site. Subsequently, AP endonuclease 1 (APE 1) cleaves the sugar-phosphate backbone at the 5′-side of the AP site resulting in 3′-hydroxyl and 5′-deoxyribose phosphate (dRP) group flanking a one-nucleotide gap. DNA pol β inserts a first nucleotide into the gap, leaving nicked DNA with a 5′-dRP flap. At this point, the repair can be accomplished via two different BER sub-pathways depending on the nature of AP site (reviewed in Ref. 40–42). In the case of regular AP sites, DNA pol β removes the 5′-deoxyribose–5′-phosphate (5′-dRP) group through its associated (dRP-lyase) activity, and the resulting nick is sealed by the DNA ligase III/XRCC1 complex. This sub-pathway is referred to as short-patch BER (SP-BER). However, if the sugar group is oxidized or reduced, DNA pol β cannot remove the 5′-dRP moiety and repair proceeds through the alternate long-patch BER (LP-BER) sub-pathway involving removal and replacement of 2–12 nucleotides. It has been found that, in the LP-BER pathway, the collaboration of DNA pols δ and ε, PCNA, RF-C and flap endonuclease 1 (Fen 1) can displace the strand 3′ to the nick and synthesize up to 12 nucleotides.[43,44] The resulting flap is cut by Fen 1 and the final nick is sealed by DNA ligase I. However, DNA pol β is also a key enzyme in LP-BER,[45] that not only initiates both sub-pathways but can also perform efficient strand displacement via the "hit and run" mechanism. In this model DNA pol β and Fen 1 act successively, followed by the action of DNA ligase I.[46]

Other DNA pols also appear to function in BER. First, dRP-lyase activities have been identified in DNA pol λ[46] and DNA pol θ,[47] suggesting a role in SP-BER. DNA pol λ has also been described as a back-up BER activity in extracts from mouse embryonic fibroblasts.[48] The cooperation of DNA pols θ and β was suggested in BER of oxidative damage.[49] The adenine misincorporated by DNA pols δ and ε opposite 7,8-dihydro-8-oxoguanine (8-oxo-G), can be removed by a specific glycosylase MUTYH, leaving the lesion on the DNA. Subsequent incorporation of C opposite 8-oxo-G on the resulting one nucleotide gapped DNA is essential for the removal of the damaged base. By using model DNA templates, purified DNA pols β and λ and knockout cell extracts, it was demonstrated that the auxiliary proteins RP-A and PCNA act as molecular switches to activate the DNA pol λ-dependent highly efficient and faithful repair of A:8-oxo-G mismatches in human cells, while suppressing the action of the less faithful DNA pol β.[50,51] Published data suggested that the switch from DNA pol β to DNA pol λ by RP-A and PCNA can be a switch from SP-BER (1 nucleotide by DNA pol β) to LP-BER (2 nucleotides by DNA pol λ) in faithful insertion of a C opposite 8-oxo-G.[52] Ubiquitination of BER proteins, including DNA pols β and λ has been reported and the functional consequences of such process discussed.[53,54] A comprehensive review dealing with post-translational modifications of BER proteins is now available.[55] Finally, a very recent review describes the relevance of several types of excision repair (BER; NER; MMMR) in telomere maintenance.[56]

Nucleotide excision repair (NER). In NER the product left for DNA repair re-synthesis is a 30-base gap. Initially it was found that PCNA is required for NER suggesting that the two replicative DNA pols δ and ε are involved in NER re-synthesis.[57] Indeed the first complete NER *in vitro* system suggested the involvement for either DNA pol δ or ε.[58,59] That both

DNA pols are involved makes sense since leading strand DNA pol ε could be borrowed for NER at the leading strand and lagging strand DNA pol δ at the lagging strand, respectively. It should be noted that the Xeroderma pigmentosum Variant gene product, identified as DNA pol η,[60] protects against UV-damage[61] (see also Section 4.4). DNA synthesis in eukaryotes requires generally two DNA pols: an inserter and an extender. DNA pol κ-deficient cells had a defect in UV damage repair that could be restored to normal sensitivity and NER, suggesting that DNA pol κ also plays a role in NER.[62] In a further study, it was shown that DNA pol κ was recruited to sites of local damage in UV-irradiated quiescent human primary fibroblasts, and was recruited to NER repair complexes in a UV and NER incision — dependent manner. In the absence of DNA pol κ, DNA pol δ or both, unscheduled DNA synthesis (UDS) was reduced by 50%, suggesting that DNA pols κ and δ were involved in the same repair synthesis pathway. In contrast, depletion of DNA pol δ and /or κ as well as DNA pol ε reduced UDS to background levels, suggesting that DNA pol ε was involved in a different repair synthesis pathway. The authors proposed that there were two modes of repair synthesis, the first requiring DNA pols δ and κ, the other employing DNA pol ε.[63]

Mismatch repair (MMR): Reconstitution of the MMR pathway *in vitro* clearly showed that DNA pol δ is the main enzyme involved in the resynthesis step after the proper recognition of the mismatch and the resection by the exonuclease 1 (EXO1).[64] An up-to-date review dealing with eukaryotic mismatch repair in relation to DNA replication is now available.[65]

Non-homologous end-joining (NHEJ). NHEJ is the process that has to repair double-stranded breaks as a consequence from DNA damage or in physiological way during V(D)J recombination (see below). DNA pols λ, DNA pol μ and TdT all have a BRCT-1 C terminal domain which was shown to interact with components of the NHEJ machineries such as Ku and XRRC4/DNA ligase IV.[66–70] Moreover, these three DNA pols were found after irradiation either in double-stranded break complexes or shown to co-localize with them. Furthermore it was found that when DNA pol λ or DNA pol μ were knocked out, recessed DNA ends decreased in NHEJ, suggesting again roles for these two DNA pols in NHEJ.[71,72] Finally, the same result was obtained in *S. cerevisiae* where DNA pol IV, the DNA pol λ-like counterpart and the only X family DNA pol from yeast (see Table 2.2 in Chapter 2), physically interacted with Dnl4-lif1 (the homologs of the human MRN and XRCC1/DNA ligase IV complexes, respectively), thus linking DNA synthesis to NHEJ.[73] Recently, it has been suggested that DNA pol θ has a prominent role in non-homologous end joining (NHEJ).[74] This observation is supported by the crystallographic structures of ternary substrate complexes of DNA pol θ, showing how the enzyme grasps the primer to bypass DNA lesions or to extend poorly annealed DNA termini to mediate end joining.[75] This primer-grip strategy, coupled with DNA pol θ's ability to dimerize, could enable the enzyme to direct DNA synthesis from unmatched primer termini during repair of double strand breaks.[76]

Recombination of immunoglobulin genes. V(D)J recombination is the process in which the variable (V), diversity (D) and joining (J) regions of the germline immunoglobulin genes are transiently broken and subsequently rejoined to allow a great variety of different immunoglobulin proteins being synthesized (reviewed in Ref. 77). First it was shown that DNA pol μ participates in kappa light chain gene rearrangement but not heavy chain gene rearrangement.[78] Next it was found that immunoglobulin heavy chain junctions from DNA pol λ-deficient animals have shorter lengths with normal N-additions, thus suggesting that DNA pol λ might be recruited during heavy chain rearrangement at a step that precedes the action of TdT.[79] In contrast to previous *in vitro* studies, analysis of animals with combined inactivation of these enzymes revealed no overlapping or compensatory activities for V(D)J recombination between these three DNA pols. This complex usage of DNA pols with distinct catalytic specificities might correspond to the specific function that the third hypervariable region assumes for each immunoglobulin chain, with DNA pol λ maintaining a large heavy chain junctional heterogeneity and DNA pol μ ensuring a restricted light chain junctional variability. The role of these three X family DNA pols was confirmed by biochemical data in which a gradient of template dependencies was proposed to have distinct biological roles for DNA pol λ, DNA pol μ and TdT[70].

Hypermutation of immunoglobulin genes. Besides the V(D)J recombination by NHEJ there is a second cycle of diversification that is called somatic hypermutation (SHM). During SHM base substitutions are introduced into a region that surrounds about 1000 bases of the antibody coding sequence.[80] SHM is induced by the enzyme activation-induced cytosine deaminase (called AID[81] and reviewed in Refs. 82 and 83). AID can deaminate cytosine to uracil in the DNA. The enzyme has a strong preference for 5′ WRC sequences (where W = A or T and R = A or G) in single-stranded DNA and it can carry out multiple C deaminations on individual DNA strands, rather than jumping from one strand to another.[84] Soon after its discovery DNA pol μ was proposed as a candidate for SHM,[85] but transgenic knock out mice for DNA pol λ and DNA pol μ were viable, fertile and did not have an abnormal SHM phenotype.[86] Later it was demonstrated that DNA pol θ plays a dominant role in SHM, likely by introducing mismatches while bypassing abasic sites generated by the uracil DNA glycosylase-mediated removal of dU residues produced by deaminase activity of AID, and by extending past such abasic sites or by synthesizing DNA during dU-dG mismatch repair.[87] Later it was found that DNA pol η and DNA pol θ function in the same pathway to generate mutation at A/T during SHM [87] and DNA pol η appears to be the limiting factor for A/T mutations in Ig genes and might thus contribute to antibody affinity maturation.[88] More recently, DNA pol ζ has also been shown to play a direct role in SHM in the absence of the dominant DNA pol η protein.[89]

Homologous recombination (HR). Two studies suggested that DNA pol η can function to extend 3′-strands that have been exchanged during homologous recombination. The action of the strand transferase, the Rad51 protein, enables a D-loop formation that can act as a primer for DNA pols. DNA synthesis from such strand invasion intermediates of HR was exclusively possible with DNA pol η suggesting an *in vitro* role for this TLS DNA pol in HR.[90] While these studies were done with human proteins, another study in chicken cell came to a similar conclusion. DT40 chicken cell can diversify their IgV genes though HR-mediated gene conversion. Cells lacking DNA pol η showed a decrease in Ig gene conversion and in double-stranded break induced HR. The defect could be rescued upon complementation with DNA pol η, suggesting that this enzyme has an important role in HR.[91] Further *in vitro* studies also demonstrated that DNA pol δ can also promote PCNA recombination associated DNA synthesis by using the 3′ OH of the invading strand in a D loop.[92] Moreover, more other data suggest that REV1/pol ζ complex and DNA pol κ can affect DNA recombination *in vivo* and *in vitro*.[93,94]

Interstrand cross link repair (ICLR). Drosophila melanogaster harbours a gene called *Mus308*, which is implicated in resistance to interstrand-crosslinking agents. A homolog of this protein has been cloned from human cells and was called DNA pol θ.[95] Moreover, with the aim to identify unconventional DNA pols from human cells, and using a special assay to fractionate HeLa extracts based on the ability to bypass DNA lesions, an aphidicolin-resistant DNA pol activity was obtained that was able to utilize either undamaged or abasic sites-containing DNA with the same efficiency. Biochemical characterization and immunoblot analysis allowed provisionally identification as a possible human homolog of DNA pol θ.[96] Further studies suggested that two DNA pols could be implicated in ICLR, the first being DNA pol θ[97] and the second DNA pol ν.[98] DNA pol κ has also been shown to process N^2-N^2 minor groove ICLs[99]). Subsequently, data have been presented showing that REV1 and DNA pol ζ are essential for TLS during ICL repair (for a review see Ref. 100).

4.4 Translesion DNA synthesis in eukaryotes generally requires two DNA polymerases: An inserter and an extender

We have introduced the translesion DNA synthesis (TLS) DNA pols in Chapter 1 (discoveries) and described in detail the properties of the eukaryotic TLS DNA pols in Chapter 3. In this chapter we will describe the functions of the TLS eukaryotic DNA pols over different DNA lesions.

Why do the TLS DNA pols have important functions in the DNA damage tolerance pathway? In contrast to the replication and the repair DNA pols, the TLS DNA pols have a much lower fidelity. Can it be of advantage when an unfaithful DNA pol acts on the DNA?

The problem: *DNA is often chemically altered*. Bases in the DNA can be chemically attacked by environmental agents such as UV-light, radioactivity, various toxic chemicals and reactive oxygen species. Moreover, many endogenous DNA damages are the consequence of hydrolysis [depurination, depyrimidation, spontaneous deamination (e.g. C leading to a U, 5-methyl-C leading to T)], oxidation (8-oxo-G, thymine glycol, cytosine hydrates, lipid peroxidation products such as etheno-A and etheno-G) and nonenzymatic methylation (7-methyl-G, 3-methyl-A and O^6-methyl-G). The amount of these attacks is very frequent: it is estimated that 10,000–100,000 alterations in the 6×10^9 bases of human genome occur per cell per day! It is a logic consequence that nature has invented different strategies to deal with all these problems.

In view of their effect on DNA synthesis by DNA pols, the lesions can be classified into "coding", "blocking" and "miscoding". First, a lesion is still "coding" when the altered base in the DNA can base pair, like the undamaged base, with the correct nucleotide (G-C and A-T). Second, a lesion is "blocking" when the alteration in the DNA base dramatically changes its shape and size so that the access of the damaged base to the active site of the DNA pol is restricted. This then results in a stop of DNA synthesis. Third, and this might be the most dangerous case, "miscoding" lesions are those that base pair with an incoming nucleotide different from the one predicted by the Watson-Crick rules. Table 4.3. shows three examples of base modifications and their consequence on eukaryotic DNA pols.

The TT dimer is representative of a large modification, 8-oxo-G as a small base modification and finally the AP site represents a complete lack of a coding base as a consequence of depurination. It is obvious that a given lesion (e.g. TT dimer) can be blocking, miscoding, or coding for different DNA pols and a given DNA pol (e.g. DNA pol η) can be blocked, miscode or code in the presence of different lesions. Table 4.3, therefore, suggests the tremendous variability that we have with the many DNA pols having different properties.

The behavior of the DNA pols does not necessary correlate with the intrinsic fidelity of DNA pols on undamaged DNA (Table 4.3). The replicative family B DNA pols α, γ, δ and ε are the least tolerant against modified DNA. This makes sense in view of their task to faithfully duplicate the genome. On the other hand the family Y DNA pols η, ι and κ together with the family X DNA pol λ are special enzymes that can act in the DNA damage tolerance pathway called TLS.[101] The TLS is an extremely important mechanism that supports DNA synthesis over lesions that could otherwise not be handled by the high-fidelity replicative DNA pols α, γ, δ and ε.

Table 4.3. Behavior of eukaryotic DNA polymerases on three different DNA lesions

DNA pol	Large modification (TT dimer)	Small modification (8-oxo-G)	Depurintion (AP site)	Error rate[a] ($\times 10^{-5}$)
DNA pol α	blocking	miscoding	blocking	10
DNA pol δ	blocking	miscoding	miscoding	1–0.1
DNA pol ε	blocking	miscoding	?	2–0.5
DNA pol γ	n.d.	miscoding	blocking	1–0.1
DNA pol β	miscoding	miscoding	miscoding	10
DNA pol λ	n.d.	miscoding[b]	miscoding	100
DNA pol μ	coding	miscoding	miscoding	10
DNA pol η	coding	miscoding[c]	blocking	3500
DNA pol ι	blocking	miscoding	blocking	72,000[d]
DNA pol κ	blocking	miscoding	blocking	580
DNA pol θ	blocking	n.d.	miscoding	240
PrimPol	miscoding	miscoding	miscoding	10

n.d. Not determined.

a: Base substitution rate.

b: DNA pol λ in the presence of PCNA and RP-A is coding.[50,51]

c: DNA pol η in the presence of PCNA and RP-A is partially coding.[50]

d: For the preferred T-G mismatch.

If a replication fork stalls at a lesion an event called S-phase checkpoint is triggered. If a fork is collapsing this might lead to DNA strand breaks, to aberrant recombination and even to programmed cell death (apoptosis). The latter might happen when the S-phase arrest is too long or if the accumulated DNA damages lead to aberrant mitosis. It is therefore important that the TLS pathways are immediately active by employing specialized DNA pols that are suited to efficiently cope with particular lesions, eventually allowing the replisome to continue DNA synthesis (Figure 4.1) Details about the DNA pol switch will be discussed in Section 4.6.

TLS occurs in two steps: first one or two bases, depending on the lesions, are inserted opposite the damage and, second, an extension from this lesion has to be carried out. We still only consider here these two fundamental TLS steps and will discuss the DNA pol switch from a

Table 4.4 Insertion and extension over DNA lesions

DNA lesion	**Inserter DNA pol**	**Extender DNA pol**	**Consequence**
cys-syn TT	DNA pol ι	DNA pol ζ or κ	error-prone
	PrimPol	PrimPol	error-prone
6–4 TT	DNA pol η or ι	DNA pol ζ	error-prone
		DNA pol θ	error-prone
6–4 TC	DNA pol η	DNA pol ζ	error-free
	DNA pol ι	DNA pol ζ	error-prone
6–4 CC	DNA pol η	DNA pol ζ	error-free
	DNA pol ι	DNA pol ζ	error-prone
AP site	DNA pol ι	DNA pol ζ	error-prone
	DNA pol η	DNA pol ζ	error-prone
	DNA pol δ	DNA pol ζ	error-prone
	PrimPol	PrimPol	error prone
	DNA pol θ	DNA pol θ	error prone
T-glycol	DNA pol δ	DNA pol ζ	error-free
	DNA pol κ	DNA pol ζ	error-free
	DNA pol θ	DNA pol θ	error-free
AAF adduct	DNA pol η	DNA pol ζ	error-free
N^2-dG acrolein adduct	DNA pol ι	DNA pol κ	error free
(+/–)BPDE-N^2dG	DNA pol κ	DNA pol κ	error-prone
8-oxo-G	DNA pol δ[a]	DNA pol δ	error-prone[a]
	DNA pol ε[a]	DNA pol ε	error-prone[a]
	DNA pol α[a]	DNA pol α	error-prone[a]
	DNA pol β[a]	DNA pol β	error-prone[a]
	DNA pol η	DNA pol η	error-prone
	DNA pol η, PCNA/RP-A	DNA pol η PrimPol	partially error-free
	PrimPol	DNA pol λ	error prone
	DNA pol λ	DNA pol λ	error-prone
	DNA pol λ/ PCNA/RP-A		error-free

a: PCNA and RP-A do not render the DNA pols error free over 8-oxo-G.

replication or a repair DNA pol in Section 4.6. Data from many laboratories suggested that many TLS pols are very proficient in incorporating nucleotides opposite certain lesions, but were unable to extend the resulting primer terminus.[102] On the other hand, DNA pols were discovered that were efficient in extending from a variety of mispaired or aberrant primer termini. For some lesions (e.g. an AP site or a T-glycol) even the faithful DNA pol δ is able to incorporate but not to elongate. The two categories were designated as TLS "inserter" DNA pols or as TLS "extender" DNA pols. While there are many "inserter" DNA pols (Table 4.4), the current knowledge is that in most cases DNA pol ζ and in rare cases DNA pol κ are "lesion extenders". Consequently, we have interplay, also called a DNA pol switch, between the "inserter" and the "extender" DNA pol. The outcome of the TLS can either be error-free or error-prone. The latter might lead to mutations in the next replication round.

From history to present days: *about error prone versus error free TLS.* UV irradiation of bacteriophages λ and ΦX174 resulted in killing but not mutagenesis of the phages after the infection of untreated host cells, whereas UV irradiation of the host cells caused mutagenesis of both untreated and irradiated phage.[103,104] Furthermore, UV irradiated single-stranded ΦX174 DNA extracted from untreated host cells was found to be replicated only to the first pyrimidine dimer.[105] Caillet–Fauquet *et al.*[106] found that UV irradiation of the host cells led to an enhancement of DNA synthesis on UV irradiated ΦX174 DNA *in vivo*, which led them to propose that pyrimidine dimers in the ΦX174 DNA can become mutagenic by causing misincorporation of deoxyribonucleotides opposite them.

These experiments laid down the foundation for the formulation by Miroslav Radman and Evelyn Witkin of the so called "SOS" (Save Our Soul) hypothesis, that postulates that a complex cellular emergency response is triggered by unrepaired DNA lesions. Implicit in the SOS hypothesis was the fact that the damaged cell could save their life at the price of increased mutagenesis and, therefore, that whatever mechanism of replication of the DNA damage was involved, it was error prone. Indeed the subsequent discovery that the SOS response in *E. coli* controlled the induction of the error prone DNA pols IV and V, belonging to the recently discovered Y family of DNA pols, reinforced this idea. This family of DNA pols are characterized by the absence of an associated $3' \rightarrow 5'$ proofreading exonuclease activity and by a relaxed structure that allows accommodation of bulky DNA lesions, such as pyrimidine dimers, in their active site, and have been named Trans Lesion Synthesis or TLS DNA pols.

However, as usual, life is not so simple and the notion, supported by numerous data, that Y family DNA pols always act as low fidelity DNA pols during translesion synthesis was challenged by the fact that XP-V patients, who bear mutations in the Rad30/pol η alleles, are prone to skin and other cancers, suggesting that the function of the DNA pol η may be antimutagenic in humans (for a review see Ref. 107). Indeed DNA pol η has been shown to replicate cys-syn cyclobutane pyrimidine dimers or CPDs, one of the major lesions resulting from UV irradiation, with high accuracy and efficiency.[108] Furthermore, DNA pol η independent bypass of CPD, which is thought to involve other TLS DNA pols, appears to be significantly more mutagenic, presumably accounting for the increased frequency of cancers in XPV patients.

The identity of some of the backup Y family DNA pols that carry out the mutagenic TLS of CPDs has been uncovered.[109] The situation is made more complicated, but even more interesting, by a finding in Livneh's group, in collaboration with others, that three main classes of TLS seem to exist in mammalian cells: two rapid and error-free and a third slow and error-prone DNA pol.[110] These results indicate that a single DNA pol, namely DNA pol ζ, plays a pivotal role in all the classes of TLS and that discrete two-polymerase combinations between ζ and other TLS DNA pols dictate error-prone or error-free TLS across the same lesion.

A novel role of DNA pol η was suggested, since it appears that DNA pol η is also required for common fragile site stability during unperturbed DNA replication.[111] When DNA pol η was depleted by short hairpin RNA–mediated depletion this affected unperturbed DNA replication. These data challenge the current idea that pol η is only responsible for TLS damage tolerance following exogenous DNA damage.

Finally, auxiliary replicative factors such as PCNA and RP-A could modulate the fidelity of TLS by a given DNA pol. One example of such behavior is the finding that PCNA and RP-A allowed the correct incorporation of dCMP opposite the very mutagenic lesion 8-oxo-guanine 1,200-fold more efficiently than the incorrect dAMP by the X family DNA pol λ.[50] This effect of PCNA and RP-A is not restricted to a particular type of DNA pol since it was also observed, although at a lower extent, for DNA pol η.

Clearly more genetic and biochemical studies are needed to begin to understand what makes a DNA pol capable, alone or in combination with others, to replicate in an error-free or error-prone mode. Nevertheless, a systematic cataloguing of how a specific DNA damage can be bypassed in a mutagenic or faithful way is needed and by which DNA pol may have considerable consequences for the prevention or even cure of some human cancers.

4.5 Expression of DNA polymerases

Regulation of the 16 DNA pols is undoubtedly a very important issue. Too much of an unfaithful DNA pol, as well as reduced levels of a faithful DNA pol, can cause genomic instability. A few examples are as follows:

1. DNA pol β, a rather inaccurate DNA pol from the X family, has been shown to confer genetic instability when up-regulated in cells, thus resulting in mutagenesis, aneuploidy and tumorigenesis.[112] Moreover, overexpression of DNA pol β strengthens the mutagenicity of oxidative damages, concomitantly with a higher cellular sensitivity and increased apoptosis. Deregulated expression of DNA pol β was proposed as a predisposition factor for mutagenic effects of oxidative stress and thus has implication in the generation and/or evolution of cancer.[113]
2. DNA pol κ, an error-prone DNA pol from the Y family is up-regulated in lung cancers, induces DNA breaks and stimulates DNA exchanges as well as aneuploidy. Excess DNA pol κ appears to favor the proliferation of competent tumor cells as observed in immunodeficient mice, suggesting that altered regulation of DNA pol κ might be related to cancer-associated genetic changes and phenotype.[114]
3. Downregulation of faithful DNA pols from the B family induces chromosome fragile site instability. This has been documented for DNA pol α and for DNA pol δ. Such findings suggested that a change in the amounts, either absolute or relative, of replicative DNA pols results in recombinogenic DNA lesions and double-stranded breaks. Eventhough these results were obtained in yeast, they are probably also valid for mammalian cells.

A study testing 68 different tumors measured the expression levels of a variety of DNA pols.[115] When the two X family DNA pols β and λ, the two Y family DNA pols ι and κ and the three replicative DNA pols α, δ and ε were tested, the data indicated that in more than 45% of these 68 tumors at least one of the X and/or Y DNA pols was over-expressed. Of particular interest was that over 30% of all tumors tested had an over-expressed DNA pol β. DNA pols λ and ι were also over-expressed but less frequently than DNA pol β. Surprisingly DNA pol κ was under-expressed in many tumor samples. Finally, a few tumors also over-expressed the three replicative DNA pols α, δ and ε while in most cases they were under-expressed. All the 68 tumors were summarized in an over-expression scale that gave the following order: DNA pol β > DNA pol ι > DNA pol λ > DNA pol α > DNA pol δ > DNA pol κ > DNA pol ε. In summary these data clearly indicated that the expression patterns of DNA pols in normal and cancer tissues are very complex and the expression levels of the DNA pols are changed in tumors and can vary from one type of a tumor to another.

A comprehensive review on DNA polymerases and cancer is now available.[116] Further details on the roles of DNA pols in human diseases can also be found in the following Chapter 5.

4.6 DNA polymerases switch between different DNA transactions

As already indicated in this and in previous Chapters the interplay of different DNA pols in the various DNA transactions are complex but so far not yet completely understood at the mechanistic level. In this Section we try to briefly address this issue in prokaryotes (4.6.1) and, more extensively, in eukaryotes (4.6.2). In the latter the regulation of DNA pols is first considered, followed by the consequences of posttranslational modifications of DNA pols and to some extent of the DNA pol clamp PCNA.

4.6.1 *Prokaryotes*

In *E. coli* the replication task is carried out by DNA pol III. It was thought for a long time that besides this replicative enzyme, DNA pol I is responsible to replace Okazaki fragments and to act in DNA repair. The three other DNA pols II, IV and V have important functions in TLS and are therefore designated as TLS DNA pols.[117] How and when are these TLS pols engaged in the bacterial cell?

The initial observation derived from studies in which DNA pol V was shown to interact on one side with the catalytic subunit of DNA pol III (the α-subunit) and with the β-clamp.[118] Moreover, it was found that mutagenesis by DNA pol IV and V depends on the interaction motif of these DNA pols with the β-clamp and a "toolbelt model" was proposed in which the β-clamp may attract several DNA pols thus allowing the DNA pol switch between the replicative DNA pol III and the TLS pols IV and V.[119] These findings were soon confirmed by structural analysis showing how a conserved motif at the carboxy-terminal end of DNA pol IV can interact with the β-clamp.[120] Competition experiments suggested that DNA pol III has the highest affinity to the β-clamp and that a second DNA pol can bind to the β-clamp. Since the three DNA pols II, IV and V are up-regulated upon SOS response, it was proposed that these three TLS DNA pols can compete for the β-clamp after DNA damage.[121] The DNA pol switch is regulated so that the action of the TLS pol IV is limited, since only under conditions that cause DNA pol III to stall on the DNA the TLS DNA pol IV can take control on the primer-template. After the stall is relieved and DNA pol III is replicating again it becomes resistant to binding by DNA pol IV, suggesting that under normal replication conditions the TLS DNA pols do not have access to the fork.[122] An elegant study addressed the question of what happens when the leading as well as the lagging strands during replication contain a block. The replication fork is uncoupled since a block in the leading strand lets lagging strand synthesis proceed, suggesting that concurrent synthesis was uncoupled.[123] When TLS was reconstituted in the presence of DNA pol III and V and in the presence of auxiliary proteins, it was found that the replicative DNA pol III is disconnected from the lesion first, then DNA pol V binds to Rec A and the β-clamp, thus allowing TLS to proceed to generate a novel DNA primer long enough for DNA pol III to continue DNA replication thereafter.[124] These findings were nicely confirmed by showing that TLS by DNA pol V is regulated by Rec A nucleoprotein filaments binding to separate single-stranded DNA molecules in *trans*.[124]

DNA pol V is regulated to properly interact with different sets of proteins. After damage the homodimeric $UmuD_2$ is over-expressed and a slow autocleavage removes 24 amino-acids from the N-terminus. This fragment $UmuD'_2$ can interact with UmuC to result in the active trimeric DNA pol V.[125,126] Published work suggested that $UmuD_2$ and $UmuD'_2$ have different interacting partners: $UmuD_2$ has a high affinity to bind DNA pol IV and the β-clamp while $UmuD'_2$ preferentially interacts with the catalytic subunit of DNA pol III (the α-subunit) and Umu C to form the active DNA pol V.[127]

It has been found that DNA pol II and IV can functionally interact with the DnaB helicase and can thus regulate the speed of unwinding by slowing it down several orders of magnitude. The two TLS DNA pols might do this by exchanging DNA pol III via the β-clamp to form "alternative" replisomes even before the DNA pol III stalls at a lesion. Since the levels of DNA pol II and IV are induced upon DNA damage the two TLS can "kick out" the replicative DNA pol III. Since they bind to the DnaB helicase they can "maintain" the architecture of the replisome, which moves now very slowly (Figure 4.5). Such a dynamic replisome action could give additional time for DNA repair even before the DNA pol III reaches the lesion on one hand and also allowing the TLS DNA pols themselves to carry out TLS.[128]

Almost 50 years ago it was proposed that UV-lesions in *E. coli* were skipped during replication resulting in discontinuities and gaps in the nascent DNA and this even in the continuous leading strand of the replication fork. The gaps were finally repaired.[129] Over three decades later replication fork reactivation downstream a blocked nascent leading strand was demonstrated *in vitro*. Biochemical reconstitution of the replication restart after replication block suggested that the block is bypassed and the downstream *de novo* priming event allows continuation of DNA replication but thus creates gaps in the nascent DNA.[130] These data clearly suggested that re-priming of DNA synthesis downstream of a blocked fork occurs not only on the lagging strand but also on the leading strand. Therefore, contrary to what we generally see in the textbooks, leading strand DNA replication does not have to be continuous (reviewed in Refs. 131 and 132).

Finally, the flexibility and the dynamics of the replisome are also documented by the fact that, after a collision with the transcribing RNA polymerase, it can displace the RNA polymerase, continuing DNA

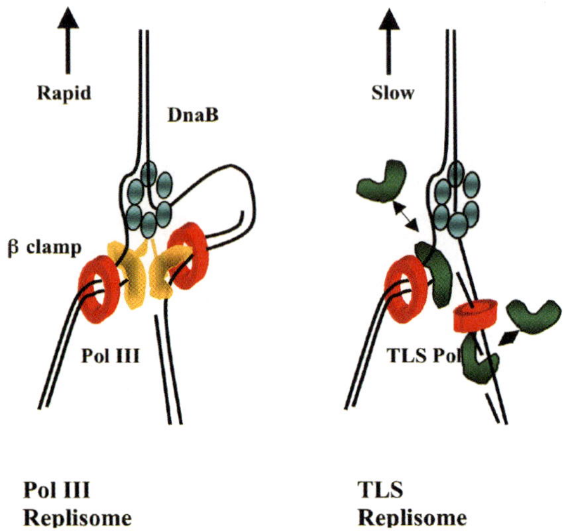

Figure 4.5 **TLS DNA polymerases can form an alternative replisome**. The left part shows the normal DNA pol III replisome. The t-subunit of the clamp loader, which is not shown, dimerizes the DNA pol III for fast coupled leading and lagging strand DNA replication. The right part documents the alternative replisome in which the TLS DNA pol II and IV replace DNA pol III at the fork, interact with the Dna B helicase thus slowing down the speed of DNA synthesis. The model predicts that leading and lagging strands are uncoupled in the TLS replisome. (Adapted from Ref. 128).

synthesis using the transcript mRNA as a primer.[133] A comprehensive review on the most recent data concerning prokaryotic translesion DNA polymerases is now available.[134]

4.6.2 *Eukaryotes*

In contrast to *E. coli* the eukaryotes have more sophisticated ways to regulate the DNA pol switches. First they possess many more DNA pols, being four from the Y family, four from the X family, four from the B family, three from the A family and one of the archeo-eukaryotic primase superfamily family. Besides this higher number of DNA pols, posttranslational modifications (PTM) of DNA pols themselves and of their processivity factor PCNA, have a great influence in regulating the proper interactions of the different DNA pols in different DNA transactions. We therefore first quickly discuss the PTM of PCNA and the known DNA pol switches.

Posttranslational modifications of PCNA: In Chapter 2 (Table 2.7) we indicated that PCNA can interact with 10 different DNA pols from the families B, X and Y. By interacting with the leading strand replicase DNA pol ε and the lagging strand replicase DNA pol δ, PCNA provides processivity to these enzymes. The interaction with the BER enzymes DNA pols β and λ allows them to perform accurate repair synthesis. The translesion DNA pol ζ, η, κ, ι, and Rev1 all can interact with PCNA. For these PTM of both PCNA and ubiquitin binding motifs in DNA pols are necessary. Here we briefly describe the PTM's of PCNA.

Phosphorylation of PCNA: The first PTM identified for PCNA was phosphorylation. Association of PCNA with nuclear chromatin in human fibroblasts is related to its phosphorylation status, suggesting that binding of PCNA to DNA synthesis sites occurs after phosphorylation.[135]

Acetylation of PCNA: The transcriptional coactivator p300 interacts with PCNA *in vitro* and *in vivo* and forms a complex with PCNA that does not depend on the S phase of cell cycle, suggesting a function in DNA repair synthesis.[136] Because p300 and histone deacetylase (HDAC1) were co-immunoprecipitated with PCNA, they are likely responsible for the acetylation and deacetylation of PCNA, respectively. Deacetylation reduced the ability of PCNA to bind to DNA pols β and δ.[137] Additional data link a mechanism of PCNA acetylation after DNA damage with DNA repair synthesis and removal of chromatin-bound PCNA, to ensure genome stability.[138]

Ubiquitination and SUMOylation of PCNA: A key element in understanding how PCNA can interact with different partners including DNA pols came with the discovery that PCNA can be ubiquinated.[139,140] Protein modification by ubiquitin is becoming important as a signal for many biological processes in eukaryotes, including regulated proteolysis, but also for non-degradative functions such as protein localization, maintenance of the genetic stability and regulation of chromatin structure. PCNA can be mono-ubiquitinated through RAD6 and RAD18 at lysine 164, modified by lysine-63-linked multi-ubiquitination, which in addition requires MMS2, UBC13 and RAD5, and it is conjugated to the small

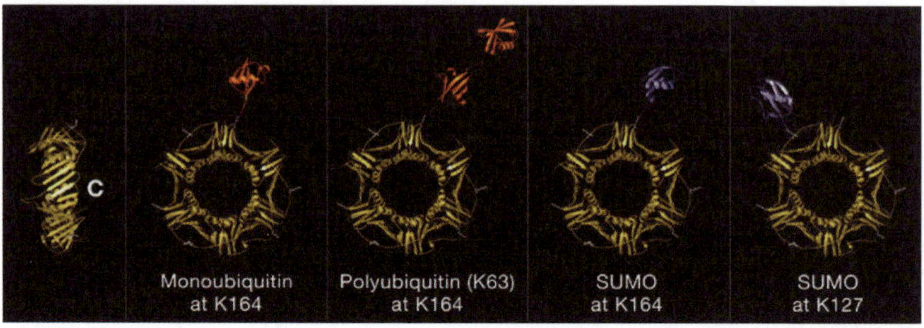

Figure 4.6 **Ubiquitination and SUMOylation of PCNA**. The structure of PCNA is shown in yellow. All three modifications are attached to lysine 164 of PCNA. The polyubiquitin chains are connected through lysine 64. PCNA from *S. cerevisiae* can be SUMOylated through lysine 127 which is in the interconnecting loop of PCNA where many proteins can bind.[3] (Reproduced with permission from Ref. 4)

ubiquitin-related modifier (SUMO) by UBC9 at lysine 164. All three modifications affect the same lysine 164 residue of PCNA, suggesting that they label PCNA for alternative functions (Figure 4.6). Recent and comprehensive reviews on the partners responding to PCNA modifications are now available.[141,142]

DNA pol switches due to PCNA ubiquitination. An extremely important issue now is the switch from the accurate replicative DNA pols ε and δ to the various TLS DNA pols. Cells can employ both transcriptional and posttranslational regulatory mechanisms to allow the TLS DNA pols under tight control to act at the replication fork when required.[143] Depending on the lesions they can be "blocking", "miscoding" or "coding" for replicative and TLS DNA pols (see Tables 4.3 and 4.4 for details). As we will see the DNA pol switch from the replicative to the TLS pols can either lead to error-prone or error-free TLS. The error-free pathway could rely on the undamaged information of the sister duplex and the mechanism here would be template switching. Therefore, two modes of switches have to be considered, first the DNA pol switches and second the DNA switch, which might be related to homologous recombination (HR).

As summarised in Refs. 134 and 143 our understanding of TLS has been transformed in the last 20 years. One of the biggest challenges mentioned in the outlook of these reviews is the ability to understand how the TLS DNA pols get access to a replication fork, a gap, or a D-loop, where they act in TLS, NER and HR. As will be discussed below it is the modification of PCNA that will likely play a key role for the DNA pol switches. How are DNA pols selected at the right time to the right place? Several findings support the hypothesis that the DNA pol best suited to perform TLS at a certain lesion will "win" the DNA pol switch. (see Ref. 144 and references therein)

These are as follows:

1. DNA pols themselves can weakly interact.
2. There are weak interactions between DNA pols and PCNA that can be increased when PCNA is ubiquinated.
3. The homotrimeric ring of PCNA could allow three DNA pols to bind simultaneously.
4. The half life of DNA Pols η and ι is very short.

In summary it appears that dynamic interactions between different DNA pols bind to the PCNA clamp and that this binding is tighter and longer if PCNA is ubiquitinated. Below are some relevant examples of PCNA/DNA pols interactions.

An archaeon PCNA is a heterotrimer. While prokaryotes have a clamp with a homodimeric structure, the eukaryotes contain a homotrimer.[3] In the archaeon *Sulfolobus solfataricus*, belonging to the third kingdom of life a heterotrimeric PCNA is present.[145] The three distinct PCNA subunits contact the DNA pol, DNA ligase and Fen1 and give a defined structure at the lagging strand of the replication fork, suggesting a preformed scanning complex at the lagging strand of the replication fork.

TLS by DNA pol η. DNA pol η is the enzyme that is altered in the disease Xeroderma pigmentosum V. It is so far one of the best studied TLS pols. First it was found that the lesion bypass by DNA pol η is a multi-DNA pol process. In contrast to the accurate bypass of a TT cyclobutane dimers by DNA pol η, other lesions appear to be bypassed in an error-prone manner (e.g. 6-4TT and G-AAF lesions), suggesting the combined action of several TLS pols depending on the lesions.[146] When cells are treated with UV, PCNA becomes monoubiquinated and this is dependent on the Rad18 protein. Specific interaction with DNA pol η is only possible after monoubiquitination of PCNA suggesting that modification of PCNA induces the DNA pol switch.[147] Rad 18 furthermore guides DNA pol η to sites where the replication fork is stalled and it does this by physical interaction with PCNA.[148] Moreover, biochemical work suggested that ubiquitination of PCNA can activate the TLS DNA pol η and REV1. Thus ubiquitinated PCNA was proposed to promote mutagenic DNA replication.[149] The DNA pol switch between the replicative DNA pol δ and the TLS DNA pol η by monoubiquitinated PCNA has to work back as well. This can only be achieved when PCNA is deubiquitinated.[150] DNA pol η has two modes to bind to PCNA: first to the interdomain connector loop of PCNA via the PCNA interacting protein (PIP) domain of DNA pol η, and second, to the lysine 164 linked ubiquitin moiety of PCNA by the Ub-binding domain (UBD) of DNA pol η. It appears that the binding to the interdomain connector loop is more important for DNA pol η to act in TLS and that a direct binding of Ub modified lysine164 to the UBD of DNA pol η appears not to be required.[151] The UBDs were generally suggested to regulate TLS for the Y family DNA pols [152] and at least for DNA pol η are responsible for correct localization.[153] TLS pols η and ι also interact among themselves and precise mapping of the two pols interacting regions in their respective ubiquitin domains demonstrated that ubiquitination of either DNA pol is a prerequisite for their physical and functional interaction.[154]

TLS by DNA pol ι. DNA pol ι can also bind to PCNA.[155] Moreover, monoubiquitination of PCNA appears to be required as well as a UBD on DNA pol ι.[152] In general, the cellular function of DNA pol ι still remains unresolved. Studies on DNA pol ι have brought insights into its possible biological roles and suggested that DNA pol ι plays important functions in protecting humans from the deleterious consequences of exposure to both oxidative and ultraviolet light-induced DNA damage (reviewed in Refs. 156 and 157). Pol ι has also been shown to prefer manganese over magnesium as divalent cation for catalysis and is highly active with low levels of manganese.[158,159].

TLS by DNA pol κ. DNA pol κ interacts with PCNA[160] and this interaction is stronger when PCNA is monoubiquitinated. The UBD is also present in DNA pol κ and is required for nuclear foci formation after

UV-damage.[161] Two PCNA interacting protein (PIP) motif domains have been identified in human DNA pol κ and shown to function redundantly in promoting TLS by DNA pol κ in human cells, even if they differ in their effects on the stimulation of PCNA-dependent DNA synthesis by DNA pol κ *in vitro*.[162] However, a recent article reports also the existence of a PCNA-Ub-independent pathway of activation of TLS by DNA pol κ and establishes that DNA pol η is an important backup polymerase in the absence of pol κ in response to DNA damage.[163]

TLS by DNA pol Rev1. The mouse Rev1 protein can interact with multiple TLS DNA pols. It interacts independently with Rev7 (a subunit of the B family TLS DNA pol ζ) and with the two Y family DNA pols η and ι, that bind to approximatively the same 100 amino acid C-terminal regions of Rev1. These data suggested that Rev1 plays a role in mediating protein–protein interactions among TLS DNA pols.[164] A UBD was also identified in Rev1 and TLS of Rev1 was enhanced upon ubiquitination of PCNA and it was suggested that this interaction was stabilized during DNA damage response.[165] In chicken DT40 cells the necessity for both the polymerase-interaction domain and the UBD was documented and the PCNA ubiquitination is essential for gap filling of postreplicative gaps. This is likely due to the fact that PCNA ubiquitination and REV1 play distinct roles in the coordination of DNA damage bypass and are temporally separated to the stalled replication fork.[166] Rev1 can promote complex formation of DNA pol ζ with the Pol 32 subunit of DNA pol δ. The Rev1 binding to pol32 of DNA pol δ might thus have a significant function for DNA pol ζ in TLS and suggests a structural role for Rev1 in modulating the binding of DNA pol ζ with Pol32 in DNA pol δ stalled at a lesion site.[167] The TLS by Rev1 appears to be mediated by two separate domains in Rev1 which are involved in distinct modes of TLS that may depend on gap filling.[168] The early mode of TLS is mediated by the BRCT domain of Rev1. At later stages Rev1, but not its BRCT domain, is required for postreplicative gap filling. An unusual site near the monomer–monomer interface of the trimeric PCNA is required for the physical interaction of PCNA with the polymerase-associated domain (PAD) of Rev1.[169]

How DNA pols work together? A complete picture of how the different DNA pol switches are regulated is far from being complete. There is still controversy in the literature concerning mechanistic details. A work published in 2006 questions some issues raised above.[170] To be able to decipher the role of PCNA monoubiquitination in the TLS process, PCNA was monoubiquitinated *in vitro* from purified *S. cerevisiae* proteins. In addition to the requirement for Rad6-Rad18, the reaction depended on the loading of the PCNA homotrimeric ring onto the DNA by the clamp loader complex RF-C. All three PCNA monomers were efficiently ubiquitinated. Next the effects of this PCNA modification on DNA synthesis by DNA pols δ, η, ζ and Rev1 were tested. Contrary to published data that postulate a role for PCNA ubiquitination in the disruption of DNA pol δ binding to PCNA or in the enhancement of the binding affinity of the TLS pols for PCNA, it was found that PCNA ubiquitination did not affect any of these processes. These observations suggested a role for PCNA monoubiquitination in disrupting the PCNA binding of a protein(s) that otherwise is (are) inhibitory to the binding of PCNA by TLS pols.

Livneh's group has shown that DNA pol ζ cooperates with DNA pols κ and ι in error-prone TLS across pyrimidine dimers in cells from XPV patients.[109] DNA pols ζ and κ but not DNA pol ι can protect XPV cells against the cytotoxic effect of UV and this is independent of NER. These results suggested that a benefit–risk balance might be inherent in TLS DNA pols, with the outcome that cells are protected against UV, but with the cost of increased mutations.

Figure 4.7 summarizes the DNA pol switch intricate dynamics. DNA pol δ and PCNA are stalled at a thymidine dimer. PCNA is monoubiquitinated after the recruitment of Rad18-Rad16 and cleavage of USP1. Monoubiquitination attracts TLS pols (here exemplified by DNA pols η and ι), which perform TLS and then dissociate and replication can restart. Such scenarios are possible with the many other TLS pols known.

Other examples to regulate DNA polymerases. DNA pol β: An elegant study suggested that DNA pol β can be polyubiquitinated by the E3 ubiquitin ligase CHIP (carboxy terminus of HSP70 interacting protein) when not bound to the DNA repair complexes and all non-employed DNA pol β molecules are subsequently rapidly degraded.[171] Moreover, the steady state level of DNA pol β might be regulated by another E3 ligase called Mule (also known as ARF-binding protein or HectH9) and its partner ARF.[172] Monoubiquitination at Lys 41, 61 and 81 renders DNA pol β as a

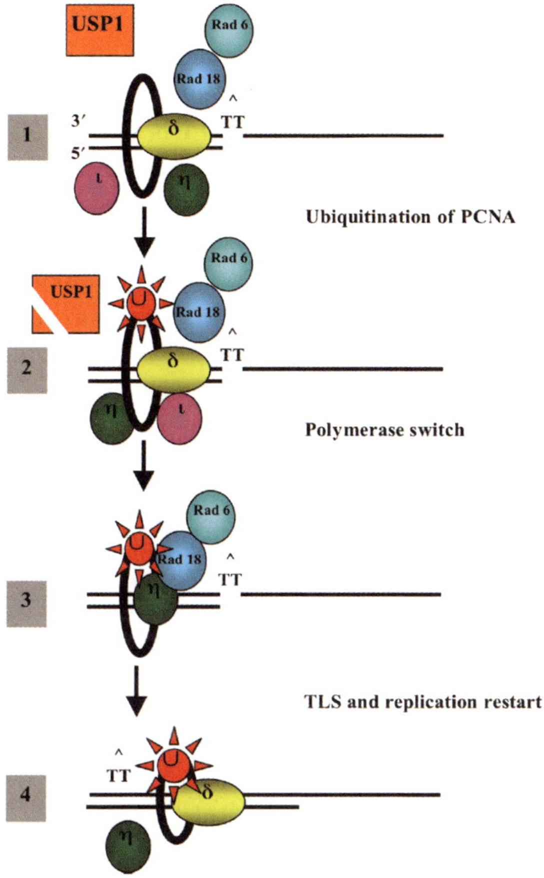

Figure 4.7 **Translesion DNA synthesis and DNA polymerase switch.** For details see text.

substrate for polyubiquitination by CHIP with subsequent degradation. If bound to ARF, Mule is inactive and thus DNA pol β cannot be ubiquitinated and is ready to perform repair in the nucleus. This is important in the view that in over 30% of the tumors tested DNA pol β is over-expressed.[115] Further modifications regulate the functions of DNA pol β. First, arginine methylation regulates the DNA pol β itself [173] and its binding to PCNA,[174] second, acetylation of DNA pol β regulates its end-trimming activity (dRP-lyase),[175] and, third, the cytoplasmic ubiquitin specific protease USP47 has been identified as the major enzyme involved in deubiquitination of DNA pol β and shown to be required for regulation of DNA repair and maintaining of genome integrity.[176] However, Quinming Fang *et al.* demonstrated that XRCC1/DNA Pol β complex formation prevents the ubiquitination and degradation of DNA Pol β. The authors proposed that DNA Pol β's stability thus depends also on binding to XRCC1.[177]

DNA pol η, ι and Rev 1: Phosphorylation can regulate the activity and the translocation of DNA pol η. After UV damage DNA pol η gets phosphorylated, and treatment with caffeine and siRNA against the ATP kinase reduced its accumulation at stalled replication forks.[178] In *Caenorhabditis elegans* during the DNA damage response, DNA pol η is degraded by the Cul4-Ddb1-Cdt2 SUMO pathway.[179] It was furthermore found that DNA pols η and ι can associate with the replication machinery in the nuclei at stalled replication forks following DNA damage. Very interestingly the two DNA pols η and ι can physically interact during this process, suggesting a coordinated function of both these TLS DNA pols.[180] This interaction is mediated by ubiquitin.[154] Finally, the interaction of DNA pol η is required for the nuclear accumulation of the Rev1 DNA pol. This interaction can suppress spontaneous mutations in human cells.[181]

DNA pol λ: Human DNA pol λ is degraded *in vivo* by the proteasome–ubiquitination pathway; however, it can be stabilized upon phosphorylation at T553. This stabilization seems to occur in particular in late S and G2 phase of the cell cycle, which probably enables the enzyme to fulfill its specific functions in DNA damage repair during these stages.[182] Similarly to other DNA repair enzymes, DNA pol λ undergoes posttranslational modifications, including ubiquitination, during cell cycle that regulates its stability and possibly its subcellular localization.[53] Further work has shown that the E3 ligase Mule mediates the degradation of DNA pol λ and that the control of DNA pol λ by Mule has functional consequences for the ability of mammalian cells to deal with 8-oxo-G lesions. Moreover, it appears that phosphorylation of DNA pol λ by Cdk2/cyclin A counteracts its Mule-mediated degradation, by promoting recruitment of DNA pol λ to chromatin as an active 8-oxo-G complex.[183]

Replicative and repair DNA pols: A human BER complex was found to be physically associated with DNA replication and cell cycle regulatory proteins, suggesting that a constitutively active DNA repair machinery is ready to be recruited when base damages occur at DNA replications forks.[184] A very recent review [185] outlines the role of Nei endonuclease VIII like 1 (NEIL1) DNA glycosylase, one of the BER-initiating enzymes in mammalian cells and whose expression is enhanced in S phase, in replicating template strand repair. Moreover, it has been shown that NER is associated with the mammalian replisome and that its efficiency depends on direct interaction between XPA and PCNA.[186]

Alternative subunit structure of DNA pol δ: DNA pol δ contains four subunits called p125, p68, p50 and p12 (for details see Chapters 2 and 3). Upon genotoxic stress treatment of human cells the four-subunit DNA pol δ was converted to a three-subunit enzyme, since the p12 was degraded,[187,188] suggesting that this might be an adaptive response to DNA damage.[189] Moreover, it was found that $CLR4^{cdt2}$, a key regulator of cell cycle progression that targets replication licensing factors, also targets the p12 subunits of DNA pol δ in response to DNA damage and on entry into S phase.[190]

In a review Alena Makarova and Burgers summarize data on eukaryotic DNA pol ζ, the enzyme responsible for the bulk of mutagenesis in eukaryotic cells in response to DNA damage.[191] They claim that new insights in mutagenesis have been derived from recent advances in the elucidation of the subunit structure of DNA pol ζ. The lagging strand DNA pol δ can share its two small Pol31 and Pol32 subunits with core DNA pol ζ assembly (being Rev3–Rev7) giving a four-subunit DNA pol ζ complex that is the active form in mutagenesis. Furthermore, DNA pol ζ forms essential interactions with the mutasome assembly factor Rev1 and with PCNA. These interactions are modulated by posttranslational modifications such as ubiquitination and phosphorylation that enhance translesion synthesis (TLS) and mutagenesis.

4.7 Functions of DNA polymerases in checkpoint control

Cell cycle events have to be coordinated in time and order and these steps are critical for a high-fidelity transmission of the genetic information and the successful cell duplication. Activation of DNA damage checkpoint and DNA replication checkpoint requires the action of DNA damage sensors and transducers.[192] Some of these factors, such as ataxia telangectasia mutated protein (ATM), ATM related protein (ATR), ATR interacting protein (ATRIP), Rad17, Rad9, Rad1 and Hus1, are thought to be involved in triggering DNA repair processes. Data in this field connecting DNA pols to checkpoint control appeared in the last 15 years. As already discussed in Section 4.2.2, the first DNA pols engaged in DNA replication are DNA pol α with its primase subunits and DNA pol ε, the leading strand DNA pol.

Functional uncoupling of the replicative DNA helicase from replicative DNA pols activates the ATR-dependent checkpoint: It was found that uncoupling of the MCM helicase and Cdc45 from the replication fork resulted in checkpoint activation.[193] Interestingly, this uncoupling also results in monoubiquitination of PCNA.[194] Successful checkpoint activation also required DNA synthesis by DNA pol α. This uncoupling renders DNA single-stranded and thus RP-A can bind. RP-A itself can activate the checkpoint [195] and does this by recruiting Rad17, a component of the alternate clamp loader [196] (see Chapter 2 for details).

DNA polymerase α and checkpoint control: The first data came from genetic experiments performed in the fission yeast *Schizosaccharomyes pombe* and these experiments suggested a role for DNA pol α in checkpoint control.[197] Later it was found that the two primase subunits of DNA pol α could affect the ChK1 activation (due to the p49 homolog of mammalian DNA pol α) and Cds1 kinase activation (due to the p58 homolog of mammalian DNA pol α).[198,199] At about the same time it was found that activation of the DNA replication checkpoint was initiated by RNA synthesis by the primase of DNA pol α.[200] This was followed by studies that elaborated the role of single-stranded DNA and DNA pol α in establishing the ATR, Hus1 DNA replication checkpoint.[201]

DNA polymerase ε and checkpoint control: Already in 1995 it was realized, thanks to genetic studies in *S. cerevisiae*, that DNA pol ε links the DNA replication machinery to the S phase checkpoint [202] and it was proposed that DNA pol ε acts as a sensor of DNA replication that coordinates the transcriptional and the cell cycle responses to replication block. Subsequently Dbp11 was found to interact with DNA pol ε and could be part of the leading strand enzyme to act in checkpoint control during the S phase of the cell cycle to sense stalled replication forks.[203] Again in

S. cerevisiae it was found that RAD9 and DNA pol ε form parallel sensory branches for transducing the DNA damage checkpoint signal.[204] The checkpoint has been identified in the C-terminal part of DNA pol ε[202] and this domain was further characterized by identifying putative zinc finger domain in the S/M phase checkpoint.[205] Next it was found that the catalytic domain of DNA pol ε was dispensable for DNA replication, DNA repair and cell viability, but the C-terminal part of DNA pol ε is the responsible part for viability.[206,207]

These data were questioned by a later study indicating that *Schizosaccharomyces pombe* DNA pol ε is required for chromosomal replication but not for the S phase checkpoint. It was proposed that the checkpoint signal operating in S phase depends on the assembly of the replication initiation complex and that this signal is generated prior to the elongation stage of DNA synthesis, e.g. through the primase action of DNA pol α.[208] Indeed the data discussed above for DNA pol α support this prediction.

Recently, work with budding yeast showed that the "alternative clamp loader" Ctf18-RFC associates with DNA pol ε via an evolutionary conserved "DNA pol ε conserved module" in the Ctf18-RFC that is produced by interaction of the carboxyl terminus of Ctf18 with the Ctf8 and Dcc1 subunits. Mutations at the end of Ctf18 disrupt the integrity of DNA pol ε binding module and block the S-phase checkpoint pathway.[209] These findings indicate that the association of Ctf18-RFC with DNA pol ε at defective replication forks is a key step in activation of the S-phase checkpoint.

DNA polymerase ζ, Rev1, λ and checkpoint control: It has been shown in *S. cerevisiae* that TLS can occur in the absence of checkpoint proteins in nucleotide excision repair (NER) proficient cells. In the absence of NER, checkpoint protein-mediated Rev1 phosphorylation by Mec1 contributed to increasing the proficiency of DNA pol ζ dependent TLS.[210] More recently, it has been shown that knockdown of DNA pol λ, but not of its close homologous DNA pol β , results in replication fork stress and activates the S phase checkpoint, slowing S phase progression in different human cancer cell lines.[211] These data present the first evidence that DNA pol λ is required for cell cycle progression and is functionally connected to S phase DNA damage response machinery in cancer cells.

PrimPol and checkpoint control: The most recently discovered human DNA pol, PrimPol, has been shown to be linked to fork progression and cell cycle completion. Chicken DT40 cells deleted for PrimPol, showed a delayed cell cycle in response to UV-C DNA damage. This delay resulted in an accumulation of PrimPol –/– cells at the G2/M border. Suppression of the checkpoint by inhibition of the kinase Chk1 in PrimPol –/– cells, resulted in an increase in mitotic cells, suggesting that the cell cycle block was, at least partially, dependent on the activation of ATR. Similarly, inhibition of the p38MAPK kinase caused a partial release of PrimPol –/– cells into M phase, but without further progression, indicating that p38 also played a role in the cell cycle arrest. The cell cycle block allowed PrimPol –/– cells to remain viable, as indicated by a decrease in apoptosis with respect to wild type cells, in response to UV-C treatment.[212]

4.8 Chapter summary

In Chapter 4 we undertook the task to look at the global functions of the several DNA pols acting in living cells. It is clear that most DNA transactions require more than one DNA pol. This back-up strategy could guarantee the insurance for keeping changes in the genome at a very low level, either during DNA replication or for the various DNA repair pathways. DNA replication in living organisms requires three DNA molecules at the replication fork. The replisome of the *E. coli* bacteriophage T7 is the simplest and so far best characterized replisome. Details of its mechanism are available mainly due to its simple component composition and to single molecule technology. The *E. coli* replisome relies on the DNA pol III, but it appears that three molecules of DNA pol III might possibly act in concert at the fork. In eukaryotes from yeast to human the three different DNA pols α, ε and δ share the workload at the DNA replication fork in initiating, leading strand and lagging strand synthesis, respectively. Proofreading is very important to minimize mistakes during DNA synthesis. We know there are proofreader DNA pols, such as the replicating DNA pol δ and ε, but there are also strategies that proofreader $3'\rightarrow5'$ exonuclease activities of DNA pol δ and ε act for non-proofreader DNA pols, such as DNA pol α.

The different DNA repair pathways have their own DNA pols, but can also borrow them from the replication machinery. Pathways to restore the integrity of DNA include BER, NER, MNR, HR, NHEJ and ICLR and they rely together on at least 10 different DNA pols. In most cases more than one DNA pol is employed for each pathway. In translesion DNA synthesis in eukaryotes generally two different DNA pols are required. One is responsible for the insertion and the other for the extension for the lesion. On an even more complex level, they all have to act in concert with the replicative DNA pols. Considering their actions in the cellular context the DNA pols have to switch between different DNA transactions.

Posttranslational modifications play an essential role in regulating the actions of many DNA pols. They include phosphorylation, acetylation, methylation and ubiquitination reactions (monoubiquitination, polyubiquitination and SUMOylation) of DNA pols as well as similar posttranslational modifications of the DNA pol clamp PCNA. Moreover, expression levels of DNA pols at the right time and the right place are essential to have DNA pols at the site of action when they are needed. Finally, regulation through checkpoint control events is dependent on "early" DNA pols in initiation of DNA replication, where DNA pol α and ε are likely involved in important control events in cell metabolism.

References

1. Hubscher, U., Maga, G. & Spadari, S. Eukaryotic DNA polymerases. *Annu Rev Biochem* **71**, 133–163 (2002).
2. Johnson, A. & O'Donnell, M. Cellular DNA replicases: components and dynamics at the replication fork. *Annu Rev Biochem* **74**, 283–315 (2005).
3. Maga, G. & Hubscher, U. Proliferating cell nuclear antigen (PCNA): a dancer with many partners. *J Cell Sci* **116**, 3051–3060 (2003).
4. Moldovan, G. L., Pfander, B. & Jentsch, S. PCNA, the Maestro of the Replication Fork. *Cell* **129**, 665–679 (2007).
5. Bergink, S. & Jentsch, S. Principles of ubiquitin and SUMO modifications in DNA repair. *Nature* **458**, 461–467 (2009).
6. Frouin, I., Montecucco, A., Spadari, S. & Maga, G. DNA replication: a complex matter. *EMBO Rep* **4**, 666–670 (2003).

7. Cotterill, S. & Kearsey, S. E. DNAReplication: a database of information and resources for the eukaryotic DNA replication community. *Nucleic Acids Res* **37**, D837–839 (2009).
8. Bell, S. D. Molecular biology: prime-time progress. *Nature* **439**, 542–543 (2006).
9. Lovett, S. T. Polymerase switching in DNA replication. *Mol Cell* **27**, 523–526 (2007).
10. Sinha, N. K., Morris, C. F. & Alberts, B. M. Efficient *in vitro* replication of double-stranded DNA templates by a purified T4 bacteriophage replication system. *J Biol Chem* **255**, 4290–4293 (1980).
11. Yuzhakov, A., Turner, J. & O'Donnell, M. Replisome assembly reveals the basis for asymmetric function in leading and lagging strand replication. *Cell* **86**, 877–886 (1996).
12. Kim, S., Dallmann, H. G., McHenry, C. S. & Marians, K. J. tau couples the leading- and lagging-strand polymerases at the Escherichia coli DNA replication fork. *J Biol Chem* **271**, 21406–21412 (1996).
13. Kim, S., Dallmann, H. G., McHenry, C. S. & Marians, K. J. Coupling of a replicative polymerase and helicase: a tau-DnaB interaction mediates rapid replication fork movement. *Cell* **84**, 643–650 (1996).
14. Pomerantz, R. T. & O'Donnell, M. Replisome mechanics: insights into a twin DNA polymerase machine. *Trends Microbiol* **15**, 156–164 (2007).
15. McInerney, P., Johnson, A., Katz, F. & O'Donnell, M. Characterization of a triple DNA polymerase replisome. *Mol Cell* **27**, 527–538 (2007).
16. Tanner, N. A. *et al.* Single-molecule studies of fork dynamics in Escherichia coli DNA replication. *Nat Struct Mol Biol* **15**, 170–176 (2008).
17. Hubscher, U., Kuenzle, C. C. & Spadari, S. Variation of DNA polymerases-alpha, -beta. and -gamma during perinatal tissue growth and differentiation. *Nucleic Acids Res* **4**, 2917–2929 (1977).
18. Conaway, R. C. & Lehman, I. R. A DNA primase activity associated with DNA polymerase alpha from Drosophila melanogaster embryos. *Proc Natl Acad Sci USA* **79**, 2523–2527 (1982).
19. Wan, L. *et al.* hPrimpol1/CCDC111 is a human DNA primase-polymerase required for the maintenance of genome integrity. *EMBO Reports* **14**, 1104–1112 (2013).
20. Garcia-Gomez, S. *et al.* PrimPol, an Archaic Primase/Polymerase Operating in Human Cells. *Molecular Cell* **52**, 541–553 (2013).
21. Bianchi, J. *et al.* PrimPol Bypasses UV Photoproducts during Eukaryotic Chromosomal DNA Replication. *Molecular Cell* **52**, 566–573 (2013).

22. Boulet, A., Simon, M., Faye, G., Bauer, G. A. & Burgers, P. M. Structure and function of the Saccharomyces cerevisiae CDC2 gene encoding the large subunit of DNA polymerase III. *EMBO J* **8**, 1849–1854 (1989).
23. Morrison, A., Araki, H., Clark, A. B., Hamatake, R. K. & Sugino, A. A third essential DNA polymerase in S. cerevisiae. *Cell* **62**, 1143–1151 (1990).
24. Garg, P., Stith, C. M., Sabouri, N., Johansson, E. & Burgers, P. M. Idling by DNA polymerase delta maintains a ligatable nick during lagging-strand DNA replication. *Genes Dev* **18**, 2764–2773 (2004).
25. Chilkova, O. *et al.* The eukaryotic leading and lagging strand DNA polymerases are loaded onto primer-ends via separate mechanisms but have comparable processivity in the presence of PCNA. *Nucleic Acids Res* **35**, 6588–6597 (2007).
26. Waga, S. & Stillman, B. The DNA replication fork in eukaryotic cells. *Annu Rev Biochem* **67**, 721–751. (1998).
27. Pursell, Z. F., Isoz, I., Lundstrom, E. B., Johansson, E. & Kunkel, T. A. Yeast DNA polymerase epsilon participates in leading-strand DNA replication. *Science* **317**, 127–130 (2007).
28. Nick McElhinny, S. A., Gordenin, D. A., Stith, C. M., Burgers, P. M. & Kunkel, T. A. Division of labor at the eukaryotic replication fork. *Mol Cell* **30**, 137–144 (2008).
29. Johnson, R. E., Klassen, R., Prakash, L. & Prakash, S. A Major Role of DNA Polymerase delta in Replication of Both the Leading and Lagging DNA Strands. *Molecular Cell* **59**, 163–175 (2015).
30. Stillman, B. Reconsidering DNA Polymerases at the Replication Fork in Eukaryotes. *Molecular Cell* **59**, 139–141 (2015).
31. Stillman, B. DNA polymerases at the replication fork in eukaryotes. *Mol Cell* **30**, 259–260 (2008).
32. Moser, B. A. *et al.* Differential arrival of leading and lagging strand DNA polymerases at fission yeast telomeres. *EMBO J* **28**, 810–820 (2009).
33. Shevelev, I. V. & Hubscher, U. The 3′-5′ exonucleases. *Nat Rev Mol Cell Biol* **3**, 364–376 (2002).
34. Kunkel, T. A. DNA replication fidelity. *J Biol Chem* **279**, 16895–16898 (2004).
35. Albertson, T. M. & Preston, B. D. DNA replication fidelity: proofreading in trans. *Curr Biol* **16**, R209–211 (2006).
36. Pavlov, Y. I. *et al.* Evidence that errors made by DNA polymerase alpha are corrected by DNA polymerase delta. *Curr Biol* **16**, 202–207 (2006).

37. Hubscher, U., Kuenzle, C. C. & Spadari, S. Functional roles of DNA polymerases beta and gamma. *Proc Natl Acad Sci USA* **76**, 2316–2320 (1979).
38. Waser, J., Hubscher, U., Kuenzle, C. C. & Spadari, S. DNA polymerase beta from brain neurons is a repair enzyme. *Eur J Biochem* **97**, 361–368 (1979).
39. Dianov, G. L. & Hubscher, U. Mammalian base excision repair: the forgotten archangel. *Nucleic Acids Research* **41**, 3483–3490 (2013).
40. Lindahl, T. & Wood, R. D. Quality control by DNA repair. *Science* **286**, 1897–1905 (1999).
41. Fortini, P. & Dogliotti, E. Base damage and single-strand break repair: mechanisms and functional significance of short- and long-patch repair subpathways. *DNA Repair (Amst)* **6**, 398–409 (2007).
42. Wallace, S. S. Base excision repair: A critical player in many games. *DNA Repair* **19**, 14–26 (2014).
43. Pascucci, B., Stucki, M., Jonsson , Z. O., Dogliotti, E. & Hübscher, U. Long patch base excision repair with purified human proteins : DNA ligase I as a patch size mediator for DNA polymerases d and e. *J Biol Chem* **274**, 33696–33702 (1999).
44. Stucki, M. *et al.* Mammalian base excision repair by DNA polymerase δ and ε. *Oncogene* **17**, 835–843 (1998).
45. Dianov, G. L., Prasad, R., Wilson, S. H. & Bohr, V. A. Role of DNA polymerase beta in the excision step of long patch mammalian base excision repair. *J Biol Chem* **274**, 13741–13743. (1999).
46. Liu, Y. *et al.* DNA polymerase beta and flap endonuclease 1 enzymatic specificities sustain DNA synthesis for long patch base excision repair. *J Biol Chem* **280**, 3665–3674 (2005).
47. Prasad, R. *et al.* Human DNA polymerase theta possesses 5′-dRP lyase activity and functions in single-nucleotide base excision repair *in vitro*. *Nucleic Acids Res* **37**, 1868–1877 (2009).
48. Braithwaite, E. K. *et al.* DNA polymerase lambda mediates a back-up base excision repair activity in extracts of mouse embryonic fibroblasts. *J Biol Chem* **280**, 18469–18475 (2005).
49. Yoshimura, M. *et al.* Vertebrate POLQ and POLbeta cooperate in base excision repair of oxidative DNA damage. *Mol Cell* **24**, 115–125 (2006).
50. Maga, G. *et al.* 8-oxo-guanine bypass by human DNA polymerases in the presence of auxiliary proteins. *Nature* **447**, 606–608 (2007).
51. Maga, G. *et al.* Replication protein A and proliferating cell nuclear antigen coordinate DNA polymerase selection in 8-oxo-guanine repair. *Proc Natl Acad Sci USA* **105**, 20689–20694 (2008).

52. van Loon, B. & Hubscher, U. An 8-oxo-guanine repair pathway coordinated by MUTYH glycosylase and DNA polymerase lambda. *Proc Natl Acad Sci USA* **106**, 18201–18206 (2009).
53. Markkanen, E., van Loon, B., Ferrari, E. & Hubscher, U. Ubiquitylation of DNA polymerase lambda. *FEBS Letters* **585**, 2826–2830 (2011).
54. Edmonds, M. J. & Parsons, J. L. Regulation of base excision repair proteins by ubiquitylation. *Experimental Cell Research* **329**, 132–138 (2014).
55. Parsons, J. L. & Dianov, G. L. Co-ordination of base excision repair and genome stability. *DNA Repair* **12**, 326–333 (2013).
56. Jia, P. P., Her, C. T. & Chai, W. H. DNA excision repair at telomeres. *DNA Repair* **36**, 137–145 (2015).
57. Shivji, K. K., Kenny, M. K. & Wood, R. D. Proliferating cell nuclear antigen is required for DNA excision repair. *Cell* **69**, 367–374 (1992).
58. Aboussekhra, A. *et al.* Mammalian DNA nucleotide excision repair reconstituted with purified protein components. *Cell* **80**, 859–868 (1995).
59. Shivji, M. K., Podust, V. N., Hubscher, U. & Wood, R. D. Nucleotide excision repair DNA synthesis by DNA polymerase epsilon in the presence of PCNA, RFC, and RPA. *Biochemistry* **34**, 5011–5017 (1995).
60. Masutani, C. *et al.* The XPV (xeroderma pigmantosum variant) gene encodes human DNA polymerase eta. *Nature* **399**, 700–704 (1999).
61. Stary, A., Kannouche, P., Lehmann, A. R. & Sarasin, A. Role of DNA polymerase eta in the UV mutation spectrum in human cells. *J Biol Chem* **278**, 18767–18775 (2003).
62. Ogi, T. & Lehmann, A. R. The Y-family DNA polymerase kappa (pol kappa) functions in mammalian nucleotide-excision repair. *Nat Cell Biol* **8**, 640–642 (2006).
63. Ogi, T. *et al.* Three DNA Polymerases, Recruited by Different Mechanisms, Carry Out NER Repair Synthesis in Human Cells. *Molecular Cell* **37**, 714–727 (2010).
64. Longley, M. J., Pierce, A. J. & Modrich, P. DNA polymerase delta is required for human mismatch repair *in vitro*. *J Biol Chem* **272**, 10917–10921 (1997).
65. Kunkel, T. A. & Erie, D. A. Eukaryotic Mismatch Repair in Relation to DNA Replication. *Annu Rev Genet* **49**, 291–313 (2015).
66. Mahajan, K. N. *et al.* Association of terminal deoxynucleotidyl transferase with Ku. *Proc Natl Acad Sci USA* **96**, 13926–13931 (1999).
67. Fan, W. & Wu, X. DNA polymerase lambda can elongate on DNA substrates mimicking non-homologous end joining and interact with

XRCC4-ligase IV complex. *Biochem Biophys Res Commun* **323**, 1328–1333 (2004).
68. Mahajan, K. N., Nick McElhinny, S. A., Mitchell, B. S. & Ramsden, D. A. Association of DNA polymerase mu (pol mu) with Ku and ligase IV: role for pol mu in end-joining double-strand break repair. *Mol Cell Biol* **22**, 5194–5202 (2002).
69. Lee, J. W. *et al.* Implication of DNA polymerase lambda in alignment-based gap filling for nonhomologous DNA end joining in human nuclear extracts. *J Biol Chem* **279**, 805–811 (2004).
70. Nick McElhinny, S. A. *et al.* A gradient of template dependence defines distinct biological roles for family X polymerases in nonhomologous end joining. *Mol Cell* **19**, 357–366 (2005).
71. Capp, J. P. *et al.* The DNA polymerase lambda is required for the repair of non-compatible DNA double strand breaks by NHEJ in mammalian cells. *Nucleic Acids Res* **34**, 2998–3007 (2006).
72. Capp, J. P. *et al.* Involvement of DNA polymerase mu in the repair of a specific subset of DNA double-strand breaks in mammalian cells. *Nucleic Acids Res* **35**, 3551–3560 (2007).
73. Tseng, H. M. & Tomkinson, A. E. A Physical and Functional Interaction between Yeast Pol4 and Dnl4-Lif1 Links DNA Synthesis and Ligation in Nonhomologous End Joining. *J Biol Chem* **277**, 45630–45637 (2002).
74. Mateos-Gomez, P. A. *et al.* Mammalian polymerase theta promotes alternative NHEJ and suppresses recombination. *Nature* **518**, 254–U285 (2015).
75. Zahn, K. E., Averill, A. M., Aller, P., Wood, R. D. & Doublie, S. Human DNA polymerase theta grasps the primer terminus to mediate DNA repair. *Nat Struct Mol Biol* **22**, 304–U351 (2015).
76. Beard, W. A. & Wilson, S. H. Structures of human DNA polymerases nu and theta expose their end game. *Nat Struct Mol Biol* **22**, 273–275 (2015).
77. Schatz, D. G. V(D)J recombination. *Immunol Rev* **200**, 5–11 (2004).
78. Bertocci, B., De Smet, A., Berek, C., Weill, J. C. & Reynaud, C. A. Immunoglobulin kappa light chain gene rearrangement is impaired in mice deficient for DNA polymerase mu. *Immunity* **19**, 203–211 (2003).
79. Bertocci, B., De Smet, A., Weill, J. C. & Reynaud, C. A. Nonoverlapping functions of DNA polymerases mu, lambda, and terminal deoxynucleotidyltransferase during immunoglobulin V(D)J recombination *in vivo*. *Immunity* **25**, 31–41 (2006).
80. Lebecque, S. G. & Gearhart, P. J. Boundaries of somatic mutation in rearranged immunoglobulin genes: 5′ boundary is near the promoter, and 3′

boundary is approximately 1 kb from V(D)J gene. *J Exp Med* **172**, 1717–1727 (1990).
81. Honjo, T., Nagaoka, H., Shinkura, R. & Muramatsu, M. AID to overcome the limitations of genomic information. *Nat Immunol* **6**, 655–661 (2005).
82. Neuberger, M. S. *et al.* Somatic hypermutation at A.T pairs: polymerase error versus dUTP incorporation. *Nat Rev Immunol* **5**, 171–178 (2005).
83. Jaszczur, M., Bertram, J. G., Pham, P., Scharff, M. D. & Goodman, M. F. AID and Apobec3G haphazard deamination and mutational diversity. *Cellular and Molecular Life Sciences* **70**, 3089–3108 (2013).
84. Pham, P., Bransteitter, R., Petruska, J. & Goodman, M. F. Processive AID-catalysed cytosine deamination on single-stranded DNA simulates somatic hypermutation. *Nature* **424**, 103–107 (2003).
85. Dominguez, O. *et al.* DNA polymerase mu (Pol mu), homologous to TdT, could act as a DNA mutator in eukaryotic cells. *Embo J* **19**, 1731–1742. (2000).
86. Bertocci, B. *et al.* Cutting edge: DNA polymerases mu and lambda are dispensable for Ig gene hypermutation. *J Immunol* **168**, 3702–3706 (2002).
87. Zan, H. *et al.* The translesion DNA polymerase theta plays a dominant role in immunoglobulin gene somatic hypermutation. *EMBO J* **24**, 3757–3769 (2005).
88. Masuda, K. *et al.* DNA polymerase eta is a limiting factor for A:T mutations in Ig genes and contributes to antibody affinity maturation. *Eur J Immunol* **38**, 2796–2805 (2008).
89. Saribasak, H. *et al.* DNA polymerase zeta generates tandem mutations in immunoglobulin variable regions. *Journal of Experimental Medicine* **209**, 1075–1081 (2012).
90. McIlwraith, M. J. *et al.* Human DNA polymerase eta promotes DNA synthesis from strand invasion intermediates of homologous recombination. *Mol Cell* **20**, 783–792 (2005).
91. Kawamoto, T. *et al.* Dual roles for DNA polymerase eta in homologous DNA recombination and translesion DNA synthesis. *Mol Cell* **20**, 793–799 (2005).
92. Li, X., Stith, C. M., Burgers, P. M. & Heyer, W. D. PCNA Is Required for Initiation of Recombination-Associated DNA Synthesis by DNA Polymerase delta. *Mol Cell* **36**, 704–713 (2009).
93. Sharma, S. *et al.* REV1 and polymerase zeta facilitate homologous recombination repair. *Nucleic acids research* **40**, 682–691 (2012).

94. Sebesta, M. *et al.* Role of PCNA and TLS polymerases in D-loop extension during homologous recombination in humans. *DNA Repair* **12**, 691–698 (2013).
95. Sharief, F. S., Vojta, P. J., Ropp, P. A. & Copeland, W. C. Cloning and chromosomal mapping of the human DNA polymerase theta (POLQ), the eighth human DNA polymerase. *Genomics* **59**, 90–96 (1999).
96. Maga, G., Shevelev, I., Ramadan, K., Spadari, S. & Hubscher, U. DNA polymerase theta purified from human cells is a high-fidelity enzyme. *J Mol Biol* **319**, 359–369 (2002).
97. Seki, M., Marini, F. & Wood, R. D. POLQ (Pol theta), a DNA polymerase and DNA-dependent ATPase in human cells. *Nucleic Acids Res* **31**, 6117–6126 (2003).
98. Marini, F., Kim, N., Schuffert, A. & Wood, R. D. POLN, a nuclear PolA family DNA polymerase homologous to the DNA cross-link sensitivity protein Mus308. *J Biol Chem* **278**, 32014–32019 (2003).
99. Minko, I. G. *et al.* Role for DNA polymerase kappa in the processing of N2-N2-guanine interstrand cross-links. *J Biol Chem* **283**, 17075–17082 (2008).
100. Sharma, S. & Canman, C. E. REV1 and DNA polymerase zeta in DNA interstrand crosslink repair. *Environ Mol Mutagen* **53**, 725–740 (2012).
101. Prakash, S., Johnson, R. E. & Prakash, L. Eukaryotic translesion synthesis DNA polymerases: specificity of structure and function. *Annu Rev Biochem* **74**, 317–353 (2005).
102. Prakash, S. & Prakash, L. Translesion DNA synthesis in eukaryotes: a one- or two-polymerase affair. *Genes Dev* **16**, 1872–1883 (2002).
103. Weigle, J. J. Induction of Mutations in a Bacterial Virus. *Proc Natl Acad Sci USA* **39**, 628–636 (1953).
104. Defais, M., Caillet-Fauquet, P., Fox, M. & Radman, M. *Mol. Gen. Genet.* **135**, 19–27 (1974).
105. Benbow, R. M., Zuccarelli, A. J. & Sinsheimer, R. L. A role for single-strand breaks in bacteriophage phi-X174 genetic recombination. *J Mol Biol* **88**, 629–651 (1974).
106. Caillet-Fauquet, P., Defais, M. & Radman, M. Molecular mechanisms of induced mutagenesis. Replication *in vivo* of bacteriophage phiX174 single-stranded, ultraviolet light-irradiated DNA in intact and irradiated host cells. *J Mol Biol* **117**, 95–110 (1977).
107. Waters, L. S. *et al.* Eukaryotic translesion polymerases and their roles and regulation in DNA damage tolerance. *Microbiol Mol Biol Rev* **73**, 134–154 (2009).

108. Johnson, R. E., Prakash, S. & Prakash, L. Efficient bypass of a thymine-thymine dimer by yeast DNA polymerase, Poleta. *Science* **283**, 1001–1004. (1999).
109. Ziv, O., Geacintov, N., Nakajima, S., Yasui, A. & Livneh, Z. DNA polymerase zeta cooperates with polymerases kappa and iota in translesion DNA synthesis across pyrimidine photodimers in cells from XPV patients. *Proc Natl Acad Sci USA* **106**, 11552–11557 (2009).
110. Shachar, S. *et al.* Two-polymerase mechanisms dictate error-free and error-prone translesion DNA synthesis in mammals. *EMBO J* **28**, 383–393 (2009).
111. Rey, L. *et al.* Human DNA Polymerase eta Is Required for Common Fragile Site Stability during Unperturbed DNA Replication. *Molecular and Cellular Biology* **29**, 3344–3354 (2009).
112. Bergoglio, V. *et al.* Enhanced expression and activity of DNA polymerase beta in human ovarian tumor cells: impact on sensitivity towards antitumor agents. *Oncogene* **20**, 6181–6187 (2001).
113. Frechet, M., Canitrot, Y., Cazaux, C. & Hoffmann, J. S. DNA polymerase beta imbalance increases apoptosis and mutagenesis induced by oxidative stress. *FEBS Lett* **505**, 229–232 (2001).
114. Bavoux, C. *et al.* Up-regulation of the error-prone DNA polymerase {kappa} promotes pleiotropic genetic alterations and tumorigenesis. *Cancer Res* **65**, 325–330 (2005).
115. Albertella, M. R., Lau, A. & O'Connor, M. J. The overexpression of specialized DNA polymerases in cancer. *DNA Repair (Amst)* **4**, 583–593 (2005).
116. Lange, S. S., Takata, K. & Wood, R. D. DNA polymerases and cancer. *Nat Rev Cancer* **11**, 96–110 (2011).
117. Wagner, J., Etienne, H., Janel-Bintz, R. & Fuchs, R. P. Genetics of mutagenesis in E. coli: various combinations of translesion polymerases (Pol II, IV and V) deal with lesion/sequence context diversity. *DNA Repair (Amst)* **1**, 159–167 (2002).
118. Sutton, M. D., Opperman, T. & Walker, G. C. The Escherichia coli SOS mutagenesis proteins UmuD and UmuD' interact physically with the replicative DNA polymerase. *Proc Natl Acad Sci USA* **96**, 12373–12378 (1999).
119. Becherel, O. J., Fuchs, R. P. & Wagner, J. Pivotal role of the beta-clamp in translesion DNA synthesis and mutagenesis in E. coli cells. *DNA Repair (Amst)* **1**, 703–708 (2002).
120. Bunting, K. A., Roe, S. M. & Pearl, L. H. Structural basis for recruitment of translesion DNA polymerase Pol IV/DinB to the beta-clamp. *EMBO J* **22**, 5883–5892 (2003).

121. Lopez de Saro, F. J., Georgescu, R. E., Goodman, M. F. & O'Donnell, M. Competitive processivity-clamp usage by DNA polymerases during DNA replication and repair. *EMBO J* **22**, 6408–6418 (2003).
122. Indiani, C., McInerney, P., Georgescu, R., Goodman, M. F. & O'Donnell, M. A sliding-clamp toolbelt binds high- and low-fidelity DNA polymerases simultaneously. *Mol Cell* **19**, 805–815 (2005).
123. Pages, V. & Fuchs, R. P. Uncoupling of leading- and lagging-strand DNA replication during lesion bypass *in vivo*. *Science* **300**, 1300–1303 (2003).
124. Fujii, S. & Fuchs, R. P. Defining the position of the switches between replicative and bypass DNA polymerases. *EMBO J* **23**, 4342–4352 (2004).
125. Reuven, N. B., Arad, G., Maor-Shoshani, A. & Livneh, Z. The mutagenesis protein UmuC is a DNA polymerase activated by UmuD', RecA, and SSB and is specialized for translesion replication. *J Biol Chem* **274**, 31763–31766 (1999).
126. Tang, M. *et al.* UmuD'(2)C is an error-prone DNA polymerase, Escherichia coli pol V. *Proc Natl Acad Sci USA* **96**, 8919–8924 (1999).
127. Simon, S. M., Sousa, F. J., Mohana-Borges, R. & Walker, G. C. Regulation of Escherichia coli SOS mutagenesis by dimeric intrinsically disordered umuD gene products. *Proc Natl Acad Sci USA* **105**, 1152–1157 (2008).
128. Indiani, C., Langston, L. D., Yurieva, O., Goodman, M. F. & O'Donnell, M. Translesion DNA polymerases remodel the replisome and alter the speed of the replicative helicase. *Proc Natl Acad Sci USA* **106**, 6031–6038 (2009).
129. Rupp, W. D. & Howard-Flanders, P. Discontinuities in the DNA synthesized in an excision-defective strain of Escherichia coli following ultraviolet irradiation. *J Mol Biol* **31**, 291–304 (1968).
130. Heller, R. C. & Marians, K. J. Replisome assembly and the direct restart of stalled replication forks. *Nat Rev Mol Cell Biol* **7**, 932–943 (2006).
131. Langston, L. D. & O'Donnell, M. DNA replication: keep moving and don't mind the gap. *Mol Cell* **23**, 155–160 (2006).
132. Lehmann, A. R. & Fuchs, R. P. Gaps and forks in DNA replication: Rediscovering old models. *DNA Repair (Amst)* (2006).
133. Pomerantz, R. T. & O'Donnell, M. The replisome uses mRNA as a primer after colliding with RNA polymerase. *Nature* **456**, 762–766 (2008).
134. Goodman, M. *et al.* Mutations for Worse or Better: Low Fidelity DNA Synthesis by SOS DNA Polymerase V is a Tightly-Regulated Double-Edged Sword. *Faseb Journal* **29** (2015).

135. Prosperi, E., Scovassi, A. I., Stivala, L. A. & Bianchi, L. Proliferating cell nuclear antigen bound to DNA synthesis sites: phosphorylation and association with cyclin D1 and cyclin A. *Exp Cell Res* **215**, 257–262 (1994).
136. Hasan, S., Hassa, P. O., Imhof, R. & Hottiger, M. O. Transcription coactivator p300 binds PCNA and may have a role in DNA repair synthesis. *Nature* **410**, 387–391 (2001).
137. Naryzhny, S. N. & Lee, H. The post-translational modifications of proliferating cell nuclear antigen: acetylation, not phosphorylation, plays an important role in the regulation of its function. *J Biol Chem* **279**, 20194–20199 (2004).
138. Cazzalini, O. *et al.* CBP and p300 acetylate PCNA to link its degradation with nucleotide excision repair synthesis. *Nucleic Acids Research* **42**, 8433-U8164 (2014).
139. Hoege, C., Pfander, B., Moldovan, G. L., Pyrowolakis, G. & Jentsch, S. RAD6-dependent DNA repair is linked to modification of PCNA by ubiquitin and SUMO. *Nature* **419**, 135–141. (2002).
140. Stelter, P. & Ulrich, H. D. Control of spontaneous and damage-induced mutagenesis by SUMO and ubiquitin conjugation. *Nature* **425**, 188–191 (2003).
141. Ulrich, H. D. Ubiquitin, SUMO, and Phosphate: How a Trio of Posttranslational Modifiers Governs Protein Fate. *Molecular cell* **47**, 335–337 (2012).
142. Ulrich, H. D. & Takahashi, T. Readers of PCNA modifications. *Chromosoma* **122**, 259–274 (2013).
143. Yang, W. & Woodgate, R. What a difference a decade makes: insights into translesion DNA synthesis. *Proc Natl Acad Sci USA* **104**, 15591–15598 (2007).
144. Lehmann, A. R. *et al.* Translesion synthesis: Y-family polymerases and the polymerase switch. *DNA Repair (Amst)* **6**, 891–899 (2007).
145. Dionne, I., Nookala, R. K., Jackson, S. P., Doherty, A. J. & Bell, S. D. A heterotrimeric PCNA in the hyperthermophilic archaeon Sulfolobus solfataricus. *Mol Cell* **11**, 275–282 (2003).
146. Bresson, A. & Fuchs, R. P. Lesion bypass in yeast cells: Pol eta participates in a multi-DNA polymerase process. *EMBO J* **21**, 3881–3887 (2002).
147. Kannouche, P. L., Wing, J. & Lehmann, A. R. Interaction of human DNA polymerase eta with monoubiquitinated PCNA: a possible mechanism for the polymerase switch in response to DNA damage. *Mol Cell* **14**, 491–500. (2004).

148. Watanabe, K. *et al.* Rad18 guides poleta to replication stalling sites through physical interaction and PCNA monoubiquitination. *EMBO J* **23**, 3886–3896. Epub 2004 Sep 3809. (2004).
149. Garg, P. & Burgers, P. M. Ubiquitinated proliferating cell nuclear antigen activates translesion DNA polymerases eta and REV1. *Proc Natl Acad Sci USA* **102**, 18361–18366 (2005).
150. Zhuang, Z. *et al.* Regulation of polymerase exchange between Poleta and Poldelta by monoubiquitination of PCNA and the movement of DNA polymerase holoenzyme. *Proc Natl Acad Sci USA* **105**, 5361–5366 (2008).
151. Acharya, N., Brahma, A., Haracska, L., Prakash, L. & Prakash, S. Mutations in the ubiquitin binding UBZ motif of DNA polymerase eta do not impair its function in translesion synthesis during replication. *Mol Cell Biol* **27**, 7266–7272 (2007).
152. Bienko, M. *et al.* Ubiquitin-binding domains in Y-family polymerases regulate translesion synthesis. *Science* **310**, 1821–1824 (2005).
153. Sabbioneda, S. *et al.* Ubiquitin-binding motif of human DNA polymerase eta is required for correct localization. *Proc Natl Acad Sci USA* **106**, E20; author reply E21 (2009).
154. McIntyre, J. *et al.* Ubiquitin mediates the physical and functional interaction between human DNA polymerases eta and iota. *Nucleic acids research* **41**, 1649–1660 (2013).
155. Haracska, L. *et al.* Targeting of human DNA polymerase iota to the replication machinery via interaction with PCNA. *Proc Natl Acad Sci USA* **98**, 14256–14261 (2001).
156. Vidal, A. E. & Woodgate, R. Insights into the cellular role of enigmatic DNA polymerase iota. *DNA Repair (Amst)* **8**, 420–423 (2009).
157. Kirouac, K. N. & Ling, H. Poli: Shining light on repair of oxidative DNA lesions and mutations. *Cell Cycle* **10**, 1520–1521 (2011).
158. Nair, D. T., Johnson, R. E., Prakash, L., Prakash, S. & Aggarwal, A. K. Human DNA polymerase iota incorporates dCTP opposite template G via a G.C + Hoogsteen base pair. *Structure* **13**, 1569–1577 (2005).
159. Frank, E. G. & Woodgate, R. Increased catalytic activity and altered fidelity of human DNA polymerase iota in the presence of manganese. *J Biol Chem* (2007).
160. Haracska, L. *et al.* Stimulation of DNA synthesis activity of human DNA polymerase kappa by PCNA. *Mol Cell Biol* **22**, 784–791 (2002).
161. Guo, C., Tang, T. S., Bienko, M., Dikic, I. & Friedberg, E. C. Requirements for the Interaction of Mouse Pol{kappa} with Ubiquitin and Its Biological Significance. *J Biol Chem* **283**, 4658–4664 (2008).

162. Yoon, J. H. *et al.* Identification of two functional PCNA-binding domains in human DNA polymerase kappa. *Genes to Cells* **19**, 594–601 (2014).
163. Wit, N. *et al.* Roles of PCNA ubiquitination and TLS polymerases kappa and eta in the bypass of methyl methanesulfonate-induced DNA damage. *Nucleic Acids Research* **43**, 282–294 (2015).
164. Guo, C. *et al.* Mouse Rev1 protein interacts with multiple DNA polymerases involved in translesion DNA synthesis. *Embo J* **22**, 6621–6630 (2003).
165. Wood, A., Garg, P. & Burgers, P. M. A Ubiquitin-binding Motif in the Translesion DNA Polymerase Rev1 Mediates Its Essential Functional Interaction with Ubiquitinated Proliferating Cell Nuclear Antigen in Response to DNA Damage. *J Biol Chem* **282**, 20256–20263 (2007).
166. Edmunds, C. E., Simpson, L. J. & Sale, J. E. PCNA ubiquitination and REV1 define temporally distinct mechanisms for controlling translesion synthesis in the avian cell line DT40. *Mol Cell* **30**, 519–529 (2008).
167. Acharya, N., Johnson, R. E., Prakash, S. & Prakash, L. Complex formation with Rev1 enhances the proficiency of Saccharomyces cerevisiae DNA polymerase zeta for mismatch extension and for extension opposite from DNA lesions. *Mol Cell Biol* **26**, 9555–9563 (2006).
168. Jansen, J. G. *et al.* Separate domains of Rev1 mediate two modes of DNA damage bypass in mammalian cells. *Mol Cell Biol* **29**, 3113–3123 (2009).
169. Sharma, N. M., Kochenova, O. V. & Shcherbakova, P. V. The Non-canonical Protein Binding Site at the Monomer-Monomer Interface of Yeast Proliferating Cell Nuclear Antigen (PCNA) Regulates the Rev1-PCNA Interaction and Pol zeta/Rev1-dependent Translesion DNA Synthesis. *Journal of Biological Chemistry* **286**, 33557–33566 (2011).
170. Haracska, L., Unk, I., Prakash, L. & Prakash, S. Ubiquitylation of yeast proliferating cell nuclear antigen and its implications for translesion DNA synthesis. *Proc Natl Acad Sci USA* **103**, 6477–6482 (2006).
171. Parsons, J. L. *et al.* CHIP-mediated degradation and DNA damage-dependent stabilization regulate base excision repair proteins. *Mol Cell* **29**, 477–487 (2008).
172. Parsons, J. L. *et al.* Ubiquitin ligase ARF-BP1/Mule modulates base excision repair. *EMBO J* **28**, 3207–3215 (2009).
173. El-Andaloussi, N. *et al.* Arginine methylation regulates DNA polymerase β. *Mol Cell* **22**, 51–62 (2006).
174. El-Andaloussi, N. *et al.* Methylation of DNA polymerase beta by protein arginine methyltransferase 1 regulates its binding to proliferating cell nuclear antigen. *Faseb J* **21**, 26–34 (2007).

175. Hasan, S. *et al.* Acetylation regulates the DNA end-trimming activity of DNA polymerase beta. *Mol Cell* **10**, 1213–1222 (2002).
176. Parsons, J. L. *et al.* USP47 is a deubiquitylating enzyme that regulates base excision repair by controlling steady-state levels of DNA polymerase beta. *Molecular cell* **41**, 609–615 (2011).
177. Fang, Q. M. *et al.* HSP90 regulates DNA repair via the interaction between XRCC1 and DNA polymerase beta. *Nat Commun* **5** (2014).
178. Chen, Y. W. *et al.* Human DNA polymerase eta activity and translocation is regulated by phosphorylation. *Proc Natl Acad Sci USA* **105**, 16578–16583 (2008).
179. Kim, S. H. & Michael, W. M. Regulated proteolysis of DNA polymerase eta during the DNA-damage response in C. elegans. *Mol Cell* **32**, 757–766 (2008).
180. Kannouche, P. *et al.* Localization of DNA polymerases eta and iota to the replication machinery is tightly co-ordinated in human cells. *EMBO J* **21**, 6246–6256 (2002).
181. Akagi, J. I. *et al.* Interaction with DNA polymerase eta is required for nuclear accumulation of REV1 and suppression of spontaneous mutations in human cells. *DNA Repair (Amst)* **8**, 585–599 (2009).
182. Wimmer, U., Ferrari, E., Hunziker, P. & Hubscher, U. Control of DNA polymerase lambda stability by phosphorylation and ubiquitination during the cell cycle. *EMBO Rep* **9**, 1027–1033 (2008).
183. Markkanen, E. *et al.* Regulation of oxidative DNA damage repair by DNA polymerase lambda and MutYH by cross-talk of phosphorylation and ubiquitination. *P Natl Acad Sci USA* **109**, 437–442 (2012).
184. Parlanti, E., Locatelli, G., Maga, G. & Dogliotti, E. Human base excision repair complex is physically associated to DNA replication and cell cycle regulatory proteins. *Nucleic Acids Res* **35**, 1569–1577 (2007).
185. Dutta, A., Yang, C. Y., Sengupta, S., Mitra, S. & Hegde, M. L. New paradigms in the repair of oxidative damage in human genome: mechanisms ensuring repair of mutagenic base lesions during replication and involvement of accessory proteins. *Cellular and Molecular Life Sciences* **72**, 1679–1698 (2015).
186. Gilljam, K. M., Muller, R., Liabakk, N. B. & Otterlei, M. Nucleotide Excision Repair Is Associated with the Replisome and Its Efficiency Depends on a Direct Interaction between XPA and PCNA. *Plos One* **7**, e49199 (2012).
187. Zhang, S. *et al.* A novel DNA damage response: rapid degradation of the p12 subunit of dna polymerase delta. *The Journal of Biological Chemistry* **282**, 15330–15340 (2007).

188. Zhang, S. *et al.* Identification of RNF8 as a Ubiquitin Ligase Involved in Targeting the p12 Subunit of DNA Polymerase delta for Degradation in Response to DNA Damage. *The Journal of Biological Chemistry* **288**, 2941–2950 (2013).
189. Meng, X. *et al.* DNA damage alters DNA polymerase delta to a form that exhibits increased discrimination against modified template bases and mismatched primers. *Nucleic Acids Research* **37**, 647–657 (2009).
190. Zhang, S. F. *et al.* A novel function of CRL4(Cdt2) regulation of the subunit structure of DNA polymerase delta in responese to DNA damage and during the S phase. *Journal of Biological Chemistry* **288**, 29550–29561 (2013).
191. Makarova, A. V. & Burgers, P. M. Eukaryotic DNA polymerase zeta. *DNA Repair* **29**, 47–55 (2015).
192. Melo, J. & Toczyski, D. A unified view of the DNA-damage checkpoint. *Curr Opin Cell Biol* **14**, 237–245 (2002).
193. Byun, T. S., Pacek, M., Yee, M. C., Walter, J. C. & Cimprich, K. A. Functional uncoupling of MCM helicase and DNA polymerase activities activates the ATR-dependent checkpoint. *Genes Dev* **19**, 1040–1052 (2005).
194. Chang, D. J., Lupardus, P. J. & Cimprich, K. A. Monoubiquitination of proliferating cell nuclear antigen induced by stalled replication requires uncoupling of DNA polymerase and mini-chromosome maintenance helicase activities. *J Biol Chem* **281**, 32081–32088 (2006).
195. Zou, L. & Elledge, S. J. Sensing DNA damage through ATRIP recognition of RPA-ssDNA complexes. *Science* **300**, 1542–1548 (2003).
196. Zou, L., Liu, D. & Elledge, S. J. Replication protein A-mediated recruitment and activation of Rad17 complexes. *Proc Natl Acad Sci USA* **100**, 13827–13832 (2003).
197. Bhaumik, D. & Wang, T. S. F. Mutational effect of fission yeast polalpha on cell cycle events. *Mol Biol Cell* **9**, 2107–2123 (1998).
198. Tan, S. & Wang, T. S. Analysis of fission yeast primase defines the checkpoint responses to aberrant S phase initiation. *Mol Cell Biol* **20**, 7853–7866 (2000).
199. Griffiths, D. J. F., Liu, V. F., Nurse, P. & Wang, T. S. F. Role of fission yeast primase catalytic subunit in the replication checkpoint. *Mol. Biol. Cell* **12**, 115–128 (2001).
200. Michael, W. M., Ott, R., Fanning, E. & Newport, J. Activation of the DNA replication checkpoint through RNA synthesis by primase. *Science* **289**, 2133–2137 (2000).
201. You, Z., Kong, L. & Newport, J. The role of single-stranded DNA and polymerase alpha in establishing the ATR, Hus1 DNA replication checkpoint. *J Biol Chem* **277**, 27088–27093 (2002).

202. Navas, T. A., Zhou, Z. & Elledge, S. J. DNA polymerase epsilon links the DNA replication machinery to the S phase checkpoint. *Cell* **80**, 29–39. (1995).
203. Araki, H., Leem, S. H., Phongdara, A. & Sugino, A. Dpb11, which interacts with DNA polymerase II(epsilon) in Saccharomyces cerevisiae, has a dual role in S-phase progression and at a cell cycle checkpoint. *Proc Natl Acad Sci USA* **92**, 11791–11795 (1995).
204. Navas, T. A., Sanchez, Y. & Elledge, S. J. RAD9 and DNA polymerase form parallel sensory branches for transducing the DNA damage checkpoint signal in *Saccharomyces cerevisiae*. *Genes Dev.* **10**, 2632–2643 (1996).
205. Dua, R., Levy, D. L. & Campbell, J. L. Role of the putative zinc finger domain of Saccharomyces cerevisiae DNA polymerase epsilon in DNA replication and the S/M checkpoint pathway. *J Biol Chem* **273**, 30046–30055 (1998).
206. Dua, R., Levy, D. L. & Campbell, J. L. Analysis of the essential functions of the C-terminal protein/protein interaction domain of Saccharomyces cerevisiae pol epsilon and its unexpected ability to support growth in the absence of the DNA polymerase domain. *J Biol Chem* **274**, 22283–22288 (1999).
207. Kesti, T., Flick, K., Keranen, S., Syvaoja, J. E. & Wittenberg, C. DNA polymerase epsilon catalytic domains are dispensible for DNA replication, DNA repair and cell viability. *Mol. Cell* **3**, 679–685 (1999).
208. D'Urso, G. & Nurse, P. Schizosaccharomyces pombe cdc20+ encodes DNA polymerase epsilon and is required for chromosomal replication but not for the S phase checkpoint. *Proc Natl Acad Sci USA* **94**, 12491–12496 (1997).
209. Garcia-Rodriguez, L. J. *et al.* A conserved Pol epsilon binding module in Ctf18-RFC is required for S-phase checkpoint activation downstream of Mec1. *Nucleic Acids Research* **43**, 8830–8838 (2015).
210. Pages, V., Santa Maria, S. R., Prakash, L. & Prakash, S. Role of DNA damage-induced replication checkpoint in promoting lesion bypass by translesion synthesis in yeast. *Genes Dev* **23**, 1438–1449 (2009).
211. Zucca, E. *et al.* Silencing of human DNA polymerase lambda causes replication stress and is synthetically lethal with an impaired S phase checkpoint. *Nucleic Acids Research* **41**, 229–241 (2013).
212. Bailey, L. J., Bianchi, J., Hegarat, N., Hochegger, H. & Doherty, A. J. PrimPol-deficient cells exhibit a pronounced G2 checkpoint response following UV damage. *Cell Cycle* **15**, 908–918 (2016).

Chapter 5

DNA Polymerases and Human Diseases

5.1 Introduction

DNA replication is a highly organized process, whose purpose is to transmit as faithfully as possible the genetic information from one generation to the next one. Many proteins participate in DNA replication, either directly at the polymerization step or indirectly by regulating the timing of DNA duplication and/or modifying the post-translational status of the proteins implicated (for more details see Chapters 3 and 4). Mutations in DNA pols or changes in their expression levels can severely perturb the overall process resulting in genetic instability that can lead to diseases, such as cancer or degenerative disorders. In this chapter we will summarize the evidence linking malfunction of DNA pols to diseases and discuss some of the most recent findings on (i) DNA pols and the maintenance of genetic stability, (ii) DNA pols and resistance to chemotherapy, and (iii) DNA pol γ and human diseases. Several reviews on this topic have been published in the last few years, suggesting that this issue is becoming of great importance in the near future.[1–4]

Table 5.1 presents an overview of DNA pols and diseases.

Table 5.1 DNA polymerases and diseases[a]

DNA polymerase	Likely physiological functions	Knock out or mutation phenotype	Diseases related to DNA polymerase
DNA pol α	-Initiation of Okazaki fragments (synthesis of short RNA primers in DNA synthesis) -Chromosomal end replication	-unknown	-N-syndrome, a form of X chromosomal breakage and T-cell leukemia?
DNA pol β	short and long patch BER	embryos are not viable Cells (MEF) = hypersensitive to alkylating agents	-overexpressed in over 30% of tumors (prostate, breast, colon, ovarian, chronic myeloid leukemia) -mutations found in gastric carcinomas and prostate cancer
PRIMPOL	Mitochondrial and nuclear DNA replication — Replication fork repriming -TLS	PRIMPOL KO mice are viable-PRIMPOL deficient mouse, embryonic fibroblast and cells display genetic instability.	Myopia; human mitochondriopathies
DNA pol δ	-DNA synthesis of the lagging strand -DNA repair (long patch BER, MMR, TLS)	-mouse pol δ exo −/− = cancer susceptibility -yeast POL3 exo −/− = mutator phenotype	-colon cancer, sporadic colorectal carcinomas -diverse cancers and reduced life-span if 3′→5′ exonuclease deficient
DNA pol ε	-DNA synthesis of the leading strand -DNA repair (long patch BER and NER)	Mouse pol ε exo −/− = cancer susceptibility	-intestine tumors -sarcomas
TdT	-V(D)J recombination and somatic hypermutation	-unknown	-immunodeficiency, and overexpressed in several acute leukemia cells, overexpression is correlated with poor prognosis

DNA pol θ	-interstrand crosslinking repair and AP site translesion synthesis, short patch BER	-mutation in Drosophila showed sensitivity to DNA crosslinking agents, elevated frequency of chromosomal aberration and altered DNA metabolism	-chromosome instability with potential cancer susceptibility
DNA pol ν	-interstrand crosslinking repair	-unknown	-unknown
DNA pol ζ	-mitosis/meiosis (spindle checkpoint) -somatic hypermutation of IgG -TLS "mismatch extender" -mouse embryonic development	-*S. cerevisiae* –/– = sensitive to various DNA damage agents - (UV-light MMS, cisplatin and IR) -apoptosis, cell proliferation and control of cell-cycle -mice –/– dye during embryonic development	-might be involved mammalian untargeted mutation
DNA pol η	-TLS insertion over TT dimers -NER -somatic hypermutation -homologous recombination -function in S-phase checkpoint	-mutation in human = xeroderma pigmentosum variant (XPV)	-skin cancer, squamous cell carcinoma
DNA pol κ	-TLS insertion and extension synthesis of various distorted lesions -NER	-overexpression causes DNA breaks and stimulates DNA exchange	-up-regulated in non-small cell lung carcinomas -reduced expression in mammary carcinoma -excess expression causes proliferation of solid tumors in immunodeficient mice

(*Continued*)

Table 5.1 (*Continued*)

DNA polymerase	Likely physiological functions	Knock out or mutation phenotype	Diseases related to DNA polymerase
DNA pol ι	TLS insertion	-overexpression in breast cancer leading to significant decrease in DNA replication fidelity and consequently contributing to genetic instability	-up- or down-regulated in different cancer cell types
REV1	-TLS insertion synthesis of AP site	-mice –/– are normal -cells –/– are sensitive to DNA damaging agents	-unknown
DNA pol λ	-DSBR (NHEJ) -long patch BER -TLS insertion and extension over 8-oxo-G -V(D)J recombination	-embryos are viable Cells (MEF) = moderate sensitivity to oxidative stress	-Overexpressed in several solid tumors. -The mutation R438W, linked to genetic instability and breast cancer.
DNA pol μ	-V(D)J recombination and somatic hypermutation	-mice –/– = defect in light chain of IgG	-up-regulation might lead to non-Hodgkin's lymphoma in B lymphocytes
telomerase	-chromosomal end replication		-overexpressed in 90% tumors -aging -Werner syndrome

[a] For detailed references see either Refs. 1–4 or in the text of Sections 5.2 to 5.4. DNA pol γ is not included here since it will be presented in a separate Table 5.4 in Section 5.4

5.2 DNA polymerases and genetic stability

As we have already indicated in previous chapters, the main task of DNA pols is to correctly incorporate bases according to the Watson-Crick base pair rules. Misincorporations during DNA replication and in DNA repair processes can lead to alterations of the DNA sequence and consequently to altered function of proteins. Along this line, Loeb's laboratory was the first to indicate the significance of multiple mutations in cancer.[5] In normal cells, most DNA damage is repaired without error leading to genetic stability. In tumor cells, on the other hand, this stability may be disturbed, leading to the accumulation of multiple mutations thus resulting in a mutator phenotype. During tumor development there is a selection for cells containing mutations that can overcome adverse conditions that limit tumor growth. Moreover, Loeb *et al.* suggested that cancers must exhibit a mutator phenotype early during their development.[6] It was found first that thousands of random mutations in individual malignant cells exist, second that during tumor progression the selection of a mutator phenotype occurs, and third that large numbers of mutations found in cancers might derive from "normal" replicative processes. The mutation frequency of cancer cells was estimated to be 210 fold higher than in normal cells.[7] In summary, one can suggest that mutation might occur via the action of "canonical" DNA pols during replication and repair as well as from the action of "specialized" DNA pols that are either up-regulated or down-regulated. Moreover, mutations in the DNA pols itself might hamper their functional accuracy again, leading to mutations and thus to genomic instability.

Not optimally regulated DNA pol levels might lead to genomic instability. In Chapter 3 we already touched the issue of the regulation of the 16 known DNA pols. Even though very little is known so far about their global regulation, a general statement is that too much of unfaithful DNA pols, e.g. from the X and Y families, might lead to genomic instability, while, on the other hand, reduced levels of faithful DNA pols can cause genomic instability as well. A nice study testing 68 different tumors measured the expression levels of a variety of DNA pols.[8] When the two X family DNA pols β and λ, the two Y family DNA pols ι and κ and the three replicative B family DNA pols α, δ and ε were tested, the data indicated that in more than 45% of these 68 tumors at least one of the X and/or Y DNA pols were over-expressed. Of particular interest was the fact that over 30% of all tumors tested had an over-expressed DNA pol β. DNA pols λ and ι were also over-expressed but less frequently than DNA pol β. Surprisingly DNA pol κ was under-expressed in many tumor samples. Finally, a few tumors also over-expressed the three replicative DNA pols α, δ and ε while in most cases they were under-expressed. All the 68 tumors were summarized in an over-expression scale that gave the following order: DNA pol β > DNA pol ι > DNA pol λ > DNA pol α > DNA pol δ > DNA pol κ > DNA pol ε. These data clearly indicated that the expression patterns of DNA pols in normal and cancer tissues are very complex and the expression levels of the DNA pols are changed in tumors and can vary from one type of a tumor to another. Another study suggested that up-regulation of DNA pol μ gene expression might contribute to the pathogenesis of a subset of non-Hodgkin's lymphoma and this might happen through DNA repair associated genomic instability.[9]

Recent studies, outlined in a number of reviews[4,10–12] highlighted how mutations in replicative DNA pols contribute to cancer. The role of each DNA pol in such a process will be detailed in the following paragraphs.

Individual DNA polymerases and their relationship to diseases. The involvement of DNA pol γ in diseases is outlined in Section 5.5.

DNA polymerase α: Even though no direct experimental evidence has implicated DNA pol α in human diseases so far, it is likely that even minor misregulations of its activity or response to checkpoint controls, might have a profound influence in the development of cancer. Indeed, altered catalytic properties have been reported for DNA pol α isolated from a hepatoma cell line. An early report suggested that DNA pol α is e.g. deficient in the N-syndrome, a form of X-chromosomal linked mental retardation associated with chromosomal breakage and a link with T-cell leukemia has been suggested but definitive studies were never published.[13]

PrimPol: PrimPol has been reported to be a potential susceptibility gene associated with development of myopia. Mutation of tyrosine 89 to aspartic acid (PrimPol Y89D) has been identified in a number of cases of high myopia, implicating it in the etiology of this disorder.[14] This finding has prompted a study aimed at examining whether this mutation resulted in any changes in the molecular and cellular activity associated with human Prim Pol.[15] This study showed that PrimPol Y89D mutation results in a striking decrease in primase and polymerase activities *in vitro*, together with reduced affinities for DNA and nucleotides. Furthermore, this mutation also reduces cell viability after DNA damage and significantly slows replication fork rates *in vivo*. Together, these findings establish that the major DNA replication defect associated with this PrimPol mutant could contribute to the onset of high myopia.

Generally speaking, it is also conceivable that mutations in PrimPol functions, affecting the balance between primase and polymerase activities, or affecting its TLS capacity, could affect cell proliferation and alter sensitivity to DNA damage agents used in human cancer therapy. In addition, since PrimPol appears to have a role in mitochondrial DNA maintenance, its deficiency could contribute to the development of some human mitochondriopathies

DNA polymerase β: As indicated above, over-expression of DNA pol β was found in about one third of all cancers tested. DNA pol β –/– mouse embryos are not viable, suggesting its essential role in embryogenesis and development.[16] DNA pol β null (–/–) embryonic cells survive in culture but are severely compromised in their ability to carry out short patch BER resulting in hypersensitivity to alkylating agents.[17] Interestingly, high levels of DNA pol β have been detected at the transcriptional and protein levels in many cancer tissues, mostly solid tumors (prostate, breast, colon, ovarian) as well as in chronic myeloid leukemia. DNA pol β up-regulation could contribute to enhancing chromosome instability and tumorigenesis even when over-expressed by only two-fold in cells, suggesting that a rigorous regulation of its expression may be essential *in vivo*.[18] DNA pol β causes chromosomal instability probably either by competing with replicative DNA pols or by producing translesion synthesis over DNA lesions. Cells over-expressing DNA pol β are much more sensitive to IR treatments by increasing apoptosis and hypermutator phenotype in surviving cells. These data indicate that up-regulation of DNA pol β strengthens both cell death and genetic changes associated with a malignant phenotype.[19] Also, over-expression of DNA pol β strengthens the mutagenicity of oxidative damages, concomitantly with a higher cellular sensitivity and

Table 5.2. Common DNA Pol β variants identified in different cancer types

DNA pol β mutation	Tumor type
K289M, E288K, S229L, R152C	Colorectal cancer
E295K, G231D, L22P, Y265C, D160N, T889C	Gastric cancer
I260M	Prostate cancer
P242R	Chromosomal aberrations in mammary cells
K167I	Oesophageal cancer

increased apoptosis.[20] Regulation of DNA pol β over-expression could be fundamental in some cancer treatment, e.g. tumor cell resistant to cisplatin drugs[21] (see Section 5.3). Mutations in DNA pol β are often identified in gastric carcinomas and cell lines thereof (see Ref. 1 and references therein). A list of the most commonly found DNA pol β variants is summarized in Table 5.2. DNA pol β mutations can increase genetic instability in several ways: some variants show lowered fidelity, such as the K289M mutant. Alternatively, other mutants, such as G231D or E295K, show very low or even no polymerase activity. In this case, these variants act as dominant negative mutants, binding the BER intermediates thus preventing access to back up DNA pols, such as DNA pol λ. This leads to accumulation of ss and ds DNA breaks. Finally, mutants with impaired dRP-lyase activity, such as L22P, fail to correctly process the BER intermediates, again leading to accumulation of DNA breaks.

Besides BER, DNA pol β functions in maintaining genomic stability also during prophase I of meiosis. Both germline and cancer-associated variants lead to aberrant DNA repair that is either error-prone or inefficient, resulting in accumulation of BER intermediates. Therefore, DNA pol β somatic and germline variants have the potential to be cancer drivers and to impact cancer therapy. Interestingly, recent results indicate that decreased DNA pol β activity during generation of immune diversity leads to lupus-like disease in mice, suggesting that decreased expression of DNA pol β in human is an underlying cause of systemic lupus erythematosus.[22]

DNA polymerases δ *and* ε: Defective DNA pol δ proofreading causes cancer susceptibility in mice with a shorter life span.[23] Single point mutation inactivates the 3′→5′ exonuclease of DNA pol δ and causes a mutator and cancer phenotype indicating that DNA pol δ proofreading suppresses spontaneous cancer development and suggesting that unrepaired DNA pol δ errors contribute to carcinogenesis.[24] Mutations in the catalytic subunit of DNA pol δ were reported in colon cancer,[25] in sporadic colorectal carcinoma[26] and in Novikoff hepatoma cells.[27] The structural basis for the high fidelity of DNA pol δ has recently been solved in *S. cerevisiae*.[28] This structure of yeast DNA pol δ is an excellent framework for understanding the effects of cancer-causing mutations in human DNA pol δ.

Recently, an in-frame deletion of a single codon occurring at Ser 605 in the catalytic domain of DNA pol δ was described, impairing its polymerase activity, but not the proofreading exonuclease. This heterozygous mutation was found in all four tested patients, out of a total of seven individuals reported so far, affected by a rare genetic syndrome (MPD, mandibular hypoplasia, deafness, and progeroid syndrome), affecting mainly teenagers and young adults (10–36 years old). So far, this deletion in DNA pol δ is the only causative mutation identified for MDP.[29]

Mouse DNA pol ε and DNA pol δ proofreading can suppress discrete mutator and cancer phenotypes. Inactivation of DNA pol ε proofreading elevates base substitution mutations and accelerate a unique spectrum of spontaneous cancers.[30] Interestingly the types of tumors were completely different from those triggered by loss of DNA pol δ proofreading. These findings distinguish DNA pol ε and δ functions *in vivo* and reveal tissue-specific requirements for DNA replication fidelity.[30]

Germline exonuclease domain mutations (EDMs) have been identified in the catalytic subunits of human DNA pols δ and ε.[31,32] such mutations predispose to "polymerase proofreading associated polyposis" (PPAP), a disease characterized by multiple colorectal adenomas and carcinoma, with high penetrance and dominant inheritance. Moreover, somatic EDMs in DNA pol ε have also been found in sporadic colorectal and endometrial cancers and also in families predisposed to cutaneous melanoma.[33]

Surprisingly, some EDMs in DNA pol ε were found to correlate with a good clinical outcome of a subset of grade 3 endometrioid carcinomas, highlighting that these mutations may have differentially impact on different tumors. Thus, caution must be taken in assigning a prognostic value to this class of mutations.[34]

Mutator variants of DNA pol ε and δ have also been characterized by two recent studies.[35,36] They both find that the mutator phenotypes of these DNA pol ε and δ variants depend upon the Dun1 effector kinase, which plays a role in the regulation of the dNTP pools and transcriptional response of DNA damage. Overall the results of these studies have been interpreted as establishing a relationship between altered dNTP pools and specific DNA pol ε or δ mutations.[36] Understanding the downstream effects of exonucleases mutations in these replicative polymerases is critical, as many of these mutations have been implicated in human cancers.

TdT: TdT is over-expressed in several acute leukemia cells, and its over-expression correlated with poor prognosis and low response to chemotherapy. It has been documented that the nucleoside analogue cordycepine in combination with the antitumor drug coformycine, inhibits TdT and displays selective toxicity against TdT^+ leukemic cells, suggesting an important functional role of TdT in the cancerous phenotype.[37]

DNA polymerase θ: A mutagenesis screen for chromosome instability in the mouse genome identified a mutation named *chaos1* (*ch*romosome *a*berration *o*ccurring *s*pontaneously 1) and mapped it to a region of chromosome 16 where DNA pol θ gene is located.[38] The identity of DNA pol θ and *chaos1* has been confirmed by direct disruption of DNA pol θ in the mouse and by correction of the phenotype with the DNA pol θ gene. This result provides evidence that DNA pol θ might potentially be involved in cancer susceptibility.

Indeed, DNA pol θ expression is higher in samples of colorectal cancer than in matched surrounding normal tissues.[39] When the expression profiles for all the human nuclear DNA pols genes were examined in patients with previously untreated primary breast cancers, DNA pol θ was the only one significantly higher in breast cancer tumor samples compared with normal breast tissues, and higher expression correlated with poor clinical outcome.[40]

Sequencing of a group of patients with hereditary breast cancer, identified a polymorphism in the promoter region of the POLQ gene, encoding for DNA pol θ. This single nucleotide substitution (c.–1060 A/G) was found to be statistically associated with the risk of developing hereditary bilateral breast cancer. This mutation disrupted the binding site for the transcription factor YingYang1 (YY1), raising the possibility that this mutation led to a reduced expression of DNA pol θ in these patients.[41]

DNA polymerase ν: No experimental evidences of a direct involvement of DNA pol ν in human diseases has been reported so far.

DNA polymerase ζ: In mouse T-lymphoma cells, stress response induced by DNA damage agents (8-methoxy-psoralen or UV-A) leads to specific, delayed and untargeted mutations.[42] It has been found that a low concentration of N-methyl-N'-nitro-N-nitrosoguanidine (MNNG), a carcinogen which can induce gastric cancer, could also induce mammalian untargeted mutations.[43] However, it is not clear which factor capable of inhibiting fidelity can be induced or activated. More recently, it was found that DNA pol ζ might be involved in the mammalian untargeted mutations induced by MNNG. The transcriptional level of REV3 gene, coding for the catalytic subunit of DNA pol ζ, is up-regulated when human cells are treated by low concentration MNNG.[44] Finally, human cells, in which the function of untargeted mutations is inhibited by antisense REV3 RNA, display characteristics of both untargeted mutations and targeted mutations.[44] The fact that DNA pol ζ is an essential gene[45,46] and that it is the main TLS extender DNA pol makes it an attractive target for chemotherapy (see Chapter 6). It has been shown that loss of DNA pol ζ provides a selective advantage in the formation of tumor cells, with enhanced spontaneous tumorigenesis in the presence and absence of p53 function.[47] Most of these tumors had REV3L–/– cells, were multifocal and were associated with mammary intraepithelial neoplasia. It has been shown that the human REV3L gene mediates resistance towards alkylating anticancer drugs.[48] More recently, it has been demonstrated that inhibition of REV3L sensitized lung cancer cells to cisplatin treatment.[49]

DNA polymerase η: DNA pol η is the prototype of a DNA pol, which is correlated without any doubt to a disease. Xeroderma Pigmentosum Variant (XPV) cells were taken to isolate a DNA pol and this DNA pol was identified as a mutant form of DNA pol η.[50] Humans with mutations in the DNA pol η gene, suffer from a disease characterised, among other phenotypes, by an increased susceptibility to sunlight-induced skin cancer.[51] DNA pol η is the primary TLS DNA pol responsible in many organisms for error-free bypasses of cis-syn cyclobutane dimers (CPDs), a major lesion resulting from UV irradiation. In fact CPDs block DNA synthesis by DNA pols like DNA pols α, δ, ε, and λ, but DNA pol η has been shown to bypass them with high accuracy and efficiency *in vitro*.[52] Consequently, mutations or deletions in DNA pol η result in a reduced efficiency to copy DNA containing CPDs. Furthermore, DNA pol η independent CPD bypass, which likely involves other TLS DNA pols such as DNA pol ζ, ι and κ is more mutagenic, presumably leading to an increased frequency of cancer in XP-V patients.[53] DNA pol η in addition to CPDs has the capacity to bypass *in vitro* a number of other DNA lesions such as 8-oxo-G,[54] acetylaminofluorene-adducted guanine[55] and adducts formed by cisplatin and oxaliplatin[56] (see also Section 5.3; DNA polymerases and resistance to chemotherapy). Moreover, DNA pol η is one of the DNA pols with the lowest fidelity on undamaged DNA *in vitro*.[57]

Analysis of DNA from lymphocytes of XP-V patients showed that DNA pol η generates hypermutation in the μ and γ switch regions of immunoglobulin genes.[58] In addition, it was shown *in vitro* that human DNA pol η and the two other Y-family enzymes DNA pols κ and ι, both of which are dispensable for somatic hypermutation, possess reverse transcriptase activity. It is thus feasible that DNA pol η may act as both a RNA-dependent and a DNA-dependent pol in somatic hypermutation.[59]

Furthermore, DNA synthesis by DNA pol η promotes fragile site stability by preventing under-replicated DNA in mitosis.[60] Recent results indicated that increased DNA damage and persistent activation of the DNA damage response induce adipocyte senescence in DNA pol η knockout mice,[61] establishing a correlation between reduced genome integrity and adipocyte senescence that leads to development of obesity and insulin resistance.

DNA polymerase κ: There is evidence that DNA pol κ is also involved in tumorigenesis. It was shown that ectopic expression of DNA pol κ promoted DNA strand breaks, aneuploidy as well as tumorigenesis in nude mice.[62] Additionally, of eight non-small cell lung carcinoma biopsies overexpressing DNA pol κ, seven displayed losses of heterozygosity compared with adjacent non-tumoral tissues. Taken together, these data suggested that misregulation of DNA pol κ can promote the emergence of a large spectrum of genetic disorders associated with a malignant phenotype. However, conflicting results have been published about DNA pol κ, as much as was the case with DNA pol ι. A reduced expression of DNA pol κ in rat mammary carcinoma cell lines and primary mammary carcinomas was found in comparison to that of the normal tissues.[63] In addition, it was reported that DNA pol κ is up-regulated by p53 in human, as well as in murine cells. The functional loss of p53 by mutation resulted in the up-regulation of DNA pol κ in human lung cancer tissues.[64] However, results from other investigators indicated that mouse but not human DNA pol κ was up-regulated in response to exposure to various DNA-damaging agents in a p53-dependent manner.[65] In summary, the available body of information suggests a complex regulation pattern of DNA pol κ. A recent study identified a series of missense mutations of DNA pol κ (E29K, G145E, F155S, E430K, L442F) in prostate cancer patients, which diminished the bypass efficiency of the enzyme at AP sites and lowered its fidelity. Thus, these mutations may contribute to prostate carcinogenesis.[66]

DNA polymerase ι: Growing evidence points to the involvement of DNA pol ι in lung cancer progression. In one study, a systematic candidate gene analysis of the pulmonary adenoma resistance 2 locus was performed. Differential gene expression in lung tissues and nucleotide polymorphisms analysis revealed that the gene encoding DNA pol ι contained 25 nucleotide polymorphisms in its coding region between A/J and BALB/cJ mice, resulting in a total of ten amino acid changes. Purified BALB/cJ and A/J DNA pol ι were shown to differ in substrate discrimination *in vitro*. Moreover, altered expression of DNA pol ι and an amino acid–changing nucleotide polymorphism were observed in human lung cancer cells, suggesting a possible role in the development of lung cancer.[67] In another case-control study, a single nucleotide polymorphism in DNA pol ι (Thr706Ala), which correlated with a significantly higher risk of lung adenocarcinoma and squamous cell carcinoma, was reported.[68] A third study provided a potential link between altered DNA pol ι expression and mutagenesis in breast cancer. DNA pol ι expression was elevated in breast cancer cells and this correlated with a significant decrease in DNA replication fidelity.[69] Interestingly, after immune depletion of DNA pol ι from nuclear extracts of these cells, a reduction in mutation frequency was measured *in vitro*. These data suggested that DNA pol ι might play a role in the high mutation frequencies observed in breast cancer cells. Another study, however, questioned the existence of a link between overexpression and DNA pol ι and cancer. The transcripts of DNA pols κ, η, ι, and ζ were measured in 131 self-paired cancerous and non-tumor samples, including 23 lung cancers, 49 stomach cancers, and 59 colorectal cancers.[63] These results indicated that, except for DNA pol η in colorectal cancers, all other DNA pols tested were significantly down-regulated in human lung, stomach, and colorectal cancers. However, this study did not rule out the possibility that the variation in the expression levels might be a function of the tumor stage or is cell-type specific. Finally, it has been shown that overexpressed DNA pol ι, regulated by JNK/c-Jun, contributes to hypermutagenesis in bladder cancer.[70]

Rev1: Rev1-deficient chicken DT40 cells display reduced viability and are sensitive to a wide range of DNA-damaging agents.[71] "Knock down" of REV1 mRNA in human cells results in a hypomutable phenotype after UV treatment.[72] In addition to its catalytic domain, Rev1 possesses a so-called BRCA1 C-terminal (BRCT) domain. Mice containing a targeted deletion of this domain are healthy, fertile and display normal somatic hypermutation. *Rev1BRCT1 –/–* cells display an elevated spontaneous frequency of intragenic deletions at HPRT locus. In addition, these cells were sensitized to exogenous DNA damages. UV-C light induced a delayed progression through late S and G2 phases of the cell cycle and many chromatid aberrations, specifically in a subset of mutant cells, but not enhanced sister chromatid exchanges. UV-C-induced mutagenesis was reduced and mutations at TT dimers were absent in *Rev1BRCT1 –/–* cells, the opposite phenotype of UV-C-exposed cells from XP-V patients, lacking DNA pol η. This suggested that the enhanced UV-induced mutagenesis in XP-V patients might depend on error-prone Rev1-dependent TLS. These data indicate a regulatory role of the Rev1 BRCT domain in TLS of a limited spectrum of endogenous and exogenous nucleotide damages during a defined phase of the cell cycle[73] (see also Section 4.6.2: DNA polymerase switches due to PCNA ubquitination). Furthermore, a recent study reveals a role of the BRCA1 C-terminal (BRCT) domain of REV1 in replication-associated gene conversion, indicating that REV1 plays multiple roles at stalled replication fork to maintain genomic integrity in response to replication stress.[74] Indeed, when endogenous REV1 transcript was reduced by 60% in bronchial epithelium, there was a significant decrease in the multiplicity of the carcinogen benzo(a)pyrene-induced lung tumors in mice.[75] These data support a central role of a TLS pathway involving REV1 in the development of lung cancer.

DNA polymerase λ: Upon treatment of mammalian cells with DNA damaging agents (UV, γ-irradiation and H_2O_2) mRNA of DNA pol λ is down-regulated[76] and mouse embryonic fibroblast (MEF) depleted of DNA pol λ are sensitive to H_2O_2 but not to alkylating agents.[77] The group of Claude-Agnès Reynaud obtained knock out DNA pol λ –/– mice in which the males were fertile, and homozygous breeding has been performed up to the third generation without a noticeable problem,[78] suggesting that this enzyme is dispensable for mouse development. When DNA pol λ levels were tested in lung epithelium of lung cancer patients it was found that DNA pol λ expression in the bronchiolar epithelia was significantly correlated with the amount of habitual smoking in the respective individuals. On the other hand, DNA pol λ expression in cancer tissues themselves did not correlate with the smoking of the patients.[79]

A cancer-related variant (R438W) of DNA pol λ has been reported, which shows lowered fidelity and impaired efficiency in NHEJ reactions. These properties correlated with an increased mutagenesis and chromosomal instability of cells expressing this mutant, suggesting a role of this DNA pol λ variant in promoting carcinogenesis.[80,81]

DNA polymerase μ: The close association of DNA pol μ with cells of the germinal centres cells, its error-prone nature and its strong expression by post germinal centres in non-Hodgkin's lymphoma B cells suggested that DNA pol μ is a candidate to be involved in this cancer. Moreover, a close association was established between DNA pol μ expression and B cells in non-Hodgkin's lymphoma.[9] It was also found that overexpression of DNA pol μ in a Burkitt's lymphoma cell line affects the somatic hyper mutation rate.[82] Interestingly, using associative learning and long-term potentiation experiments, it has been shown that DNA pol μ –/– mice maintain the ability to learn at ages when wild-type mice do not.[83] Expression and biochemical analyses suggest that brain aging is delayed in DNA pol μ –/–, being associated with reduced error-prone DNA oxidative repair activity and more efficient mitochondrial function. This appears to be the first example in which the genetic ablation of a DNA-repair function results in a substantially better maintenance of learning ability, together with fewer signs of brain aging, in old mice.

Table 5.3 Human DNA polymerases, genes, chromosomal localizations and single nucleotides polymorphisms[a]

DNA polymerase	Gene of the catalytic subunit	Chromosomal localization[b]	SNPS[c]	Frequency (%)[4d]
DNA pol α	POLA1	Xp22.1-21.3	0	0
PrimPOL	CCDC111	4q35.1	1[e]	1.5[e]
DNA pol β	POLB	8p11.2	3	0.6–2.1
DNA pol γ	POLG1	15q25	9	3–5
DNA pol δ	POLD1	19q13.3	8	0.6–37
DNA pol ε	POLE	12q5.3	11	1–13
DNA pol θ	POLQ (chaos1)	3q13.33	14	0.6–48
DNA pol ν	POLN	4q16.3	8	2.3–50
DNA pol ζ	REV3L	6q21	25	0.6–50
DNA pol η	POLH	6q21.1	6	0.6–8.4
DNA pol κ	POLK (hDinB)	5q13	6	0.6–5.2
DNA pol ι	POLI	18q21.1	8	0.6–38
REV1	REV1L	2q11.1–11.2	16	0.6–40
DNA pol λ	POLL	10q23	2	2–14
DNA pol μ	POLM	7p13	5	5–10

a: Data from Ref. 1 where more details concerning SNPs are documented. b: Taken from the NIEHS Environmental Genome Project. c: Number of non-synonymous codon changes within the sequence of the POL gene. d: Allele frequencies in individuals. e: Data from Ref. 14

Single nucleotide polymorphism in DNA polymerases: Single nucleotide polymorphisms (SNPs) have been identified in most DNA pols (Table 5.3). So far over 120 SNPs were found in human populations in the different DNA pols and they were predicted to result in non-synonymous amino-acid substitutions of DNA pols (reviewed in Ref. 1 and see also NCBI data base dbSNP). Very little is known about the functional consequences of these variations. These are examples where a deleterious effect led to haploinsufficiency. This was exemplified for DNA pol $\beta^{+/-}$ mice, which are tumor prone.[84] Furthermore DNA pol SNP variants might also affect DNA pol gene expression[85] and as a consequence the deleterious effect could be the amount of a particular DNA pol, which is up- and/or down-regulated as a consequence of SNP variants (see also above).

5.3 DNA polymerases and resistance to chemotherapy

Drug resistance is a major obstacle for the successful chemotherapy treatment of cancer. Although high response rates are often initially observed in patients, resistance frequently occurs, rendering subsequent therapy largely ineffective. An example of such a problem is the cellular resistance induced by the antitumor drug cisplatin, a platinum compound which is one of the most effective and broadly used anticancer drug so far, particularly useful for treatment of testicular cancer.[86] Cisplatin is believed to exert its cytotoxic effects by interacting with DNA where it inhibits both replication and transcription and induces programmed cell death.[87]

Cell lines exhibit different mechanisms that account for their acquired resistance to cisplatin. These mechanisms include a) decreased platinum accumulation; b) elevated levels of proteins such as glutathione transferase or metallothionein, which can sequester cisplatin before it reaches its pharmacological target; c) enhanced repair capacity to remove Pt-DNA lesions; d) alteration in the type of Pt-DNA lesions. An additional mechanism of resistance is the consequence of an increased capacity of the cell to tolerate platinum-DNA lesions. This has been shown to be the primary mechanism that caused decreased cisplatin sensitivity in a series of cell lines.[88] The mechanisms of replicative bypass of cisplatin-DNA lesions that take place *in vivo* are still not completely understood but a number of eukaryotic DNA pols have been shown capable of mutagenic bypass *in vitro* of the major cisplatin-DNA adduct, the Pt-(GpG) intrastrand cross-link. Among these DNA pols are the X family DNA pol β[89,90] and DNA pol μ,[91] the Y family DNA pol η[56] and the B family DNA pol ζ.[92] DNA pol β could be a particularly good candidate for participating in the *in vivo* TLS of cisplatin lesions because this DNA pol has been found to be overexpressed in cisplatin resistant cell lines.[93] Moreover, it has been shown that the frequency of mutations induced by cisplatin increased significantly in CHO cells over expressing DNA pol β compared to control cells displaying a normal level of enzyme, and the use of cell extracts supported the hypothesis that error-prone translesion replication was one of the key determinants of tolerance phenotype.[94] Therefore, it can reasonably be postulated that overexpression of DNA pol β contributes to replicative bypass of cisplatin-DNA lesions *in vivo*. Of course, in addition to DNA pol β, other DNA pols could participate, together with other proteins implicated in DNA replication, in TLS processes leading to cisplatin resistance.

Survival through S phase as a result of replicative bypass should give the cell an additional period of time to repair DNA adducts in arrested G2 phase, suggesting a link between TLS and resistance as a consequence of augmented capacity of the cell to repair DNA lesions. An additional link between replicative bypass and cisplatin resistance may be enforced by the finding that over-expression of several proto-oncogenes is correlated with cisplatin resistance following drug exposure.[95] Although the molecular mechanisms that could relate proto-oncogene induction to cisplatin resistance have not yet been explored, *in vivo* mutagenic replication of a single-stranded vector bearing the major cisplatin lesion Pt-(GdG) placed on codon 13 within the human H-ras gene has been reported.[96] Interestingly, the mutations observed with the highest frequency in this system result in the amino-acid substitution that is known to be key step in the activation of the H-ras proto-oncogene.

In conclusion, as far as resistance to cisplatin is concerned, an essential strategy would be to inhibit DNA pols involved in TLS in order to increase the replication inhibiting capacity of the drug. In the case of DNA pol β it should be noted that this enzyme is implicated in DNA repair so that inefficient DNA repair synthesis could strengthen the cytotoxicity of potentially lethal repair intermediates by, for instance, increasing the chances of formation of lethal double strand breaks.

5.4 DNA polymerase γ and human diseases

Mitochondrial DNA polymerase: Mitochondrial DNA is replicated by the DNA pol γ encoded by the nuclear gene POLG. Pol γ is a heterotrimer composed of a 140-KDa catalytic subunit and two 55-kDa accessory subunits that form themselves a dimer.[97] The crystal structures of the human heterotrimeric DNA pol γ holoenzyme and of a variant of its processivity factor have been determined. The holoenzyme structure reveals an unexpected assembly of the mitochondrial DNA replicase where the catalytic subunit interacts with its processivity factor primarily via a domain absent in all other DNA pols. This domain provides a structural model for supporting both the intrinsic processivity of the catalytic subunit alone and the enhanced processivity of the three-subunits holoenzyme. The DNA pol γ structure also provides a context for interpreting the phenotypes of diseases-related mutations in the polymerase.[98,99] The catalytic subunit is synthesized as a precursor polypeptide with a 25 amino-terminal leader sequence that functions to target the polypeptide to the mitochondria. In addition to DNA pol γ, the Twinkle DNA helicase and the mitochondrial single-stranded binding protein (mtSSB) are required for replication and repair of the mitochondrial genome. The catalytic subunit of DNA pol γ contains DNA polymerase, 3′→5′proofreading exonuclease, and dRP-lyase activities. The domain of the 140-kDa subunit comprises an exonuclease domain and a polymerase domain separated by a linker region. The accessory subunits facilitate tighter dNTPs and double-stranded DNA binding and are required for highly processive DNA synthesis. DNA pol γ accurately replicates DNA *in vitro*, with a measured error frequency of 2 to 4×10^{-6} per nucleotide. Reviews on the role of DNA pol γ in mitochondrial DNA replication and repair are summarized in Refs. 100 and 101.

Several human mitochondrial diseases have been associated with mutations in DNA pol γ and the most important ones are listed in Table 5.4 and briefly discussed below. Comprehensive reviews on DNA pol γ forms harboring disease alterations have been published.[102–104]

Progressive External Ophthalmoplegia (PEO): PEO is a mitochondrial disorder associated with mtDNA mutations and deletions and is usually transmitted in an autosomal dominant trait. PEO is characterized by

Table 5.4 DNA polymerase γ and human Diseases[a]

Physiological functions	Diseases related to DNA polymerase γ
Mitochondrial DNA replication	Progressive external ophthalmoplegia (PEO)
	Ataxia neuropathy
Mitochondrial DNA repair	Alpers syndrome
	Premature aging (?)
	Parkinson disease
	Ataxia neuropathy

a: see text for details and references

progressive weakening of external eye muscle leading to blepharoptosis and ophtalmoparesis. The disease is often accompanied by cataract and hearing loss and may result in development of neuromuscular problems. PEO has been linked to four different chromosomal loci, including genes encoding for the mitochondrial DNA helicase Twinkle,[105] a gene that encodes adenine nucleotide translocator (ANTI),[106] the catalytic subunit of the mitochondrial DNA pol[107] and in the gene encoding the p55 accessory subunits.[108] All the dominant Pol mutations leading to PEO have been mapped to the polymerase domain[109] and four amino acids substitutions, G923D, R943H, Y955C and A957S, were analyzed biochemically.[110] All these substitutions showed low catalytic activity, ranging from 30% to 1% of the wt DNA pol and, for two of them, namely R943H and Y955C, also decreased processivity. The clinical severity of PEO correlated with the extent of reduction of DNA synthesis of the mutations was examined. The single mutation G451E found in the p55 subunit disrupts interaction between the accessory and the catalytic subunit, ultimately causing deletions in the mtDNA of this patient.[108]

Alpers syndrome: Alpers syndrome is a rare but severe, heritable autosomal recessive disease that affects young children. The disease may manifest within the first year of life and the patients show progressive cerebral degeneration leading to mental deterioration, cortical blindness and deafness eventually leading to death. To date 45 different point mutations in DNA pol γ have been associated with Alpers syndrome, but only two, A467T and W748S have been characterized so far. The A467T mutant enzyme possessed only 4% of the wt polymerase activity and failed to interact with the accessory subunit p55.[111] Similarly, W748S protein exhibited low polymerase activity, low processivity and severe DNA binding defect.[112] Sequence analysis of the C-terminal polymerase region of DNA pol γ revealed a cluster of four Alpers mutations at highly conserved residues in the thumb subdomain and two in the adjacent palm subdomain.[113] Biochemical characterization of purified recombinant mutant proteins revealed that, although retaining their binding capacity with the accessory subunit p55, the proteins mutated in the thumb domain had only 1% of the wt polymerase activity while proteins mutated in the palm subdomain retained 50–70% of wt polymerase activity. Mutations S305R and P1073L have been also associated with Alpers syndrome.

These mutations severely reduced the catalytic activity of DNA pol γ in *in vitro* reconstituted replisome assays. Consistently, yeast cells heterozygous for these mutations, showed a 75% reduction of mtDNA.[114]

Premature aging: Mice harboring genetic defects in the proofreading exonuclease activity of the mitochondrial DNA pol γ ($PolG^{mut/mut}$) displayed both increased mitochondrial DNA mutations and multiple symptoms of premature aging, providing a causative link between mtDNA mutations and ageing phenotypes in mammals.[115] However, subsequent work from another group, using an alternative method to measure mutations, came to the conclusion that mitochondrial single-base mutations do not limit the lifespan of mice.[116] By contrast, the frequency of deletion mutations has been found to correlate with the aging phenotype.[117]

However, these data still appear to be controversial, as documented by two publications. First, it has been reported that the absolute level of deletions in mutator mice is quite low, especially when compared with the level of point mutations in the same mice.[118] The authors argued that the available data are insufficient to conclude that mtDNA mutations drive premature aging in mtDNA mutator mice. Second, molecular analyses were carried out to determine the mechanism whereby the mtDNA mutations impair respiratory chain function.[119] These authors found that mitochondrial protein synthesis is unimpaired in mtDNA mutator mice, consistent with the observed minor alterations of steady-state levels of mitochondrial transcripts. Their findings consequently refute arguments that circular mtDNA molecules with large deletions are driving the premature aging phenotype. Clearly the question concerning a causative role for accumulating mtDNA mutations in aging needs further investigation.

Parkinson disease and ataxia neuropathy: Parkinson disease is a frequent neurodegenerative disease and a significant co-segregation of Parkinsonism with mutations in the POLG gene was described.[120,121] Mutations in POLG are associated with a form of ataxia neuropathy called mitochondrial associated ataxia syndrome (MIRAS).[122] Symptoms of ataxia involving mutations in POLG include peripheral neuropathy, involuntary movements and epileptic seizures. One mutation, the A467T, is located in the linker region of the catalytic subunit of DNA pol γ and has been also found in PEO and Alpers syndromes.[123,124] Biochemical analysis indicated that the A467T mutant had strongly reduced polymerase activity but possessed almost normal proofreading exonuclease activity and failed to interact with the p55 subunit.[111]

5.5 Chapter summary

In this chapter we highlighted the possible connections between DNA pols and diseases. In addition, we also attempted to critically evaluate those cases where the experimental data are not yet fully convincing.

In the last ten years numerous novel DNA pols have been revealed, but their exact cellular functions still await clarification. Table 5.1 summarises the known eukaryotic DNA pols and their relationships with diseases. Since DNA pols are the working horses to faithfully copy the DNA in all DNA transactions such as DNA replication, DNA repair, DNA recombination and TLS, one is not surprised that slight changes in their behaviour might lead to genomic instability and therefore to cancer. The literature suggests that mutations might occur via the action of "canonical" DNA pols during replication and repair as well as from "specialized" DNA pols, of which there is too much (up-regulated) or not enough (down-regulated) in the cell. Moreover, mutations in the DNA pols themselves might also hamper their functional accuracy again leading to mutations and thus to genomic instability. So far over 120 single nucleotide polymorphisms (SNPs) were identified in human populations of different DNA pols and they were predicted to result in non-synonymous amino-acid substitutions within their sequences but very little is known about the functional consequences of these variations. Since DNA pols are targets for chemotherapy (for details see Chapter 6) it is important to discuss their potential role in chemotherapy resistance. DNA pol γ is the prototype of a DNA pol that has been directly linked to a variety of diseases other than cancer, such as Progressive External Ophthalmoplegia, Alpers syndrome, premature aging, Parkinson disease and ataxia neuropathy.

Genetic or even epigenetic factors involved in mis-regulation of the DNA pols could represent novel prognostic, diagnostic and therapeutic tools in medicine, and understanding the molecular details of the processes linking DNA pols properties to diseases will be a big challenge for the future.

References

1. Sweasy, J. B., Lauper, J. M. & Eckert, K. A. DNA polymerases and human diseases. *Radiat Res* **166**, 693–714 (2006).
2. Loeb, L. A. & Monnat, R. J., Jr. DNA polymerases and human disease. *Nat Rev Genet* **9**, 594–604 (2008).
3. Ramadan, K., Maga, G. & Hubscher, U. in *Genome Integrity: Facets and Perspectives* (ed D.-H. Lankenau) (Springer Verlag, 2007).
4. Lange, S. S., Takata, K. & Wood, R. D. DNA polymerases and cancer. *Nat Rev Cancer* **11**, 96–110 (2011).
5. Loeb, K. R. & Loeb, L. A. Significance of multiple mutations in cancer. *Carcinogenesis* **21**, 379–385 (2000).
6. Loeb, L. A., Loeb, K. R. & Anderson, J. P. Multiple mutations and cancer. *Proc Natl Acad Sci USA* **100**, 776–781 (2003).
7. Bielas, J. H., Loeb, K. R., Rubin, B. P., True, L. D. & Loeb, L. A. Human cancers express a mutator phenotype. *Proc Natl Acad Sci USA* **103**, 18238–18242 (2006).
8. Albertella, M. R., Lau, A. & O'Connor, M. J. The overexpression of specialized DNA polymerases in cancer. *DNA Repair (Amst)* **4**, 583–593 (2005).
9. Chiu, A. *et al.* DNA polymerase mu gene expression in B-cell non-Hodgkin's lymphomas — An analysis utilizing in situ hybridization. *Am J Pathol* **161**, 1349–1355 (2002).
10. Sohl, C. D., Ray, S. & Sweasy, J. B. Pools and Pols: Mechanism of a mutator phenotype. *P Natl Acad Sci USA* **112**, 5864–5865 (2015).
11. Waterfall, J. J. & Meltzer, P. S. Avalanching mutations in biallelic mismatch repair deficiency syndrome. *Nature Genetics* **47**, 194–196 (2015).
12. Ganai, R. A. & Johansson, E. DNA Replication-A Matter of Fidelity. *Molecular Cell* **62**, 745–755 (2016).
13. Floy, K. M., Hess, R. O. & Meisner, L. F. DNA polymerase alpha defect in the N syndrome. *Am J Med Genet* **35**, 301–305 (1990).
14. Zhao, F. X. *et al.* Exome sequencing reveals CCDC111 mutation associated with high myopia. *Human Genetics* **132**, 913–921 (2013).
15. Keen, B. A., Bailey, L. J., Jozwiakowski, S. K. & Doherty, A. J. Human PrimPol mutation associated with high myopia has a DNA replication defect. *Nucleic Acids Research* **42**, 12102–12111 (2014).
16. Gu, H., Marth, J. D., Orban, P. C., Mossmann, H. & Rajewsky, K. Deletion of a DNA polymerase beta gene segment in T cells using cell type-specific gene targeting. *Science* **265**, 103–106 (1994).

17. Sobol, R. W. *et al.* Requirement of mammalian DNA polymerase-beta in base-excision repair. *Nature* **379**, 183–186 (1996).
18. Bergoglio, V. *et al.* Evidence of finely tuned expression of DNA polymerase beta in vivo using transgenic mice. *FEBS Lett* **566**, 147–150 (2004).
19. Frechet, M. *et al.* Deregulated DNA polymerase beta strengthens ionizing radiation-induced nucleotidic and chromosomal instabilities. *Oncogene* **21**, 2320–2327 (2002).
20. Frechet, M., Canitrot, Y., Cazaux, C. & Hoffmann, J. S. DNA polymerase beta imbalance increases apoptosis and mutagenesis induced by oxidative stress. *FEBS Lett* **505**, 229–232 (2001).
21. Louat, T. *et al.* Antitumor activity of 2′,3′-dideoxycytidine nucleotide analog against tumors up-regulating DNA polymerase beta. *Mol Pharmacol* **60**, 553–558 (2001).
22. Senejani, A. G. *et al.* Mutation of POLB Causes Lupus in Mice. *Cell Rep* **6**, 1–8 (2014).
23. Goldsby, R. E. *et al.* Defective DNA polymerase-delta proofreading causes cancer susceptibility in mice. *Nat Med* **7**, 638–639 (2001).
24. Goldsby, R. E. *et al.* High incidence of epithelial cancers in mice deficient for DNA polymerase delta proofreading. *Proc Natl Acad Sci USA* **99**, 15560–15565. Epub 12002 Nov 15512. (2002).
25. Da Costa, L. T. *et al.* Polymerase delta variants in RER colorectal tumors. *Nat Genet* **9**, 11–14 (1995).
26. Flohr, T. *et al.* Detection of mutations in the DNA polymerase delta gene of human sporadic colorectal cancers and colon cancer cell lines. *Int J Cancer* **80**, 919–929 (1999).
27. Popanda, O., Flohr, T., Fox, G. & Thielmann, H. W. A mutation detected in DNA polymerase delta cDNA from Novikoff hepatoma cells correlates with abnormal catalytic properties of the enzyme. *J Cancer Res Clin Oncol* **125**, 598–608 (1999).
28. Swan, M. K., Johnson, R. E., Prakash, L., Prakash, S. & Aggarwal, A. K. Structural basis of high-fidelity DNA synthesis by yeast DNA polymerase delta. *Nat Struct Mol Biol* **16**, 979–986 (2009).
29. Weedon, M. N. *et al.* An in-frame deletion at the polymerase active site of POLD1 causes a multisystem disorder with lipodystrophy. *Nature Genetics* **45**, 947–950 (2013).
30. Albertson, T. M. *et al.* DNA polymerase {varepsilon} and {delta} proofreading suppress discrete mutator and cancer phenotypes in mice. *Proc Natl Acad Sci USA* **106**, 17101–17104 (2009).

31. Palles, C. *et al.* Germline mutations affecting the proofreading domains of POLE and POLD1 predispose to colorectal adenomas and carcinomas. *Nature Genetics* **45**, 136–144 (2013).
32. Rayner, E. *et al.* A panoply of errors: polymerase proofreading domain mutations in cancer. *Nat Rev Cancer* **16**, 71–81 (2016).
33. Aoude, L. G. *et al.* POLE mutations in families predisposed to cutaneous melanoma. *Fam Cancer* **14**, 621–628 (2015).
34. Meng, B. *et al.* POLE exonuclease domain mutation predicts long progression-free survival in grade 3 endometrioid carcinoma of the endometrium. *Gynecol Oncol* **134**, 15–19 (2014).
35. Williams, L. N. *et al.* dNTP pool levels modulate mutator phenotypes of error-prone DNA polymerase epsilon variants. *P Natl Acad Sci USA* **112**, E2457–E2466 (2015).
36. Mertz, T. M., Sharma, S., Chabes, A. & Shcherbakova, P. V. Colon cancer-associated mutator DNA polymerase delta variant causes expansion of dNTP pools increasing its own infidelity. *P Natl Acad Sci USA* **112**, E2467–E2476 (2015).
37. Kodama, E. N., McCaffrey, R. P., Yusa, K. & Mitsuya, H. Antileukemic activity and mechanism of action of cordycepin against terminal deoxynucleotidyl transferase-positive (TdT+) leukemic cells. *Biochem Pharmacol* **59**, 273–281 (2000).
38. Shima, N. *et al.* Phenotype-based identification of mouse chromosome instability mutants. *Genetics* **163**, 1031–1040 (2003).
39. Pillaire, M. J. *et al.* A 'DNA replication' signature of progression and negative outcome in colorectal cancer. *Oncogene* **29**, 876–887 (2010).
40. Lemee, F. *et al.* DNA polymerase theta up-regulation is associated with poor survival in breast cancer, perturbs DNA replication, and promotes genetic instability. *P Natl Acad Sci USA* **107**, 13390–13395 (2010).
41. Brandalize, A. P. C. *et al.* A DNA repair variant in POLQ (c.-1060A > G) is associated to hereditary breast cancer patients: a case-control study. *Bmc Cancer* **14** (2014).
42. Boesen, J. J. B. *et al.* Stress Response Induced by DNA Damage Leads to Specific, Delayed and Untargeted Mutations. *Molecular & General Genetics* **234**, 217–227 (1992).
43. Zhang, X. S., Yu, Y. N. & Chen, X. R. Evidence for Nontargeted Mutagenesis in a Monkey Kidney-Cell Line and Analysis of Its Sequence Specificity Using a Shuttle-Vector Plasmid. *Mutation Research* **323**, 105–112 (1994).

44. Zhu, F. & Zhang, M. DNA polymerase zeta: new insight into eukaryotic mutagenesis and mammalian embryonic development. *World J Gastroenterol* **9**, 1165–1169 (2003).
45. Wittschieben, J. *et al.* Disruption of the developmentally regulated Rev3I gene causes embryonic lethality. *Curr Biol* **10**, 1217–1220 (2000).
46. Lange, S. S., Wittschieben, J. P. & Wood, R. D. DNA polymerase zeta is required for proliferation of normal mammalian cells. *Nucleic Acids Research* **40**, 4473–4482 (2012).
47. Wittschieben, J. P. *et al.* Loss of DNA Polymerase zeta Enhances Spontaneous Tumorigenesis. *Cancer Research* **70**, 2770–2778 (2010).
48. Roos, W. P. *et al.* The Translesion Polymerase Rev3L in the Tolerance of Alkylating Anticancer Drugs. *Molecular Pharmacology* **76**, 927–934 (2009).
49. Wang, W. J. *et al.* REV3L modulates cisplatin sensitivity of non-small cell lung cancer H1299 cells. *Oncol Rep* **34**, 1460–1468 (2015).
50. Masutani, C. *et al.* The XPV (xeroderma pigmantosum variant) gene encodes human DNA polymerase eta. *Nature* **399**, 700–704 (1999).
51. Kunkel, T. A. Considering the cancer consequences of altered DNA polymerase function. *Cancer Cell* **3**, 105–110 (2003).
52. Johnson, R. E., Prakash, S. & Prakash, L. Efficient bypass of a thymine-thymine dimer by yeast DNA polymerase, Poleta. *Science* **283**, 1001–1004. (1999).
53. Wang, Y. *et al.* Evidence that in xeroderma pigmentosum variant cells, which lack DNA polymerase eta, DNA polymerase iota causes the very high frequency and unique spectrum of UV-induced mutations. *Cancer Res* **67**, 3018–3026 (2007).
54. Maga, G. *et al.* 8-oxo-guanine bypass by human DNA polymerases in the presence of auxiliary proteins. *Nature* **447**, 606–608 (2007).
55. Yuan, F. e. a. Specificity of DNA lesion bypass by the yeast DNA polymerase eta. *J Biol Chem* **275**, 8233–8239 (2000).
56. Vaisman, A., Masutani, C., Hanaoka, F. & Chaney, S. G. Efficient translesion replication past oxaliplatin and cisplatin GpG adducts by human DNA polymerase eta. *Biochemistry* **39**, 4575–4580 (2000).
57. Prakash, S., Johnson, R. E. & Prakash, L. Eukaryotic translesion synthesis DNA polymerases: specificity of structure and function. *Annu Rev Biochem* **74**, 317–353 (2005).
58. Zeng, X., Negrete, G. A., Kasmer, C., Yang, W. W. & Gearhart, P. J. Absence of DNA polymerase eta reveals targeting of C mutations on the

nontranscribed strand in immunoglobulin switch regions. *J Exp Med* **199**, 917–924. Epub 2004 Mar 2029. (2004).
59. Franklin, A., Milburn, P. J., Blanden, R. V. & Steele, E. J. Human DNA polymerase-eta, an A-T mutator in somatic hypermutation of rearranged immunoglobulin genes, is a reverse transcriptase. *Immunol Cell Biol* **82**, 219–225 (2004).
60. Bergoglio, V. *et al.* DNA synthesis by Pol eta promotes fragile site stability by preventing under-replicated DNA in mitosis. *The Journal of Cell Biology* **201**, 395–408 (2013).
61. Chen, Y. W., Harris, R. A., Hatahet, Z. & Chou, K. M. Ablation of XP-V gene causes adipose tissue senescence and metabolic abnormalities. *P Natl Acad Sci USA* **112**, E4556–E4564 (2015).
62. Bavoux, C. *et al.* Up-regulation of the error-prone DNA polymerase {kappa} promotes pleiotropic genetic alterations and tumorigenesis. *Cancer Res* **65**, 325–330 (2005).
63. Pan, Q., Fang, Y., Xu, Y., Zhang, K. & Hu, X. Down-regulation of DNA polymerases kappa, eta, iota, and zeta in human lung, stomach, and colorectal cancers. *Cancer Lett* **217**, 139–147 (2005).
64. Wang, Y. *et al.* Elevated expression of DNA polymerase kappa in human lung cancer is associated with p53 inactivation: Negative regulation of POLK promoter activity by p53. *Int J Oncol* **25**, 161–165 (2004).
65. Velasco-Miguel, S. *et al.* Constitutive and regulated expression of the mouse Dinb (Polkappa) gene encoding DNA polymerase kappa. *DNA Repair (Amst)* **2**, 91–106 (2003).
66. Yadav, S., Mukhopadhyay, S., Anbalagan, M. & Makridakis, N. Somatic Mutations in Catalytic Core of POLK Reported in Prostate Cancer Alter Translesion DNA Synthesis. *Hum Mutat* **36**, 873–880 (2015).
67. Wang, M. *et al.* Pol iota is a candidate for the mouse pulmonary adenoma resistance 2 locus, a major modifier of chemically induced lung neoplasia. *Cancer Res* **64**, 1924–1931 (2004).
68. Lee, G. H. *et al.* Analysis of lung tumorigenesis in chimeric mice indicates the Pulmonary adenoma resistance 2 (Par2) locus to operate in the tumor-initiation stage in a cell-autonomous manner: detection of polymorphisms in the Poli gene as a candidate for Par2. *Oncogene* **22**, 2374–2382 (2003).
69. Yang, J., Chen, Z., Liu, Y., Hickey, R. J. & Malkas, L. H. Altered DNA polymerase iota expression in breast cancer cells leads to a reduction in DNA replication fidelity and a higher rate of mutagenesis. *Cancer Res* **64**, 5597–5607 (2004).

70. Yuan, F. *et al.* Overexpressed DNA Polymerase Iota Regulated by JNK/c-Jun Contributes to Hypermutagenesis in Bladder Cancer. *Plos One* **8** (2013).
71. Simpson, L. J. & Sale, J. E. Rev1 is essential for DNA damage tolerance and non-templated immunoglobulin gene mutation in a vertebrate cell line. *EMBO J* **22**, 1654–1664 (2003).
72. Clark, D. R., Zacharias, W., Panaitescu, L. & McGregor, W. G. Ribozyme-mediated REV1 inhibition reduces the frequency of UV-induced mutations in the human HPRT gene. *Nucleic Acids Res* **31**, 4981–4988 (2003).
73. Jansen, J. G. *et al.* The BRCT domain of mammalian Rev1 is involved in regulating DNA translesion synthesis. *Nucleic Acids Res* **33**, 356–365. Print 2005. (2005).
74. Yang, Y. R. *et al.* FANCD2 and REV1 cooperate in the protection of nascent DNA strands in response to replication stress. *Nucleic Acids Research* **43**, 8325–8339 (2015).
75. Dumstorf, C. A., Mukhopadhyay, S., Krishnan, E., Haribabu, B. & McGregor, W. G. REV1 Is Implicated in the Development of Carcinogen-Induced Lung Cancer. *Mol Cancer Res* **7**, 247–254 (2009).
76. Aoufouchi, S. *et al.* Two novel human and mouse DNA polymerases of the polX family. *Nucl Acids Res* **28**, 3684–3693 (2000).
77. Braithwaite, E. K. *et al.* DNA polymerase lambda protects mouse fibroblasts against oxidative DNA damage and is recruited to sites of DNA damage/repair. *J Biol Chem* **280**, 31641–31647 (2005).
78. Bertocci, B. *et al.* Cutting edge: DNA polymerases mu and lambda are dispensable for Ig gene hypermutation. *J Immunol* **168**, 3702–3706 (2002).
79. Ohba, T. *et al.* Expression of an X-family DNA polymerase, pol lambda, in the respiratory epithelium of non-small cell lung cancer patients with habitual smoking. *Mutat Res* **677**, 66–71 (2009).
80. Capp, J. P. *et al.* The R438W polymorphism of human DNA polymerase lambda triggers cellular sensitivity to camptothecin by compromising the homologous recombination repair pathway. *Carcinogenesis* **31**, 1742–1747 (2010).
81. Nemec, A. A. *et al.* Estrogen Drives Cellular Transformation and Mutagenesis in Cells Expressing the Breast Cancer-Associated R438W DNA Polymerase Lambda Protein. *Mol Cancer Res* **14**, 1068–1077 (2016).
82. Ruiz, J. F. *et al.* Overexpression of human DNA polymerase mu (Pol mu) in a Burkitt's lymphoma cell line affects the somatic hypermutation rate. *Nucleic Acids Res* **32**, 5861–5873 (2004).
83. Lucas, D. *et al.* Increased Learning and Brain Long-Term Potentiation in Aged Mice Lacking DNA Polymerase mu. *Plos One* **8** (2013).

84. Cabelof, D. C. *et al.* Haploinsufficiency in DNA polymerase beta increases cancer risk with age and alters mortality rate. *Cancer Res* **66**, 7460–7465 (2006).
85. Pastinen, T., Ge, B. & Hudson, T. J. Influence of human genome polymorphism on gene expression. *Hum Mol Genet* **15 Spec No 1**, R9–16 (2006).
86. Wang, D. & Lippard, S. J. Cellular processing of platinum anticancer drugs. *Nat Rev Drug Discov* **4**, 307–320 (2005).
87. Kelland, L. The resurgence of platinum-based cancer chemotherapy. *Nat Rev Cancer* **7**, 573–584 (2007).
88. Johnson, S. W., Laub, P. B., Beesley, J. S., Ozols, R. F. & Hamilton, T. C. Increased platinum-DNA damage tolerance is associated with cisplatin resistance and cross-resistance to various chemotherapeutic agents in unrelated human ovarian cancer cell lines. *Cancer Res* **57**, 850–856 (1997).
89. Hoffmann, J. S., Pillaire, M. J., Garcia-Estefania, D., Lapalu, S. & Villani, G. In vitro bypass replication of the cisplatin-d(GpG) lesion by calf thymus DNA polymerase beta and human immunodeficiency virus type I reverse transcriptase is highly mutagenic. *J Biol Chem* **271**, 15386–15392 (1996).
90. Hoffmann, J. S. *et al.* DNA polymerase beta bypasses in vitro a single d(GpG)-cisplatin adduct placed on codon 13 of the HRAS gene. *Proc Natl Acad Sci USA* **92**, 5356–5360 (1995).
91. Havener, J. M. *et al.* Translesion synthesis past platinum DNA adducts by human DNA polymerase mu. *Biochemistry* **42**, 1777–1788 (2003).
92. Shachar, S. *et al.* Two-polymerase mechanisms dictate error-free and error-prone translesion DNA synthesis in mammals. *Embo Journal* **28**, 383–393 (2009).
93. Ali-Osman, F., Berger, M. S., Rairkar, A. & Stein, D. E. Enhanced repair of a cisplatin-damaged reporter chloramphenicol-O-acetyltransferase gene and altered activities of DNA polymerases alpha and beta, and DNA ligase in cells of a human malignant glioma following in vivo cisplatin therapy. *J Cell Biochem* **54**, 11–19 (1994).
94. Canitrot, Y. *et al.* Overexpression of DNA polymerase beta in cell results in a mutator phenotype and a decreased sensitivity to anticancer drugs. *Proc Natl Acad Sci USA* **95**, 12586–12590 (1998).
95. Scanlon, K. J., Kashani-Sabet, M., Tone, T. & Funato, T. Cisplatin resistance in human cancers. *Pharmacol Ther* **52**, 385–406 (1991).
96. Pillaire, M. J. *et al.* Mutagenesis in monkey cells of a vector containing a single d(GPG) cis-diamminedichloroplatinum(II) adduct placed on codon 13 of the human H-ras proto-oncogene. *Nucleic Acids Res* **22**, 2519–2524 (1994).

97. Yakubovskaya, E., Chen, Z., Carrodeguas, J. A., Kisker, C. & Bogenhagen, D. F. Functional human mitochondrial DNA polymerase gamma forms a heterotrimer. *J Biol Chem* **281**, 374–382 (2006).
98. Falkenberg, M. & Larsson, N. G. Structure casts light on mtDNA replication. *Cell* **139**, 231–233 (2009).
99. Lee, Y. S., Kennedy, W. D. & Yin, Y. W. Structural insight into processive human mitochondrial DNA synthesis and disease-related polymerase mutations. *Cell* **139**, 312–324 (2009).
100. Kaguni, L. S. DNA polymerase gamma, the mitochondrial replicase. *Annu Rev Biochem* **73**, 293–320 (2004).
101. Graziewicz, M. A., Longley, M. J. & Copeland, W. C. DNA polymerase gamma in mitochondrial DNA replication and repair. *Chem Rev* **106**, 383–405 (2006).
102. Chan, S. S. & Copeland, W. C. DNA polymerase gamma and mitochondrial disease: Understanding the consequence of POLG mutations. *Biochim Biophys Acta* **1787**, 312–319 (2009).
103. Copeland, W. C. Defects in mitochondrial DNA replication and human disease. *Crit Rev Biochem Mol* **47**, 64–74 (2012).
104. Copeland, W. C. & Longley, M. J. Mitochondrial genome maintenance in health and disease. *DNA Repair* **19**, 190–198 (2014).
105. Spelbrink, J. N. *et al.* Human mitochondrial DNA deletions associated with mutations in the gene encoding Twinkle, a phage T7 gene 4-like protein localized in mitochondria. *Nat Genet* **28**, 223–231 (2001).
106. Kaukonen, J. *et al.* Role of adenine nucleotide translocator 1 in mtDNA maintenance. *Science* **289**, 782–785 (2000).
107. Van Goethem, G., Dermaut, B., Lofgren, A., Martin, J. J. & Van Broeckhoven, C. Mutation of POLG is associated with progressive external ophthalmoplegia characterized by mtDNA deletions. *Nat Genet* **28**, 211–212 (2001).
108. Longley, M. J. *et al.* Mutant POLG2 disrupts DNA polymerase gamma subunits and causes progressive external ophthalmoplegia. *Am J Hum Genet* **78**, 1026–1034 (2006).
109. Longley, M. J., Graziewicz, M. A., Bienstock, R. J. & Copeland, W. C. Consequences of mutations in human DNA polymerase gamma. *Gene* **354**, 125–131 (2005).
110. Graziewicz, M. A., Longley, M. J., Bienstock, R. J., Zeviani, M. & Copeland, W. C. Structure-function defects of human mitochondrial DNA polymerase in autosomal dominant progressive external ophthalmoplegia. *Nat Struct Mol Biol* **11**, 770–776. Epub 2004 Jul 2018. (2004).
111. Chan, S. S., Longley, M. J. & Copeland, W. C. The common A467T mutation in the human mitochondrial DNA polymerase (POLG) compromises

catalytic efficiency and interaction with the accessory subunit. *J Biol Chem* **280**, 31341–31346 (2005).

112. Chan, S. S., Longley, M. J. & Copeland, W. C. Modulation of the W748S mutation in DNA polymerase gamma by the E1143G polymorphismin mitochondrial disorders. *Hum Mol Genet* **15**, 3473–3483 (2006).
113. Kasiviswanathan, R., Longley, M. J., Chan, S. S. & Copeland, W. C. Disease mutations in the human mitochondrial DNA polymerase thumb subdomain impart severe defects in mitochondrial DNA replication. *J Biol Chem* **284**, 19501–19510 (2009).
114. Qian, Y. F., Ziehr, J. L. & Johnson, K. A. Alpers disease mutations in human DNA polymerase gamma cause catalytic defects in mitochondrial DNA replication by distinct mechanisms. *Front Genet* **6** (2015).
115. Trifunovic, A. *et al.* Premature ageing in mice expressing defective mitochondrial DNA polymerase. *Nature* **429**, 417–423 (2004).
116. Vermulst, M. *et al.* Mitochondrial point mutations do not limit the natural lifespan of mice. *Nat Genet* **39**, 540–543 (2007).
117. Vermulst, M. *et al.* DNA deletions and clonal mutations drive premature aging in mitochondrial mutator mice. *Nat Genet* **40**, 392–394 (2008).
118. Kraytsberg, Y., Simon, D. K., Turnbull, D. M. & Khrapko, K. Do mtDNA deletions drive premature aging in mtDNA mutator mice? *Aging Cell* **8**, 502–506 (2009).
119. Edgar, D. *et al.* Random point mutations with major effects on protein-coding genes are the driving force behind premature aging in mtDNA mutator mice. *Cell Metab* **10**, 131–138 (2009).
120. Luoma, P. *et al.* Parkinsonism, premature menopause, and mitochondrial DNA polymerase gamma mutations: clinical and molecular genetic study. *Lancet* **364**, 875–882 (2004).
121. Luoma, P. T. *et al.* Mitochondrial DNA polymerase gamma variants in idiopathic sporadic Parkinson disease. *Neurology* **69**, 1152–1159 (2007).
122. Hakonen, A. H. *et al.* Mitochondrial DNA polymerase W748S mutation: a common cause of autosomal recessive ataxia with ancient European origin. *Am J Hum Genet* **77**, 430–441 (2005).
123. Van Goethem, G. *et al.* Recessive POLG mutations presenting with sensory and ataxic neuropathy in compound heterozygote patients with progressive external ophthalmoplegia. *Neuromuscul Disord* **13**, 133–142 (2003).
124. Naviaux, R. K. & Nguyen, K. V. POLG mutations associated with Alpers' syndrome and mitochondrial DNA depletion. *Ann Neurol* **55**, 706–712 (2004).

Chapter 6

Human DNA Polymerases and Chemotherapy

6.1 DNA polymerases are important chemotherapeutic targets

Uncontrolled DNA replication certainly contributes to pathological states such as cancer, autoimmune diseases, and many viral and bacterial infections. Therefore, the DNA pols of human/animal cells, viruses and bacteria are potential targets for chemotherapy of these pathological states in humans and animals.

Antivirals compounds targeting DNA pols and reverse transcriptases have assumed an unexpected importance only a few decades ago, since until then, it appeared unconceivable to specifically interfere with viral replication without affecting normal cell metabolism.

Inhibitors have been developed that successfully target the DNA pols of human Herpesviruses and the Hepatitis B virus, as well as the reverse transcriptase (RT) of HIV-1.

Albeit no anticancer drugs selectively targeting human DNA pols have been approved so far, the plethora of different DNA pols existing in human cells and their differential expression in normal versus tumoral tissues, offer a great deal of opportunities to develop novel agents.[1–4] Several inhibitors are being thus developed, specifically targeting DNA repair enzymes. This approach is justified by the fact that often cancer cells rely on a narrower range of DNA repair pathways than normal cells.

In this Chapter 6 we will first illustrate the strategies and the rationale behind the design of novel compounds targeting human DNA pols. Subsequently we will provide examples of inhibitors of human DNA pols.

6.2 Strategies and problems for the design of inhibitors of DNA polymerases

6.2.1 *Substrate analogs*

The most straightforward approach to inhibit an enzymatic activity is to develop substrate analogs. These are molecules which retain chemical and structural features of the natural substrates that are important for their interaction with the enzyme's active site, but lack one or more reactive groups essential for the chemical step. Thus, substrate analogs (or "false substrates" as they are sometimes named) will compete with the natural substrates of the enzyme, resulting in a decrease of the overall efficiency of the enzymatic reaction.

The obvious limitation of such an approach is the need to know in detail the structure of the substrate, its mode of interaction with the enzyme and the nature and mechanism of the chemical reaction.

Since all these limitations have been overcome in the case of the DNA pols, as detailed in the previous chapters, it is of no surprise that the first class of DNA pol inhibitors ever developed was based on analogs of one of the natural substrates of DNA pols: the nucleotides. Already in the 1970's it was evident that by using nucleotides, which lacked the reactive 3′-hydroxyl group on the deoxyribose ring, essential for the polymerization of the growing DNA chain, it was possible to inhibit the reaction by DNA pols.[5,6] In fact, these analogs are incorporated in the growing DNA chain by DNA pols, then preventing further elongation (Figure 6.1). Due to their mechanism of action, compounds of this class, whose prototype are the 2′,3′-dideoxy-nucleotides, are called chain terminators and still represent the majority of the DNA pol and RT inhibitors used in antiviral chemotherapy.

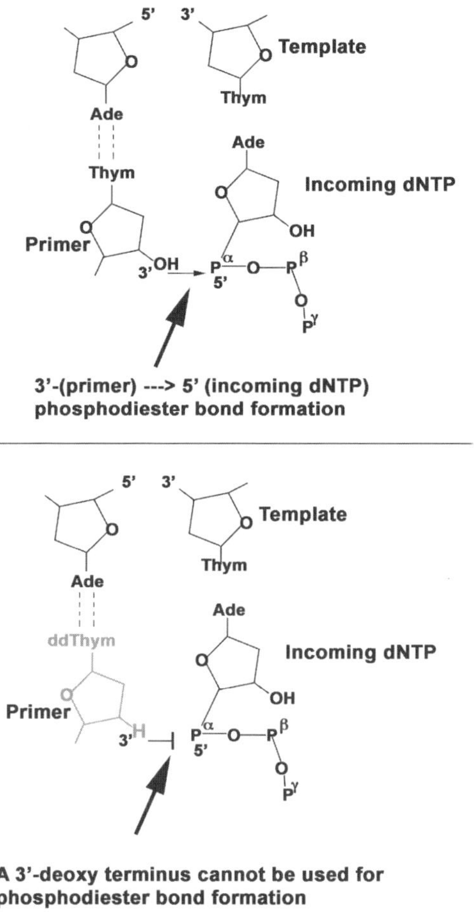

Figure 6.1 **Mechanism of action of dideoxynucleotides.** Upper panel: the hydroxyl group at the 3′ position of the deoxyribose carries on the nucleophylic attack to the α-phosphate group at the 5′-position of the incoming deoxynucleotide triphosphate to catalyze chain elongation. Bottom panel: when a dideoxynucleotide is incorporated, the resulting 3′ of the dideoxyribose lacks a suitable hydroxyl group to start the chemical bond formation, thus preventing chain elongation.

However, one major obstacle in the use of nucleotide analogs is that they must carry three phosphate groups at the 5′-position of the sugar ring, in order to be functional. These deoxyribonucleoside triphosphates (dNTPs), however, do not easily traverse the cellular membrane, due to the negative charges of their phosphate groups. Thus, while *in vitro* reactions with purified enzymes can be easily inhibited by purified dNTPs, *in vivo* these analogs have to be administered in their nucleosidic precursor form (i.e. without phosphate groups). These prodrugs then need to be activated intracellularly by specific nucleoside kinases up to their triphosphate form. The need of additional activation steps for nucleoside analogs has several important implications. For example, modifications that might improve the binding of the dNTPs to the target enzyme can decrease their ability to interact with the activating kinases in their nucleoside form. Moreover, while deoxynucleosides triphosphates are selective substrates of DNA pols, unphosphorylated nucleosides are recognized by a number of different cellular enzymes, thus resulting in lower stability (due to the action of nucleoside hydrolases) and/or unwanted off-targeting effects and cytoxicity (as a result of their interference with other cellular metabolic pathways).

Thus, in designing nucleoside analogs, it is necessary to try not only to optimize their interaction with the target enzyme (DNA pols), but also their ability to be activated in the infected cell without being rapidly degraded and/or being toxic to the host organism. As will become clear later in this chapter, in discussing the different analogs developed as antiviral agents, a number of strategies have been devised in order to balance efficacy and safety.

6.2.2 *Non-substrate analogs*

As outlined in Chapters 2 and 3, we now have a detailed knowledge of the different steps of the DNA polymerization reaction and the thermodynamic and kinetic barriers responsible for the exact sequential order of molecular events leading to the incorporation of nucleotides into a nucleic acid lattice. In principle, any of these steps can be targeted by inhibitors that do not necessarily need to resemble natural substrates, but rather interact with aminoacid residues of the enzyme, which are important for the different conformational changes that drive the reaction (see also Section 4.2). The best known examples of such non-substrate inhibitors are probably the non-nucleoside inhibitors of HIV-1 RT (NNRTIs). These molecules bind to a hydrophobic pocket that is close to, but distinct from, the active site. Contrary to nucleoside analogs, they do not take any interaction with catalytic amino-acids, but rather induce short- and long-range structural rearrangements that reduce the rate of the "open-to-close" conformational change of the enzyme, that is essential for catalysis. Thus, they act as a sort of allosteric inhibitors of RT. Clearly, the structural features which are important for the interaction of NNRTIs with their binding site are totally different from those governing the binding of the substrates to the enzyme and indeed NNRTIs have chemical structures completely unrelated to nucleosides. These non-substrate inhibitors have important advantages over nucleoside analogs: i) they are generally very selective for the target enzyme and ii) they do not require intracellular activation in the form of prodrugs. Indeed, NNRTIs have proven safer and more tolerated by patients than nucleoside analogs in clinical practice.

One major disadvantage of non-substrate analogs is that it is not intuitive, as it is for substrate analogs, to figure out which kind of chemical scaffold might be effective against a given target enzyme. Indeed, the first NNRTIs have been discovered in late 1980's by high-throughput random screening of hundreds of thousands different synthetic molecules. In recent years, however, more sophisticated technologies have been developed to allow the rational design of non-substrate inhibitors.

6.2.3 *Novel in silico technologies for designing inhibitors of DNA polymerases*

The availability of several crystal structures of proteins which are also relevant pharmaceutical targets, both in their unligated state and in complex with substrates/inhibitors, coupled to the ever increasing power of modern computational methods, both in terms of hardware and software, opens new possibilities for drug design. *In silico* techniques allow to build pharmacophores, to screen virtual libraries and to identify specifically tailored inhibitors even in such cases when no previous inhibitors were known for that particular molecular target. Molecular dynamics, on the other hand, is starting to help to clarify the molecular details of ligand–protein interactions, hence providing a powerful tool for mechanism of action studies.

Virtual screening. Molecules specifically interacting with a given target can be identified through a combination of structural biology,

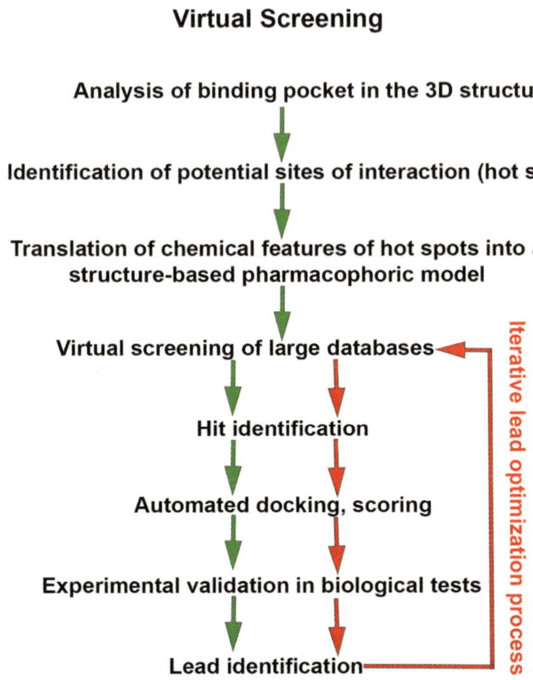

Figure 6.2 **The virtual screening (VS) technique.** After the first round (green arrows), the process is iterated to optimize the lead compound (red arrows).

screening of chemical libraries and rational drug design. The classical approach of experimental high-throughput screening (HTS) of large compound libraries is now being integrated, and in some instances replaced, by the more sophisticated knowledge-based computer-aided virtual screening (VS) approach (Figure 6.2).

The VS approach has several advantages over the classical HTS approach (Table 6.1). VS, also referred to as *in silico* screening, makes use of computational models to evaluate a specific biological activity of compounds in order to filter either existing databases or virtual libraries leading to the identification of molecules with activity against the target of interest.[7] By means of virtual screening of small molecules databases it is possible to identify new potential inhibitors against a target of interest.

Table 6.1 A comparison of HTS and VS approaches

Main difference	**High-throughput screening (HTS)** — **Technology-driven**	**Virtual screening (VS)** — **Knowledge-driven**
Advantages	1) Provides immediately the experimental confirmation for the hit compound 2) No structural information required for the target 3) No binding criteria required	1) Low cost 2) No need for large-scale synthesis of compounds libraries 3) Provides mechanistic information on ligand binding 4) Selectivity criteria of the desired stringency can be easily implemented 5) Generates criteria for compound optimization 6) Generates information easily expandable to other targets 7) Diverse chemistries can be exploited simultaneously
Disadvantages	1) High costs 2) Requires large-scale synthesis of chemical libraries 3) No mechanistic information for ligand binding 4) Limited to the chemistry of the tested library 5) Information not expandable to novel targets	1) Structural information required 2) Provides only virtual hits (need to be experimentally confirmed) 3) Highly dependent on the scoring/filter function 4) Requires precise binding criteria

Homology modeling and structure-based design. A pharmacophore is the description of the three-dimensional (3D) arrangements of essential features enabling a molecule to exert a particular biological effect. Structure-based pharmacophores constitute a very useful tool in drug design, both in the discovery and optimization processes.[8] The structure-based pharmacophore methodologies require the knowledge of the 3D-structure of the ligand–protein target complex. Briefly, the protocol involves the computation of points of minimum energy in the binding site of interest followed by the conversion of such points into pharmacophoric features. With this technology it is theoretically possible to determine all the physical–chemical features that a putative ligand (for example an inhibitor) should possess in order to selectively bind to any given target. Hence, specific pharmacophore models will result from this application.

Molecular dynamics. Another powerful methodology is certainly the molecular dynamics (MD).[9] MD is a computational tool that permits to explore the concerted movements of both small molecules (ligands) and large macromolecular (proteins) entities. MD simulations can be applied to selected ligand candidates in order to accurately calculate binding free energies and to optimize the ligand structure by taking into account the protein flexibility. MD is becoming an ideal companion tool to crystallography for the mechanism of action studies. In fact, while a crystal is a sort of snapshot of the most stable conformation of a protein–ligand complex, MD studies allow investigating the microscopic conformational rearrangements during the process of binding of the inhibitor to the target enzyme.

6.3 The rationale behind the inhibition of DNA polymerases

Tumor cells are characterized by genetic changes, which in turn drive genomic instability. How cancer cells respond to DNA damages has thus important implications for the development of a cancer phenotype. Key players in all DNA damage response and repair pathways are the DNA pols, which are essential to replace lost or damaged pieces of information (DNA sequences). The mistakes made by DNA pols during DNA synthesis, can contribute to the genetic changes driving normal cells towards a tumor phenotype. Thus repair DNA pols are promising targets for innovative chemotherapy.[1]

6.3.1 *The synthetic lethality approach*

Chemotherapy can be defined as the science of selective toxicity. That means that, ideally, the target of any chemotherapeutic approach should be unique to the cell population meant to be killed, while absent in the healthy cells. This goal, however, is not always easily achieved, especially in the context of cancer chemotherapy. In fact, a cancer cell relies on the same enzymatic activities as a normal cell, while differences are mostly due to aberrant expression (up- or down-regulation) of specific enzymes and proteins. Thus, the traditional chemotherapeutic approach aiming at killing proliferating cells, has serious drawbacks. Firstly, not all cancer cells have a high proliferative index; secondly, also healthy cell populations are normally proliferating in the body, for example in bone marrow, and thus are potentially being targeted by chemotherapy with serious toxic side effects.

Novel molecular targeted chemotherapeutic approaches are being developed, with the aim of selectively killing cancer cells, without affecting normal cell populations. In this context, the synthetic lethality approach has recently attracted much attention.

The "synthetic lethality" concept was defined in the context of genetic studies in Drosophila.[10,11] It refers to the fact that loss of one gene function may be tolerated by the cell (or organism) through reliance on another redundant pathway. Compromising this latter by ablating a second gene function, results in cell lethality.

Several years ago, Lee Hartwell and colleagues observed that one of the steps towards tumor transformation, is the loss of an oncosuppressor gene. They realized that cancer cells compensate the loss through redundant gene functions, but blocking the gene product required to compensate for the tumor suppressor loss, could kill cancer cells.[12] This "synthetic lethality" would be specific for the tumor cells, since often in cancer cells both alleles of the oncosuppressor are lost, while they are functional in healthy tissues.

The proof-of-principle for the applicability of the synthetic lethality approach in cancer, was obtained through the development of inhibitors of the of poly(ADP-ribose) polymerase 1 (PARP-1, ARTD1).[13,14] This enzyme plays a key role in initiating the cascade leading to repair of damaged DNA bases. Small nicotinamide analogues inhibitors of ARTD1 were developed.[15,16] Targeting ARTD1 with such inhibitors in tumor cells deficient of the BRCA1 and BRCA2 genes, resulted in selective killing for the cancer cells, with minimal toxicity on wild-type cells.[17] The BRCA1 and BRCA2 proteins are involved in HR-directed repair of DNA double strand breaks (DSBs). These can result from replication forks stalling at interstrand crosslinks lesions, generated by classic anticancer drugs such as cis-Pt. Defects in HR can be compensated by NHEJ, the second main pathway for repair of DSBs. Activation of this process, however, requires the action of ARTD1. Thus, inhibition of ARTD1 in a BRCA1 or BRCA2 genetic null background, concomitantly with cis-Pt treatment, resulted in synthetic lethality of cancer cells.[18,19] Several ARTD1 inhibitors are now in clinical trials for the treatment of BRCA-null ovarian cancer.

Along this line, ideally any protein which is essential in some DNA damage response pathway can be exploited in search of synthetic lethality effects. Since specialized DNA pols are involved in virtually every DNA repair pathway, as we have seen in Chapter 4, they represent an attractive target for synthetic lethality approaches (see below).

Again, proof-of-principle studies have shown that inhibition of DNA pol β or DNA pol γ was synthetically lethal in the context of MSH2- or MLH1-deficiency, respectively.[20] MSH2 and MLH1 are key components of the mismatch-repair (MMR) pathway and involved in the tolerance of oxidative DNA damages. The majority of hereditary nonpolyposis colorectal carcinomas are defective in either gene.

6.3.2 *DNA polymerases in cancer progression and chemoresistance*

As outlined in Chapters 2 and 3, several specialized DNA pols, particularly those of the Y family, are endowed with a very low fidelity on undamaged templates. Conversely, they are capable of performing error-free synthesis over damaged DNA. Thus, careful balance of their levels of expression and tight regulation of their activity is required in order to avoid unwanted effects. Given these facts, it is not surprising that several studies showed that altered expression of these error-prone DNA pols may increase genetic instability, mutagenesis and, as a consequence, accelerate neoplastic transformation.[21] Very often, tumor cells show a complex pattern of deregulation of these DNA pols, with both increase and reduction in the expression of the different enzymes,[22] again indicating the importance of a critical regulation for proper cell homeostasis. As mentioned in Chapter 4, overexpression of DNA pol η, DNA pol β and DNA pol θ has been strongly correlated with enhanced malignancy or poor response to therapy in several cancers (reviewed in Refs. 1 and 3). Additionally, tumor specific genetic variants of these DNA pols have been identified, which may result in increased mutagenesis such as the DNA pol β T889C[23] or R152C[24] variants.

Thus, inhibition of promutagenic DNA pols may represent a useful strategy to counteract genetic instability and prevent or slow down cancer transformation.

Besides promoting genetic instability, TLS DNA pols play a key role in resistance to chemotherapeutic agents damaging DNA. For example, a clear role of DNA pol η and DNA pol β in enhancing tumor chemoresistance to cis-Pt has been documented.[25–28] Similarly, DNA pol ζ has been implicated in resistance to alkylating agents and radiation therapy.[29] Again, targeting these enzymes may be a good strategy to increase tumor sensitivity towards other treatments.

6.4 Inhibitors of family X DNA polymerases

As outlined in Chapters 3 and 4, DNA pol λ and TdT are playing major roles in error-prone pathways of NHEJ of broken DNA ends. TdT has been found overexpressed in several forms of acute leukemia and its overexpression correlates with poor prognosis and low response to chemotherapy.[30] Similarly, DNA pol λ has been found overexpressed in different solid tumors[31] and DNA pol β overexpression led to chromosome instability and hypermutator phenotype in cells.[32] DNA pol β is found overexpressed in many tumor types.[31] Thus, all these three DNA pols are potentially interesting targets for anticancer chemotherapy. Several classes of small molecule inhibitors have been developed for TdT, while identification of synthetic inhibitors of DNA pols λ and β is still at an early stage.

6.4.1 *Terminal deoxynucleotidyl transferase inhibitors*

5-alkylated arabinofuranosyl nucleotides. 1-beta-D-arabinofuranosyluracil 5′-triphosphate (ara-UTP) and its 5-alkylated derivatives 1-beta-D-arabino-furanosylthymine 5′-triphosphate (ara-TTP): 1-beta-D-arabinofuranosyl-5-ethyluracil 5′-triphosphate (ara-EtUTP); 1-beta-D-arabinofuranosyl-

5′-propyluracil 5′-triphosphate (ara-PrUTP); 1-beta-D-arabinofuranosyl-5-butyluracil 5′-triphosphate (ara-BuUTP), are competitive inhibitors of TdT, with respect to the nucleotide dTTP. Inhibition was dependent on the length of the alkyl group at position 5, with long chain substituents (propyl and butyl groups) being the most effective.[33]

2′,3′-dideoxynucleoside 5′-triphosphates. Nucleoside triphosphates lacking the 3′-OH group are the classical chain-terminating dideoxy-nucleotides (ddNTPs), routinely employed in Sanger sequencing reactions (see also Chapter 8). They all inhibit competitively TdT, with respect to the natural dNTPs, blocking DNA elongation. Among 3′-deoxy nucleotides, it has to be mentioned also cordycepin (3′-deoxy adenosine), a derivative of the ribonucleoside adenosine, which once triphosphorylated is effectively incorporated by TdT. In combination with the adenosine deaminase inhibitor deoxycoformycin, it proved to be selectively toxic towards TdT+ leukemic cells.[34]

Carbocyclic nucleoside analogs. Substitution of the ribose moiety with a carbocyclic ring led to the synthesis of the nucleoside analog 1-(6-amino-purin-9-yl)-2-(hydroxymethyl)cyclopentane. Its 1S,2R enantiomer was more active as a TdT inhibitor than the corresponding 1R,2S isomer. Even though the compounds were active in their triphosphate form, no incorporation was detected, suggesting that they inhibited the enzyme forming a dead-end complex at a step preceding chemistry.[35]

5-*nitroindole-2′-deoxyribose-5′-triphosphate. The* 5-nitroindole-2′-deoxyriboside triphosphate analog, has been shown to act as an effective chain terminator for TdT. By generating specific antibodies against its nitroindole moiety, a novel method for DNA detection was developed, based on 3′-labeling of DNA by TdT with this unnatural nucleoside.[36]

Biotinylated nucleotides. Another very useful and generally used method for *in vivo* labeling of broken DNA ends is the TUNEL (TdT-mediated dUTP-biotin nick end-labeling) assay. It is based on the ability of TdT of incorporating nucleotide analogues carrying bulky substituents, such as a biotin moiety. Biotinylated dUTP is normally used as substrate and labelled DNA can be then revealed by interaction of biotin with fluorescinated avidin.[37]

Aryl diketo hexenoic acid. Non-nucleoside inhibitors that do not have a structure resembling a natural nucleoside, have potentially several advantages over nucleoside analogs. They do not require intracellular phosphorylation for activation, thus bypassing one of the rate-limiting step of the intracellular metabolism. In addition, they do not bind to the active site but to specific pockets which are unique to the target enzymes. Also, bearing no resemblance with physiological molecules, they have fewer non-specific interactions with cellular enzymes. These properties make non-nucleoside inhibitors generally more selective for the target and less toxic than nucleoside analogs.

The aryl diketo acids (6-alkyl 2-hydroxy-4-oxo-2,5-hexadienoic acid) derivatives, were found to be potent and selective inhibitors of TdT. Some of them also showed activity against the terminal transferase-like activity of DNA pol λ, but not for template-dependent DNA synthesis.[38] A subsequent series of pyrrolyl- and indolyl- diketo hexenoic acid derivatives were found to be extremely selective for TdT only. Kinetic studies showed an unusual competitive inhibitory mechanism with respect to the dNTPs. Crystal structures of TdT in complex with two of these inhibitors, showed that they bind close to the dNTP binding site, with their aromatic rings occluding the position normally occupied by the nucleotide base moiety. Thus, they represent the first nucleotide-competing non-nucleoside inhibitors of TdT. These compounds were toxic to TdT+ MOLT-4 leukemic cells, but not towards TdT- HeLa cells. They induced cell cycle arrest at the G1/G2 border and apoptosis.[39]

6.4.2 *DNA polymerase λ and β inhibitors*

Non-nucleoside analogs. Until recently, known non-nucleoside inhibitors of DNA pols β and λ were only of natural origin (see Section 6.6) and were usually not very selective, being able to target also other DNA pols. A study identified the monophosphate form of the immunosuppressive drug bredinin as an inhibitor of DNA pols α and β, able to compete with both the nucleotide and nucleic acid substrates.[40] In an effort for finding specific DNA pol λ and β inhibitors, a high-throughput screening of a chemical library of about 9,000 synthetic small molecules, allowed the identification of 11 hit compounds inhibiting DNA pols β and/or λ, belonging to three different classes: rhodanines, carbohydrazides and 2,4-pentadiones. Rhodanine derivatives were found to be the most potent and selective non-nucleoside inhibitors of DNA pol λ, with respect to DNA pol β.[41] A subsequent expansion of this work led to the design and synthesis of additional rhodanine-based inhibitors of DNA pol λ and/or β, with affinities in the low micromolar range. These compounds were also able to sensitize cancer cells to the approved cancer drug temozolomide.[42] *In silico* structure-based screening of a library of 140,000 small molecules, led to the identification of compound NSC-124854, a methoxyphosphonic acid derivative, as a potent inhibitor of DNA pol β able to sensitize cancer cells to temozolomide treatment.[43] Recently, *in silico* screening of a commercial database of 35 millions compounds led to the identification of 5-methoxyflavones as DNA pol β inhibitors, which were able to protect cultured neuronal cells from beta-amyloid toxicity.[44]

Nucleoside analogs. No selective nucleoside inhibitors for human repair DNA pols have been developed so far. However, several nucleoside analogs with anticancer and antiviral activity can be incorporated by human DNA pols β and λ. For example, the triphosphate form of the anticancer drug cytarabine (AraCTP) can be efficiently incorporated by DNA pol λ[45] and DNA pol β.[46] Also the triphosphate forms of several anti-HIV nucleoside analogs can be incorporated by DNA pols λ and β.[47,48]

Overall, it seems that DNA pols β and λ may be responsible for a significant part of the cytotoxic effects of these anticancer and antiviral drugs.

6.5 Inhibitors of family Y DNA polymerases

Due to the important role of TLS in the DNA damage tolerance and the potentially mutagenic outcome of the bypass events, TLS-specialized enzymes are being regarded as potential chemotherapy targets.[49] Unfortunately, development of specific inhibitors for Y-family DNA pols η, ι, κ and Rev1 is still in its infancy. The methanocarba-2′-deoxy nucleoside analog MC-dATP, has been found to act as a substrate for DNA pol ι, blocking further chain elongation. This compound has been proposed as a potential drug for the treatment of tumors where DNA pol ι is overexpressed such as breast cancer and esophageal carcinoma.[50] Y-family DNA pols have been shown also to incorporate the triphosphate forms of anti-HIV nucleoside analogs with high efficiency.[47,48] The synthetic nucleoside triphosphate analogs Benzi-TP and BIMTP were incorporated by DNA pol η opposite normal G or a cis-Pt-1,2-d(GpG) intrastrand crosslink.[51] Screening of a library of about 320 small molecules, led to the identification of nonnucleoside indole-thiobarbituric acid derivatives (ITBA) as selective inhibitors of DNA pol η. These compounds were allosteric mixed-type competitors of nucleotide binding.[52]

6.6 Natural inhibitors of DNA polymerases

Given the extraordinary diversity of chemical entities present in nature, intensive efforts have been devoted to the identification of molecules of natural origins, either from plants or from various microrganisms, able to inhibit cellular DNA pols. These efforts led to the isolation of dozens of natural compounds with anti-proliferative activity and with DNA pols inhibition capability in *in vitro* assays. However, one major shortcoming of these compounds is the lack of absolute specificity, which enables them to act upon DNA pols from different families. Some examples of the most recently identified and selective natural inhibitors of specialized DNA pols are given below.

Petasiphenol. This polyhydroxylated aromatic compound is derived from the Japanese plant *Petasites japonicus*. It has been shown to inhibit DNA pol λ, but not DNA pols α, δ, ε, β and TdT. Its mechanism of action is non-competitive with respect to both primer/template and dNTPs substrates.[53]

Curcumin. Curcumin (diferuloylmethane) is a phenolic derivative isolated from the plant *Curcuma longa* with anti-inflammatory properties. It has been shown to inhibit specifically DNA pol λ in a non-competitive manner, albeit with low potency (ID50 = 70 μM).[54]

Phytosterols. Screening of plant crude extracts allowed the identification of stigmasterol, β-sitosterol and β-sitosterol-β-D-glucoside as potent inhibitors of the dRP-lyase activity of DNA pol β. These natural compounds, extracted from *Cladogynus orientalis*, *Hymenache donacifolia*, *Heteropsis integerrima*, and *Acacia pilispina,* were also found to potentiate the anti-cancer effects of bleomycin on cultured cells.[55]

Triterpenoid derivatives. Isomalabaricane triterpenoids isolated from the sponge *Rhabdastrella globostellata*, were found to inhibit DNA pol β by inducing a dead-end complex of the enzyme with the DNA substrate. These compounds showed cytotoxic activity towards an ovarian cancer cell line.[56]

Funicone derivative. Screening of extracts from the mycelia of different marine fungal strains from Australia, allowed the identification of 3-O-methylfunicone as a selective inhibitor of mammalian Y-family DNA pols. The most sensitive was DNA pol κ, while no effects were found against family A, B or X DNA pols. This inhibitor also suppressed proliferation of colon carcinoma cells.[57]

Acylated steryl glycosides. From extracts of soybean (*Glycine max L.*), the acylated steryl glycoside β-sitosteryl (6′-O-linoleoyl)-glucoside was isolated and shown to inhibit Y-family DNA pols, without effects on family A, B or X enzymes. DNA pol κ was the most affected. The inhibition was non-competitive with respect to both the nucleotide and primer/template substrates.[58]

Pinophilins and Penicilliols. Cultures from two sea-water fungi (*Penicillium daleae* and *Penicillium pinophilum*) led to the identification of 5-methoxy-furanones and azaphilones compounds with DNA pols inhibitory activity. The 5-methoxy furanones were found selective inhibitors of Y-family DNA pols, while the azaphilones were found specific for family A, B and Y enzymes, with no activity against family X DNA pols. Azaphilones were competitive inhibitors with respect to the nucleotide substrate, while 5-methoxy-furanones were non-competitive inhibitors.[59,60]

6.7 Chapter summary

Beside a role in DNA replication, DNA pols play essential roles in DNA repair and DNA damage tolerance repair pathways. Many specialized DNA pols acting in such pathways are potentially mutagenic. We know that deregulation of the activity or expression levels of these DNA pols have a role in promoting tumorigenesis. In addition, specialized DNA pols are important for the tolerance by cancer cells of the DNA damages induced by conventional chemotherapeutic agents. Due to genetic mutations, many cancer types are dependent on a fewer number of DNA repair pathways than normal cells. Disabling a DNA pol acting in these pathways, may thus result in higher sensitivity to DNA damaging agents or even in synthetic lethality. Thus, selective inhibition of repair DNA pols constitutes a promising approach to develop anticancer drugs.

Random screenings of large libraries of molecules in enzymatic assays have allowed, so far, the identification of a relatively small number of selective DNA pols inhibitors. However, in the near future, computer-aided pharmacophore identification starting from structural information, together with *in silico* screening and automated docking procedures, is expected to foster the identification of small molecules inhibitors, specifically tailored to a particular DNA pol.

References

1. Maga, G. & Hubscher, U. Repair and translesion DNA polymerases as anticancer drug targets. *Anti-Cancer Agent Me* **8**, 431–447 (2008).
2. Crespan, E., Garbelli, A., Amoroso, A. & Maga, G. Exploiting the Nucleotide Substrate Specificity of Repair DNA Polymerases To Develop Novel Anticancer Agents. *Molecules* **16**, 7994–8019 (2011).
3. Lange, S. S., Takata, K. & Wood, R. D. DNA polymerases and cancer. *Nat Rev Cancer* **11**, 96–110 (2011).
4. Mentegari, E., Kissova, M., Bavagnoli, L., Maga, G. & Crespan, E. DNA Polymerases λ and β: The Double-Edged Swords of DNA Repair. *Genes-Basel* **7**, E57 (2016).
5. Cozzarelli, N. R. The mechanism of action of inhibitors of DNA synthesis. *Annual review of biochemistry* **46**, 641–668 (1977).
6. Sanger, F., Nicklen, S. & Coulson, A. R. DNA sequencing with chain-terminating inhibitors. *P Natl Acad Sci USA* **74**, 5463–5467 (1977).
7. Kirchmair, J. *et al.* Enhancing drug discovery through in silico screening: Strategies to increase true positives retrieval rates. *Current Medicinal Chemistry* **15**, 2040–2053 (2008).
8. Dror, O., Shulman-Peleg, A., Nussinov, R. & Wolfson, H. J. Predicting molecular interactions in silico: I. A guide to pharmacophore identification and its applications to drug design. *Current Medicinal Chemistry* **11**, 71–90 (2004).
9. Klepeis, J. L., Lindorff-Larsen, K., Dror, R. O. & Shaw, D. E. Long-timescale molecular dynamics simulations of protein structure and function. *Current Opinion in Structural Biology* **19**, 120–127 (2009).

10. Dobzhansky, T. Genetics of natural populations; recombination and variability in populations of Drosophila pseudoobscura. *Genetics* **31**, 269–290 (1946).
11. Lucchesi, J. C. Synthetic lethality and semi-lethality among functionally related mutants of Drosophila melanfgaster. *Genetics* **59**, 37–44 (1968).
12. Hartwell, L. H., Szankasi, P., Roberts, C. J., Murray, A. W. & Friend, S. H. Integrating genetic approaches into the discovery of anticancer drugs. *Science (New York, NY)* **278**, 1064–1068 (1997).
13. Audeh, M. W. *et al.* Oral poly(ADP-ribose) polymerase inhibitor olaparib in patients with BRCA1 or BRCA2 mutations and recurrent ovarian cancer: a proof-of-concept trial. *Lancet (London, England)* **376**, 245–251 (2010).
14. Tutt, A. *et al.* Oral poly(ADP-ribose) polymerase inhibitor olaparib in patients with BRCA1 or BRCA2 mutations and advanced breast cancer: a proof-of-concept trial. *Lancet (London, England)* **376**, 235–244 (2010).
15. Virag, L. & Szabo, C. The therapeutic potential of poly(ADP-ribose) polymerase inhibitors. *Pharmacol Rev* **54**, 375–429 (2002).
16. Southan, G. J. & Szabo, C. Poly(ADP-ribose) polymerase inhibitors. *Current medicinal chemistry* **10**, 321–340 (2003).
17. Bryant, H. E. & Helleday, T. Poly(ADP-ribose) polymerase inhibitors as potential chemotherapeutic agents. *Biochem Soc Trans* **32**, 959–961 (2004).
18. Ashworth, A. A synthetic lethal therapeutic approach: poly(ADP) ribose polymerase inhibitors for the treatment of cancers deficient in DNA double-strand break repair. *Journal of clinical oncology: official journal of the American Society of Clinical Oncology* **26**, 3785–3790 (2008).
19. Lord, C. J. & Ashworth, A. Targeted therapy for cancer using PARP inhibitors. *Current opinion in pharmacology* **8**, 363–369 (2008).
20. Martin, S. A. *et al.* DNA Polymerases as Potential Therapeutic Targets for Cancers Deficient in the DNA Mismatch Repair Proteins MSH2 or MLH1. *Cancer Cell* **17**, 235–248 (2010).
21. Kunkel, T. A. Considering the cancer consequences of altered DNA polymerase function. *Cancer Cell* **3**, 105–110 (2003).
22. Hoffmann, J. S. & Cazaux, C. Aberrant expression of alternative DNA polymerases: A source of mutator phenotype as well as replicative stress in cancer. *Semin Cancer Biol* **20**, 312–319 (2010).
23. Tan, X. *et al.* Clinical significance of a point mutation in DNA polymerase beta (POLB) gene in gastric cancer. *Int J Biol Sci* **11**, 144–155 (2015).
24. Zhou, T. *et al.* R152C DNA Pol beta mutation impairs base excision repair and induces cellular transformation. *Oncotarget* **7**, 6902–6915 (2016).

25. Canitrot, Y. *et al.* Overexpression of DNA polymerase beta in cell results in a mutator phenotype and a decreased sensitivity to anticancer drugs. *Proc Natl Acad Sci USA* **95**, 12586–12590 (1998).
26. Yang, J. *et al.* Cells deficient in the base excision repair protein, DNA polymerase beta, are hypersensitive to oxaliplatin chemotherapy. *Oncogene* **29**, 463–468 (2010).
27. Srivastava, A. K. *et al.* Enhanced expression of DNA polymerase eta contributes to cisplatin resistance of ovarian cancer stem cells. *P Natl Acad Sci USA* **112**, 4411–4416 (2015).
28. Teng, K.-Y. *et al.* DNA polymerase eta protein expression predicts treatment response and survival of metastatic gastric adenocarcinoma patients treated with oxaliplatin-based chemotherapy. *J Transl Med* **8**, 126 (2010).
29. Shi, T.-Y. *et al.* DNA polymerase zeta as a potential biomarker of chemoradiation resistance and poor prognosis for cervical cancer. *Med Oncol* **30**, 500 (2013).
30. Di Santo, R. & Maga, G. Human terminal deoxynucleotidyl transferases as novel targets for anticancer chemotherapy. *Curr Med Chem* **13**, 2353–2368 (2006).
31. Albertella, M. R., Lau, A. & O'Connor, M. J. The overexpression of specialized DNA polymerases in cancer. *DNA Repair (Amst)* **4**, 583–593 (2005).
32. Frechet, M. *et al.* Deregulated DNA polymerase beta strengthens ionizing radiation-induced nucleotidic and chromosomal instabilities. *Oncogene* **21**, 2320–2327 (2002).
33. Ono, K., Ogasawara, M. & Nakane, H. Inhibition of terminal deoxynucleotidyltransferase by various 5-alkylated derivatives of 1-beta-arabinofuranosyluracil 5′-triphosphate: substituent effects on inhibitory action. *Biomed Pharmacother* **37**, 288–292 (1983).
34. McCaffrey, R. *et al.* Selective killing of leukemia cells by inhibition of TdT. *Haematol Blood Transfus* **28**, 24–27 (1983).
35. Theil, F., Ballschuh, S., Flatau, S., von Janta-Lipinski, M. & Matthes, E. Chemo-enzymatic synthesis of a new type of enantiomerically pure carbocyclic nucleoside analogues with strong inhibitory effects on terminal deoxynucleotidyl transferase. *Bioorg Med Chem* **6**, 701–706 (1998).
36. Smith, C. L. *et al.* Use of 5-nitroindole-2′-deoxyribose-5′-triphosphate for labelling and detection of oligonucleotides. *Nucleosides Nucleotides* **17**, 555–564 (1998).
37. Gorczyca, W., Gong, J. & Darzynkiewicz, Z. Detection of DNA strand breaks in individual apoptotic cells by the in situ terminal deoxynucleotidyl transferase and nick translation assays. *Cancer Res* **53**, 1945–1951 (1993).

38. Locatelli, G. A. *et al.* Diketo hexenoic acid derivatives are novel selective non-nucleoside inhibitors of mammalian terminal deoxynucleotidyl transferases, with potent cytotoxic effect against leukemic cells. *Mol Pharmacol* **68**, 538–550 (2005).
39. Costi, R. *et al.* 6-aryl-2,4-dioxo-5-hexenoic acids, novel integrase inhibitors active against HIV-1 multiplication in cell-based assays. *Bioorg Med Chem Lett* **14**, 1745–1749 (2004).
40. Horie, T. *et al.* A 5′-monophosphate form of bredinin selectively inhibits the activities of mammalian DNA polymerases in vitro. *Int J Mol Med* **1**, 83–90 (1998).
41. Strittmatter, T. *et al.* Small Molecule Inhibitors of Human DNA Polymerase lambda. *ACS Chem Biol* **6**, 314–319 (2011).
42. Strittmatter, T. *et al.* Expanding the scope of human DNA polymerase lambda and beta inhibitors. *ACS Chem Biol* **9**, 282–290 (2014).
43. Jaiswal, A. S. *et al.* DNA polymerase beta as a novel target for chemotherapeutic intervention of colorectal cancer. *Plos One* **6**, e16691 (2011).
44. Merlo, S. *et al.* Identification of 5-Methoxyflavone as a Novel DNA Polymerase-Beta Inhibitor and Neuroprotective Agent against Beta-Amyloid Toxicity. *J Nat Prod* **78**, 2704–2711 (2015).
45. Garcia-Diaz, M., Murray, M. S., Kunkel, T. A. & Chou, K. m. Interaction between DNA Polymerase and Anticancer Nucleoside Analogs. *Journal of Biological Chemistry* **285**, 16874–16879 (2010).
46. Prakasha Gowda, A. S., Polizzi, J. M., Eckert, K. A. & Spratt, T. E. Incorporation of Gemcitabine and Cytarabine into DNA by DNA Polymerase β and Ligase III/XRCC1. *Biochemistry* **49**, 4833–4840 (2010).
47. Brown, J. A., Pack, L. R., Fowler, J. D. & Suo, Z. Pre-steady-state kinetic analysis of the incorporation of anti-HIV nucleotide analogs catalyzed by human X- and Y-family DNA polymerases. *Antimicrob Agents Ch* **55**, 276–283 (2011).
48. Vyas, R., Zahurancik, W. J. & Suo, Z. Structural basis for the binding and incorporation of nucleotide analogs with L-stereochemistry by human DNA polymerase lambda. *P Natl Acad Sci USA* **111**, E3033–3042 (2014).
49. Tomasso, A., Casari, G. & Maga, G. What makes y family pols potential candidates for molecular targeted therapies and novel biotechnological applications. *Curr Mol Med* **14**, 96–114 (2014).
50. Eoff, R. L. *et al.* Selective modulation of DNA polymerase activity by fixed-conformation nucleoside analogues. *Angew Chem Int Ed Engl* **49**, 7481–7485 (2010).

51. Nilforoushan, A., Furrer, A., Wyss, L. A., van Loon, B. & Sturla, S. J. Nucleotides with altered hydrogen bonding capacities impede human DNA polymerase eta by reducing synthesis in the presence of the major cisplatin DNA adduct. *Journal of the American Chemical Society* **137**, 4728–4734 (2015).
52. Coggins, G. E. *et al.* N-Aroyl indole thiobarbituric acids as inhibitors of DNA repair and replication stress response polymerases. *ACS Chem Biol* **8**, 1722–1729 (2013).
53. Mizushina, Y. *et al.* Petasiphenol: a DNA polymerase lambda inhibitor. *Biochemistry* **41**, 14463–14471 (2002).
54. Mizushina, Y. *et al.* Monoacetylcurcumin: a new inhibitor of eukaryotic DNA polymerase lambda and a new ligand for inhibitor-affinity chromatography. *Biochem Biophys Res Commun* **337**, 1288–1295 (2005).
55. Li, S. S., Gao, Z., Feng, X., Jones, S. H. & Hecht, S. M. Plant sterols as selective DNA polymerase beta lyase inhibitors and potentiators of bleomycin cytotoxicity. *Bioorg Med Chem* **12**, 4253–4258 (2004).
56. Clement, J. A., Li, M., Hecht, S. M. & Kingston, D. G. Bioactive isomalabaricane triterpenoids from Rhabdastrella globostellata that stabilize the binding of DNA polymerase beta to DNA. *J Nat Prod* **69**, 373–376 (2006).
57. Mizushina, Y. *et al.* 3-O-methylfunicone, a selective inhibitor of mammalian Y-family DNA polymerases from an Australian sea salt fungal strain. *Mar Drugs* **7**, 624–639 (2009).
58. Horie, S. *et al.* beta-Sitosteryl (6′-O-linoleoyl)-glucoside of soybean (Glycine max L.) crude extract inhibits Y-family DNA polymerases. *J Oleo Sci* **59**, 621–630 (2010).
59. Myobatake, Y. *et al.* Pinophilins A and B, inhibitors of mammalian A-, B-, and Y-family DNA polymerases and human cancer cell proliferation. *J Nat Prod* **75**, 135–141 (2012).
60. Kimura, T. *et al.* Penicilliols A and B, novel inhibitors specific to mammalian Y-family DNA polymerases. *Bioorganic & Medicinal Chemistry* **17**, 1811–1816 (2009).

Chapter 7

Polymerase Chain Reaction and Heat-Stable DNA Polymerases: The History and the Potential of Evolved DNA Polymerases

It was in 1985 when, thanks to the enzymatic properties of DNA pols, the paramount technology of polymerase chain reaction[1] was developed in a Californian company named Cetus Inc.

Hindsight — basic science with DNA pols at a medically oriented faculty of Zürich in 1975–1980. When Spadari and Hübscher were investigating the function of mammalian DNA polymerases (pols) α, β, and γ between 1974–1978 (for details see Chapter 1), the latter, who was a graduate student at the Department of Veterinary Pharmacology and Biochemistry of the University of Zürich, was often told by a number of pathologists, microbiologists, parasitologists, physiologists and nutritionists not to lose his time on such "sub-molecular nonsense". Basically the comments, not made, however, by clinicians, were as follows: "It is impossible that a simple biochemical molecule such as an enzyme called DNA pol can ever be useful for a practical application". Here is the story in Hübscher's words: "At this time (1974) I tried to promote my modest career (almost a decade before PCR was discovered) as a simple veterinarian interested in basic science and in particular in an enzyme named DNA pol. I realized that I felt in love with this enzyme. With this idea I was immediately attacked by some considered as opinion leaders at that time at our faculty saying: 'Hübscher, are you crazy to work on an issue that will never be of any practical use?' Fortunately, my supervisors let me do my work. Forty years later, as "DNA pols scientists", we do not have to apologize anymore! The practical development of the use of a DNA pol in PCR documents in a convincing way that the scientific community has always to be open to issues that are not immediately obvious to everyone. In other words the DNA pol studies led not only to the polymerase chain reaction (PCR), but also to the DNA sequencing technology invented by Fred Sanger some years earlier. Nowadays the DNA pols serve millions of people and animals as diagnostic and prophylactic tools and have certainly saved thousands of lives. Moreover, I am sure that the successors of the mentioned Departments are now daily using PCR and DNA sequencing!"

But let's now go back to the paramount issue of PCR.

Concept of PCR: When the concept of PCR was created by Kary Mullis in 1983, the basic properties of DNA pols were known. A DNA pol requires a template, a primer to start DNA synthesis, the four deoxyribonucleoside triphosphates, magnesium, and certain additives such as buffer, salt and others. The concept of PCR was not to invent novel properties of DNA pols but rather to use them in a reaction that does not generally occur in nature, namely amplification. When Mullis first brought the PCR idea in 1983 to Cetus, a biotech company founded in 1971 in the San Francisco Bay area, the enthusiasm was not overwhelming. Nevertheless, in collaboration with the hard working scientists and technicians such as Henry Ehrlich, Norman Arnheim, Randall K. Saiki, Glen Horn, Corey Levenson, Steven Scharf, Fred Faloona and Tom White, the company achieved within two years reproducible amplification of DNA. This led to the first patent application on March 28th, 1985. On September 20th, 1985 a paper was sent to *Science* and appeared on 20th December 1985 entitled "Enzymatic amplification of beta-globin genomic sequences and restriction site analysis for diagnosis of sickle cell anaemia" by R K. Saiki, S. Scharf, F. Faloona, K.B. Mullis, G. Horn, H.A. Ehrlich and N. Arnheim.[1] In this paper an application but not the technique of PCR *per se* was the main topic. Mullis tried to publish at the same time the PCR method in

Nature and *Science* and both journals rejected his work as insufficient for a top journal. Under the title "Specific synthesis of DNA *in vitro* via polymerase-catalysed chain reaction" Mullis and Faloona published their results in 1987 in *Methods in Enzymology*,[2] meaning that one of the most revolutionary technology concepts in medical science and biotechnology was not published via a "regular" referee process! Nevertheless, the inventor of the PCR concept Kary Mullis was eventually awarded the Noble Prize for Chemistry in 1993.

The *three steps* of the PCR cycle are as follows: first, denature the DNA to be amplified at around 95°C; second, anneal an excess of primers between 50–70°C depending on their sequences and, third, synthesize the DNA by the DNA pol. However, here came the initial problem: Because *E. coli* DNA pol I was used in the early days of PCR, fresh enzyme had to be added after each cycle, because it was inactivated upon denaturation and hybridization of DNA. This was not only laborious and time-consuming, but also enhanced the danger of carryover contamination, a problem that in the early days of PCR was common to many laboratories.

Around 1985 the idea came up at Cetus to purify a heat-thermostable DNA pol. Kary Mullis apparently refused to isolate this DNA pol, but David Gelfand and his technical assistant Susanne Stoffel purified in a very short time a DNA pol of about 90,000 daltons that was heat-stable and free of contaminating endo- and exonucleases. In addition, the enzyme was unexpectedly very efficient in adding dNTP's at 70–75°C when non-ionic detergents were added. With this enzyme from the organism *Thermus aquaticus*, now called *Taq* DNA pol, they were quickly in a position to develop and to optimize single tube PCR for over 30 cycles. Moreover, Gelfand and Stoffel soon cloned and successfully over-expressed *Taq* DNA pol in *E. coli* so that large amounts of extremely pure enzyme became available to the scientific community. This made the PCR technology spread over the globe in a very short time.[3,4] It was this great discovery of heat-stable DNA pol for PCR that exponentially increased the scientific publications using PCR after 1990. The last December 1989 issue of *Science* praised "*DNA polymerase as the molecule of the year*"[5] and this attribution was due to the spreading of the PCR technology as a result of its ease thanks to heat-stable DNA pols.

PCR has deeply changed the daily practices in many laboratories. Thousands of scientists very quickly began using PCR for various purposes. Many novel applications appeared, such as nested PCR, single-molecule amplification, inverse PCR, multiplex amplification, direct DNA sequencing, quantification of virus and bacterial load, single gamete typing, reverse-transcriptase PCR (known as single-enzyme RT-PCR), dUMP incorporation to prevent carry over, combinatorial libraries, aptamers, amplification from ancient DNA, *in situ* PCR, sequence-tagged sites, long PCR, accurate PCR by using a combination of DNA pol and 3′→5′ exonuclease. These applications led to a "golden age" of research on thermophiles, because many DNA pols appeared with different properties. Moreover, site-directed mutagenesis allowed engineering of DNA pols with desired properties (see also Chapter 8: Synthetic evolution of DNA polymerases for novel properties). In conclusion, PCR has changed the daily work not only in scientific laboratories of topics from basic to applied, but also in practical terms, because many diagnostic and forensic laboratories now use it routinely.

Development in the last six years: many novel and very useful applications of DNA pols in biotechnology appeared (for reviews see Refs. 6–8 and Chapter 8). In short, these reviews describe the combination of novel properties of DNA pols in the fields of PCR, next-generation sequencing (NGS), DNA cloning, single nucleotide polymorphism (SNP) detection, whole genome amplification (WGA), synthetic biology as well as many fields of molecular diagnostics (e.g. in presence of inhibitors present in specimens to be analysed).

Moreover, directed evolution of DNA pols has provided a revolution in biotechnology and in medical research.[9] As will be outlined in Chapter 9 in more detail, screening in combination with sophisticated selection methods,[9] such as genetic complementation and autogene selection, activity-based phage display, compartmentalised self-replication (CSR) and compartmentalised self-tagging (CST), will generate DNA pols with novel properties. Evolved DNA pol mutants have interesting properties for many applications.[9] (and they will be discussed in more details with reference to their authors in Chapter 8): altered fidelity, increased thermal stability, enhanced inhibitor tolerance, altered repertoires for substrates (modified dNTPs, modified sugars, dNTPs with modified nucleobases, modified template DNA), and the capacity to perform PCR of damaged and ancient DNA. Finally, mutants within one DNA pol are in many occasions combined with mutants from another DNA pol and are either blends of DNA pols or even heterodimeric enzymes.

In a remarkable work the Philipp Holliger group has synthesized genetic polymers capable of heredity and evolution.[10] Briefly, they showed that genetic information can be stored in and recovered from six alternative genetic polymers based on simple nucleic acid architectures not found in nature. These polymers are called xeno-nucleic acids (XNA's). Selected XNA aptamers have been shown to have, beyond heredity, also the capacity for Darwinian evolution and folding into defined novel structures. Thus, it is argued that heredity and evolution, two hallmarks of life, are not limited to DNA and RNA but are likely to be emergent properties of polymers capable of information storage.

Finally, a technique that has been initially described in 2000 by Tsugunori Notomi *et al.*,[11] has been named loop-mediated isothermal amplification (LAMP) of DNA. As will be discussed in Chapters 8 and 9 this technology has been introduced in a variety of applications. The main advantage is that DNA can be amplified under *isothermal conditions* with high selectivity, high efficiency and high speed. In addition, as we will also see in Chapters 8 and 9, further variations will include rolling circle amplification (RCA), recombinase polymerase amplification (RPA), nicking enzyme amplification reaction (NEAR) and helicase-dependent amplification (HDA).

References

1. Saiki, R. K. *et al.* Enzymatic amplification of beta-globin genomic sequences and restriction site analysis for diagnosis of sickle cell anemia. *Science* **230**, 1350–1354 (1985).
2. Mullis, K. B. & Faloona, F. A. Specific synthesis of DNA in vitro via a polymerase-catalyzed chain reaction. *Methods Enzymol* **155**, 335–350 (1987).
3. Saiki, R. K. *et al.* Primer-directed enzymatic amplification of DNA with a thermostable DNA polymerase. *Science* **239**, 487–491 (1988).

4. Lawyer, F. C. *et al.* Isolation, characterization, and expression in Escherichia coli of the DNA polymerase gene from Thermus aquaticus. *J Biol Chem* **264**, 6427–6437 (1989).
5. Guyer, R. L. & Koshland, D. E., Jr. The Molecule of the Year. *Science* **246**, 1543–1546 (1989).
6. Gardner, A. F. & Kelman, Z. DNA polymerases in biotechnology. *Frontiers in Microbiology* **5**, 659, doi:10.3389/fmicb.2014.00659 (2014).
7. Zhang, L., Kang, M., Xu, J. & Huang, Y. Archaeal DNA polymerases in biotechnology. *Applied Microbiology and Biotechnology* **99**, 6585–6597, doi:10.1007/s00253-015-6781-0 (2015).
8. Ishino, S. & Ishino, Y. DNA polymerases as useful reagents for biotechnology — the history of developmental research in the field. *Frontiers in Microbiology* **5**, 465, doi:10.3389/fmicb.2014.00465 (2014).
9. Chen, T. & Romesberg, F. E. Directed polymerase evolution. *FEBS Letters* **588**, 219–229, doi:10.1016/j.febslet.2013.10.040 (2014).
10. Pinheiro, V. B. *et al.* Synthetic genetic polymers capable of heredity and evolution. *Science* **336**, 341–344, doi:10.1126/science.1217622 (2012).
11. Notomi, T. *et al.* Loop-mediated isothermal amplification of DNA. *Nucleic Acids Research* **28**, E63 (2000).

Chapter 8

Synthetic Evolution of DNA Polymerases for Novel Properties

8.1 Why design enzymes with novel properties?

Most chemical reactions in a cell are catalyzed by enzymes possessing a high substrate specificity that has evolved over billions of years. The versatility of the thousands of enzymes is enormous not only in their specificity, but also in their adaptive behavior to altered environmental and intracellular conditions. As indicated in a News and Views note in *Nature*[1] "*the ability of designing new enzymes on the basis of our knowledge of protein structure and reaction mechanism can charitably be described as primitive*". The first approaches to use enzymes came from industrial applications with the purpose to reduce high temperatures for catalytic chemical processes and to circumvent extremes of pH. Moreover, reaction specificity can be increased, a more pure product can be obtained and last but not least the environment can be protected upon applying "mild" enzymatic conditions compared to the harsh chemical reactions with many side products. The growing use of industrial enzymes is dependent on innovation to improve performance and to reduce costs.[2]

As described in detail in Chapter 7 the polymerase chain reaction (PCR) has immediately brought the DNA pols into the very center of biotechnical applications. They include many novel applications for thousands of different purposes, such as nested PCR, single-molecule amplification, inverse PCR, multiplex amplification, direct DNA sequencing, quantification of virus and bacterial load, single gamete typing, reverse transcriptase PCR (known as single enzyme RT-PCR), dUMP incorporation to prevent carry over, combinatorial libraries, SELEX to generate aptamers and ribozymes, amplification from ancient DNA, *in situ* PCR, sequence tagged sites, long PCR, accurate PCR by using a combination of DNA polymerase and 3′→5′ exonuclease. Moreover, the research on thermophiles came into its golden age, since with the introduction of heat stable DNA pols in PCR many thermostable DNA pols appeared with certain different properties. Even though we have learned a lot by structural studies as well as biochemical analyses on how DNA pols function, synthetic evolutionary approaches have given insight into the importance of certain amino-acids and motifs in the active site of DNA pols.

In this Chapter 8 we discuss:

1. the tight active site of DNA pols to which the substrates fit,
2. methods to evolve DNA pols with novel and improved properties,
3. applications of DNA pols with novel properties and
4. DNA pols with at least 36 novel properties (for details see Section 8.5 and references therein) have been generated.

8.2 DNA polymerases have a tight active site to which the substrates fit

As already discussed in Chapter 2 DNA pols have a form of a human right hand. The structures of many DNA pols have been solved in great details (reviewed in Refs. 3–5). On the other hand DNA pols are complex enzymes that have to adopt in their active site a DNA template, a primer and the four dNTPs. In view of this complexity one wonders how the active site of DNA pols is geometrically constructed to adopt all four dNTPs. It was argued that the Watson-Crick hydrogen bonds themselves are not necessary for efficient and accurate copy of a base pair by the DNA pol (reviewed in Ref. 6). Important issues in this connection appear to be the minor groove hydrogen bonding, base stacking, water solvation and steric effects. Many different approaches were tested to address this issue (see Ref. 6 and references therein). The first direction is the testing of nucleoside analogs as substrate probes for DNA pols. Such analogs include methylated bases, construction of bases with altered shapes and sizes that do not contain any hydrogen bonding, base analogs that still contain the hydrogen bonding but with an increased size of the base and sterically augmented sugars. Eric Kool concluded that "*among the non-covalent interaction in the active site, steric effects are the most likely candidates for the lion's share of fidelity during DNA synthesis. The steric effects depend on the complementarity of size and shape of the two bases being paired, and on the tightness of the polymerase active site under study. Watson-Crick hydrogen bonding groups may also contribute to fidelity, but may do so in part because of the steric effect of water bound to them.*"[6] Since it is known that different DNA pols have different active site geometries and thus different accuracies, one can expect that different modified bases have different effects on different DNA pols.

In general it is evident from Figure 8.1, that the four combinations of Watson-Crick base pairing have only little variation in the *overall* shape.

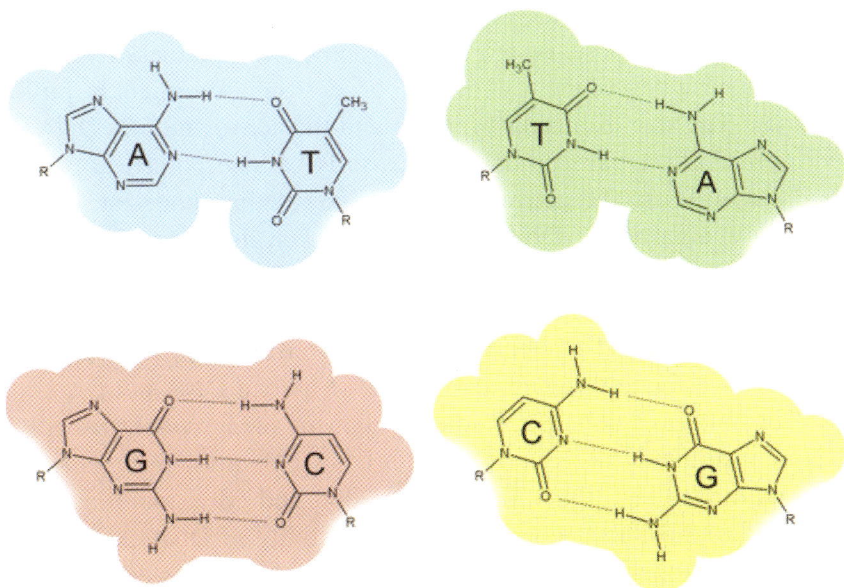

Figure 8.1 **Space filling shapes of the four different naturally occurring base pairs.** For details see text and Ref. 6.

Figure 8.2 **Overlay form of the four base pair shapes from Figure 8.1**. Red: G-C; yellow: C-G; blue: A-T and green: T-A. For details see text and Ref. 6.

If, however, the four shapes are overlayed one can see from Figure 8.2 that there is considerable variation in the minor groove and at the edges of the major groove. It was suggested that the active site tightness hypothesis predicts to what degree the wall of this consensus pocket might be pushed outwards.[6] The size exclusion hypothesis in the active site of a DNA pol predicts that a base pair can only successfully be incorporated if the predicted shape fits in the active site. In summary, this model clearly can predict how a particular DNA pol, depending on its shape in the active site, can adapt different types of base pairs.

Francesca Di Pasquale *et al.* used 4′alkylated nucleotides and primers containing 4′alkylated nucleotides at the 3′-terminal position as steric probes to investigate different active sites of human DNA pol β and the 3′→5′ exonuclease deficient Klenow fragment of *E. coli* DNA pol I.[7] Kinetic experiments suggested that both DNA pols must vary considerably in the active site tightness. Small methyl and ethyl modification in nucleoside triphosphates compromised the activity of DNA pol β, while the extension from the modified primer is only slightly disturbed. When, however, these experiments were conducted with the 3′→5′ exonuclease deficient Klenow fragment of *E. coli* DNA pol I, incorporation of the modified nucleotide was only slightly reduced, while modified primers, depending on the size of the modification, reduced the catalytic efficiency by several orders of magnitude. These results clearly suggested that there is indeed a profound influence of steric effects of different DNA pols on their accuracy of DNA synthesis. To support their hypothesis[7] a graphic representation of the distances of the van der Waals radii between DNA and the two enzymes DNA pol β and the DNA pol from *Bacillus stearothermophilus* (Bst DNA pol) was made (Figure 8.3). Bst DNA pol is highly homologous to the 3′→5′ exonuclease deficient Klenow fragment of *E. coli* DNA pol I,[8–10] of which there is not enough structural information available so far.[7]

From Figure 8.3. it is evident that the shapes are different between DNA pol β and Bst DNA pol when either the incoming dNTP was compared to the coding template nucleotide [compare Figure 8.3 (a) to (b)] or when the 3′primer terminus was compared to the template nucleotide

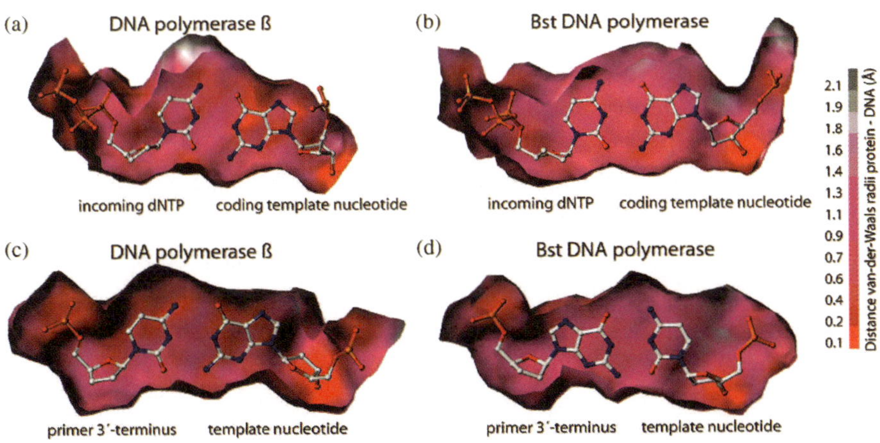

Figure 8.3 **Comparison of human DNA polymerase β to *Bacillus stearothermophilus* DNA polymerase**. The graphics represent the distances of the van der Waals radii between the DNA and the two DNA pols. For details see text. Reproduced with permission of Ref. 7.

[compare Figure 8.3 (c) to (d)]. In summary, these experimental data supported nicely the size exclusion hypothesis proposed by Kool.[6]

8.3 Methods to evolve DNA polymerases with novel properties

Several techniques have been developed to artificially evolve novel properties of DNA pols (reviewed in Refs. 11 and 12). Five of them are discussed here in more details. The advantages and disadvantages of some of them are indicated in the recent review by Tingjian Chen and Floyd Romesberg[12] (see Table 1 therein).

8.3.1 *Detection and characterization of DNA polymerases and mutants thereof by functional complementation in Escherichia coli*

The first *in vivo* screens were developed in the laboratory of Lawrence Loeb in 1993, in which a bacterial complementation system to identify and characterize mammalian DNA pol β mutants was designed and employed.[13] In this complementation system, DNA pol β was able to replace both the replicative and repair functions of DNA pol I in the *E. coli* recA718 polA12 double mutant. It furthermore allowed isolation and evaluation of mutants of a DNA pol in *E. coli* and to characterize them for a better understanding of their functions. Later, this technique was expanded to the Taq DNA pol I.[14] Expression of Taq DNA pol I in *E. coli* complemented the growth defect caused by a temperature-sensitive mutation in the host DNA pol I. The nucleotide sequence encoding the amino acids 659–671 of the O-helix of Taq DNA pol I, corresponding to the substrate binding site, was replaced with oligonucleotides containing random nucleotides. With this functional Taq DNA pol I, mutants were selected based on colony formation at the non-permissive temperature. Finally, catalytically active mutants of HIV RT were generated by random sequence mutagenesis and selected in *E. coli* for their ability to complement the temperature-sensitive phenotype of a DNA pol I mutant.[15] The mutations recovered included most of those associated with drug resistance as well as so far unidentified mutations.

Further studies with DNA pol I from *E. coli* revealed that several recently identified mutants with abilities to incorporate nucleotides with bulky adducts have mutations that are not located within conserved regions. Analysis of these mutants might be useful for our understanding of how DNA pols select bases with high fidelity.[16] Moreover, DNA pols with increased capacity to incorporate rNTPs[17] (see also below: A DNA pol becomes an RNA pol) or to increase its fidelity[18] were evolved. Mutations that produced an antimutator phenotype were found throughout the entire polymerase domain, with 12% clustered in the so called M-helix. A single mutation in the M-helix resulted in increased base discrimination suggesting that interaction with the M-helix is a determinant of fidelity.

8.3.2 *DNA polymerase evolution by random point mutagenesis*

By using 96- and/or 386-well plates, a primer/template is attached to their surfaces and individual clones, that have previously been established (e.g by error-prone PCR of a DNA pol of choice), and tested in primer extension assays (reviewed in Ref. 12). The probe can also be a fluorophore or an affinity tag such as biotin or an antibody.[19] Finally, high-throughput ELISA-screening can be applied by using oligonucleotide arrays.[20]

The method includes the construction of a large library and creates random mutations in the DNA pol (reviewed in Ref. 21 and in Figure 8.4.). For the DNA pol of choice primers are designed so the entire enzyme can be amplified by PCR. PCR is performed under error-prone conditions that have to be optimized in each case.

This can be achieved by varying the relative dNTP concentrations (non-balanced nucleotides), by replacing $MgCl_2$ with $MnCl_2$, which is known to render DNA pols less faithful or by mutagenic dNTPs (e.g. 8-oxo-dGTP). Next, with these DNA molecules a library is constructed in *E. coli* and this library is sorted in microtiter plates (e.g. 384 wells). After growth of the bacteria, extracts are prepared and DNA pol activities are measured with an

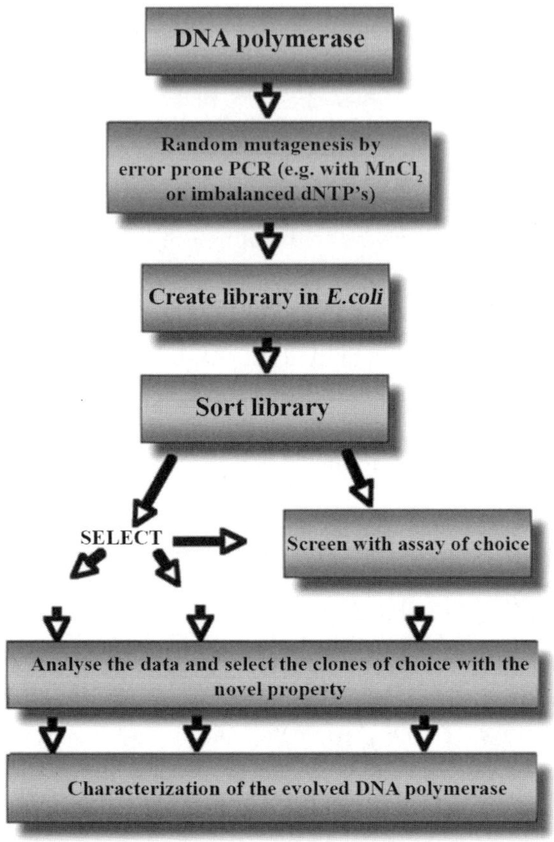

Figure 8.4 **DNA polymerase evolution by random point mutagenesis**. For details see text and references therein.

assay of design. This is done to select for the property of choice (see below). These experiments can now be carried out by an automated robotic system. As an example Katharina Sauter and Andreas Marx took Taq DNA pol I (Klentaq) performed error-prone PCR and screened the library clones for activity in the micro titer plates directly after heat inactivation of the host proteins (among others also the *E. coli* DNA pols I, II, III, IV and V)). They performed the subsequent step in automated liquid handling. DNA pol activities were monitored by using SYBR green I for the optical quantification of the different clones. With this a Taq DNA pol I was evolved that possesses reverse transcriptase activity.[22]

8.3.3 *DNA polymerase evolution by compartmentalized self-replication (CSR) and by compartmentalized self-tagging (CST)*

The compartmentalized self-replication (CSR) technology was pioneered by the Holliger group for the evolution of new DNA pol properties. This method exploits the water-in-oil emulsion technology.[23] It is based on the self-replication onto a separate compartment that cannot exchange proteins and DNA anymore and ensures the linkage between phenotype and genotype. First, a library of clones is cloned and expressed in *E. coli* [Figure 8.5. (a)]. Active Taq DNA pol I molecules were selected by incubating the bacteria in a solution containing flanking primers for the Taq DNA pol I and the dNTPs. Then the DNA pols are segregated into the aqueous solution to allow self-replication of the own polymerase gene by PCR. The generated DNA pols variants are released and can be re-cloned for another round of selection until the DNA pol molecules of choice are finally selected. With this technology a Taq DNA pol I with 11-fold higher

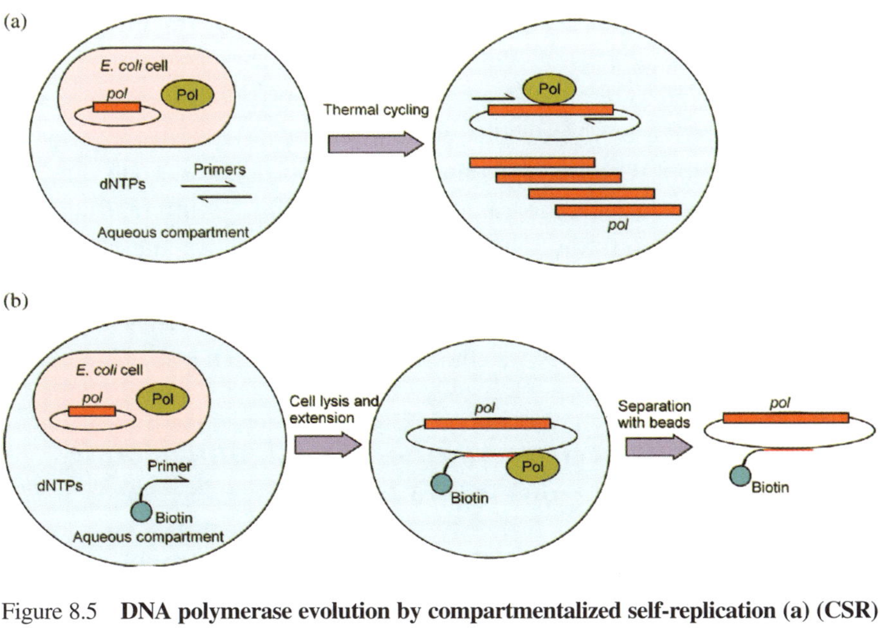

Figure 8.5 **DNA polymerase evolution by compartmentalized self-replication (a) (CSR) and by complementarized self-tagging (b) (CST)**. For details see text. Adapted from Ref. 12.

thermostability and a 130-fold higher resistance to the DNA pol inhibitor heparin was engineered.[24] Moreover, Taq DNA pol I molecules were constructed by this method that contain in addition to the DNA pol, also RNA pol and reverse transcriptase activities.

A variant of CSR is called compartmentalized self-tagging (CST). Here a biotinylated primer is taken that can hybridize to the DNA pol gene, thus allowing to recover plasmids with only a limited amount of extension.[25]

8.3.4 *DNA polymerase evolution by phage display*

Phage display is a powerful technique to covalently attach protein to a filamentous bacteriophage (reviewed in Ref. 26). DNA pols can now be covalently attached and expressed as a fusion protein. The DNA pols of choice are selected in several steps: first, a DNA substrate is cross-linked to the phage; second, the active DNA pols synthesize DNA and generate a product containing biotin and third, the product is purified by affinity chromatography. This cycle can be repeated and the final DNA pol can then be characterized after recovery. This occurs after extensive washing, when the desired phage is recovered by treatment with DNase I and reinfection of *E. coli* by the DNA pol gene (Figure 8.6.). The Stoffel fragment of Taq DNA pol I was made as a p3-Jun/Fos-Stoffel fragment fusion[27] and could thereby be evolved to be an efficient RNA polymerase.[28] Romesberg *et al.*[29] created mutants Taq DNA pols that could efficiently synthesize an unnatural polymer from 2′-O-methyl ribonucleoside triphosphates. Their directed evolution resulted in relocating a critical side chain to a different position in the DNA pol, therefore re-engineering the overall active site while preserving critical protein–DNA interactions. One DNA pol was identified that could incorporate modified nucleotides with the same efficiency and accuracy compared to normal DNA as the wild-type DNA pol (see also below).

8.3.5 *DNA polymerase evolution by oligonucleotide addressed enzyme assay (OAEA)*

A novel approach to characterize DNA pol mutants was invented by Robert Kranaster and Marx by developing a method for DNA pol profiling on DNA arrays[20] and was named oligonucleotide-addressed enzyme

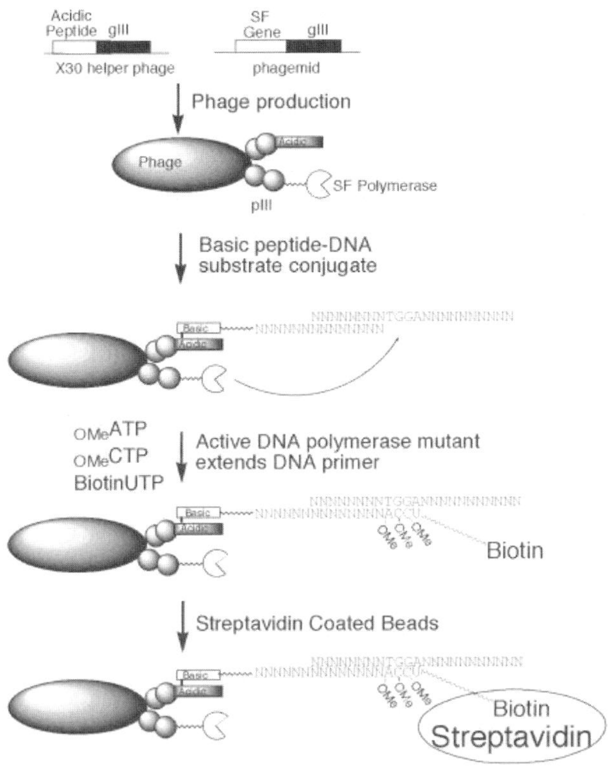

Figure 8.6 **DNA polymerase evolution by phage display**. For details see text. Adapted from Ref. 29.

assay (OAEA). The principle consists of one spotting and incubation step. In the first spotting step 5′NH_2-$(CH)_6$ modified oligonucleotides are covalently attached to activated glass slides.[30] The oligonucleotides are attached in defined spot rows (Figure 8.7). In the second spotting step DNA pol mutants to be characterized in the presence of dATP, dGTP, dCTP, biotin-dUTP and different DNA template to be tested will be applied to the oligonucleotides in nanoliter quantities. Interestingly, even though the nanodrops dried out they can be rehydrated without loss of DNA pol function (exemplified here with Taq DNA pol I, Klentaq). The drops are then rehydrated in a humidity-saturated chamber at 50°C. In case of DNA synthesis the incorporated biotin binds with an alexa546-streptavidine conjugate that will create a fluorescence signal. In summary,

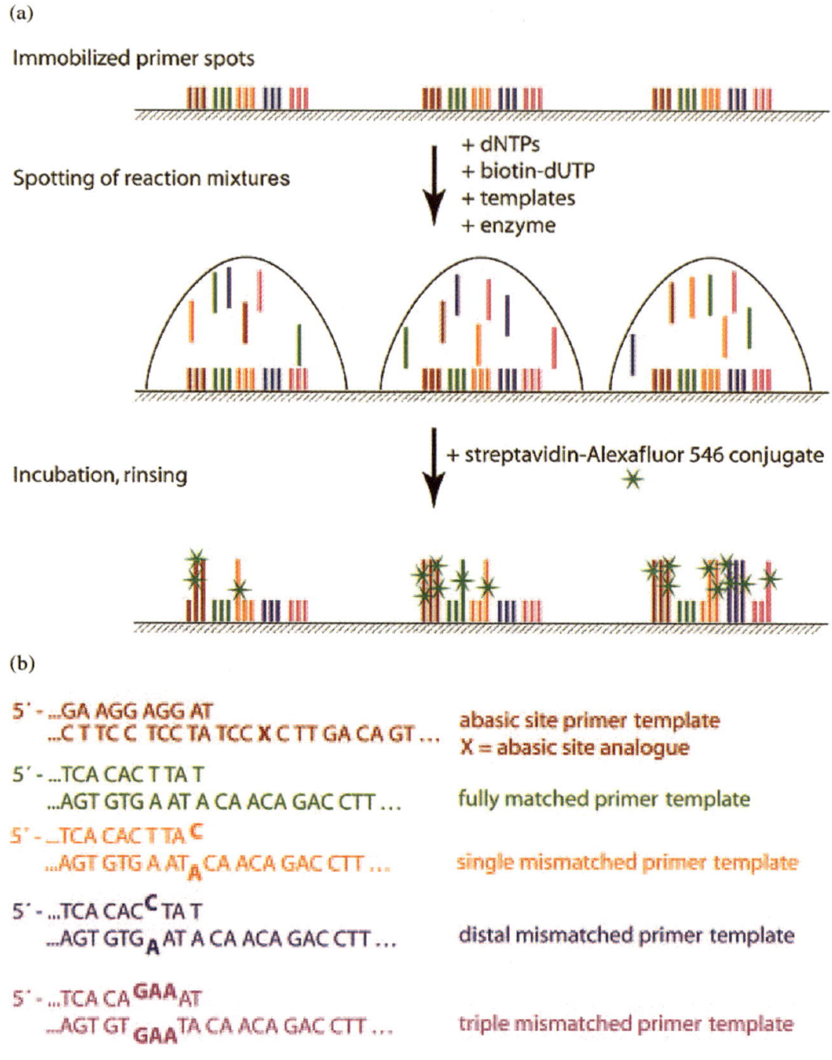

Figure 8.7 **DNA polymerase evolution by the oligonucleotide-addressed enzyme assay (OAEA).** (a) Top: the short colored strips are immobilized primers; middle: the longer stripes represent DNA template that hybridize to the corresponding primers; bottom: after successful DNA synthesis the Alexa 546-streptavidin can bind to the incorporated biotin and this gives a green signal (stars). (b) Sequence of the different templates according to colors. Reproduced with permission of Ref. 20.

this method is a microarray assay that can be used not only to characterize DNA pol mutants but can also be employed for the evolution of DNA pols with novel properties. Finally, this method, if combined in a single experiment with the already powerful microarrays containing several hundreds of thousands of different oligonucleotides, might be extremely informative. However, one must bear in mind that evolving heat-labile DNA pols might need laborious optimization to prevent drying and rehydration that may irreversibly inactivate labile DNA pols.

8.4 Applications of DNA polymerases with novel properties

A few general remarks are indicated here for a quick overview:

(1) The most successful approaches were performed by using Taq DNA pol I. This is evident from its biotechnical applications mainly in PCR. Improvements were made in thermostability, fidelity, processivity, RT-PCR, decreased pausing at GC-rich regions, resistance to heparin, an anticoagulant found in patient samples.
(2) Vent and Taq DNA pols I were generated that increase the incorporation of dideoxy chain terminators and they are used for improved DNA sequencing and genotyping.
(3) *E. coli* DNA pol I was evolved into RNA pols, with enhanced processivity, increased fidelity and higher extension fidelity.
(4) Several Thermus species (Taq, Tth and Tfl) were evolved to carry out PCR from different damaged DNA. This might be of great practical use to forensically analyze DNA from dead bodies that were decaying over a long time period.
(5) A cocktail of seven Taq pols with multiple mutations was able to PCR amplify from a 47,000-year-old DNA from the cave bear *Ursus spelaeus*.
(6) A Pfu DNA pol with increase in base selectivity was rationally designed substituting two amino acids in the active site into hydrophobic counterparts (D514L and K593M).

DNA pols can be mutated, mostly in their active sites, to adapt novel properties (see e.g. Refs. 11, 12 as reviews). Thirty-six novel and/or improved properties of DNA pols including their references are listed below. Some of their applications are described in more details in the text in Sections 8.5.1 to 8.5.10.

1. increased processivity,[31–33]
2. increased thermostability,[24]
3. increased thermostability in a reverse transcriptase,[34]
4. increased fidelity,[18,32,35]
5. increased substrate selectivity,[36]
6. increased selectivity for mismatch extension,[35,37]
7. increased elongation,[38]
8. increased strand displacement activity,[39]
9. increased tolerance against uracil,[40]
10. cold sensitive mutants,[41,42]
11. improved RT-PCR enzymes,[22,34,43–48]
12. resistance to inhibitors (e.g. heparin,[24] blood,[49,50] crude clinical samples,[51] crude soil,[49] environmental inhibitors[52]),
13. damaged DNA amplification and ancient DNA amplification,[53–55]
14. increased translesion synthesis,[56,57]
15. PCR from 5′methyl cytosine allele specific amplification,[19,58,59]
16. DNA pols active on surfaces,[60]
17. synthesis of unnatural polymers from 2′O-methyl rNTP,[29]
18. expansion of the genetic code: synthesis of DNA containing PICS self-pair,[61]
19. reversible terminators,[62]
20. a DNA pol becomes an RNA pol,[19,47,63]
21. evolution from non-standard h bond patterns,[29,64]
22. DNA pol hybrids,[65]
23. heterodimeric DNA pols,[66,67]
24. increased DNA binding,[50]
25. DNA synthesis with Cy-dyes, incorporation of non-standard dNTPs,[68]
26. expansion of the genetic code,[64]
27. sequencing evolution[69] (e.g. NGS),
28. incorporation of two C5-modified nucleotides,[70]
29. detection of DNA adducts,[71]
30. synthesis of branched DNA,[72]

31. incorporation of C5-modified pyrimidine and C7-modified 7-deazapurine NTP,[73]
32. SNP detection,[30,62]
33. loop-mediated isothermal amplification of DNA,[74]
34. evolving hydrophobic analogs,[75]
35. ultrafast PCRs (http://news.berkeley.edu/2015/07/31/light-speeds-up-pcr-test/ and http://www.gna-bio.com/technology/),
36. recombinase polymerase amplification.[76]

8.5 DNA polymerases with novel and/or improved properties

8.5.1 *DNA polymerases with improved properties for PCR*

Increased processivity[31–33]: Work from the Richardson lab has shown almost 20 years ago that the high processivity of T7 DNA polymerase is achieved through its tight binding to a heterogeneous processivity factor of *E. coli*, called thioredoxin.[31]

Another innovative approach was to create chimeric DNA pols. The thioredoxin binding domain (TBD) of the T3 DNA pol was inserted into thermostable Taq DNA pol I at an analogous position in the thumb domain.[32] This converts the Taq DNA pol from a low processive to a highly processive enzyme. The enhancement in processivity was 20–50-fold when compared with the wild-type Taq DNA pol I. The chimeric thioredoxin/Taq DNA pol I copied six to seven times more faithfully than Taq DNA pol I alone and synthesized 2–3-fold fewer AT→GC transition mutations. Wang *et al.*[33] reported a protein engineering–based approach to enhance the processivity of DNA pols by covalently linking the polymerase domain to a non-specific sequence dsDNA binding protein. By using the Sso7d protein from *Sulfolobus solfataricus* they showed that the processivity of both family A and family B DNA pols can be significantly enhanced. Two conclusions were evident from this work: (i) fusion of a heterologous dsDNA binding protein to a DNA pol can increase its processivity and (ii) the fusion enzyme exhibited advantages in PCR compared to unmodified enzymes.

Increased thermostability[24]: By using the compartmentalized self-replication (CSR) system (see methodology above) the Holliger group identified variants of Taq DNA pol with a 11-fold higher thermostability than the wild-type enzyme.[24] Veronica Barrioluengo *et al.* constructed thermostable HIV-1 group O reverse transcriptase variants with the same fidelity as murine leukaemia virus reverse transcriptase.[34] For RT-PCR DNA pol enzymes see Section 8.5.6. "Improved reverse-transcriptase PCR enzymes".

Increased fidelity[18,32,35]: The accuracy of DNA pols is essential for their ability to copy DNA faithfully. The fidelity of those DNA pols, that contain a 3′→5′ exonuclease activity can be changed by the ratio of polymerase versus exonuclease activites. The Loeb laboratory has, as already indicated above, used a bacterial complementation system to identify and characterize mammalian DNA pol β mutants.[13] DNA pols of the family A are most well characterized in the motifs A and B, which regulate the fidelity of replication and the incorporation of nucleotide analogs such as dideoxyribonucleotides. Mutants were identified with abilities to incorporate nucleotides with bulky adducts and the analysis of these mutants will help to understand how DNA pols select bases with high fidelity.[16] The tolerance of *E. coli* DNA pol I for amino acid substitutions in the active site and in different segments of DNA pol I was determined and the effects of these substitutions on the fidelity of DNA synthesis were tested.[18] The DNA pol I mutant library, with random substitutions throughout the polymerase domain, was evolved for activity. It was found that two-thirds of single-base substitutions were tolerated without loss of activity, and this plasticity occurred at evolutionarily conserved regions. Four-hundred-and-eight mutants of the active library were screened for alterations in fidelity *in vivo* in *E. coli*. The mutation frequencies varied from 10^6-fold compared with wild type (e.g. 1000-fold lower up to 1000-fold higher fidelity). Mutations that produced a high fidelity and thus an anti-mutator phenotype were distributed throughout the polymerase domain, with 12% clustered in the M-helix[18] suggesting that this domain is a determinant of fidelity. As shown in the processivity section above, the thioredoxin binding domain (TBD) of the T3 DNA pol was inserted into thermostable Taq DNA pol I at an analogous position in the thumb domain,[32] leading not only to a higher processivity but also to a higher fidelity.[32] Daniel Summerer *et al.* described an efficient automated high-throughput setup for rapid parallel screening of DNA pol mutant libraries. The readout is based on enzyme activity through detection by fluorescence. By randomizing a gene cassette of the *E. coli* DNA pol I Klenow fragment that was 3′→5′ exonuclease deficient, they identified mutants that had a significant higher extension fidelity.[35]

Increased substrate selectivity[36]: These findings of Ref. 35 were then transferred to thermostable Taq DNA pol I as a tool to evolve a better PCR-based genotyping. In view of the fact that the hydrogen bonding network between the DNA pol active site and its primer and template has an influence on the selectivity of a DNA pol, a Pfu DNA pol was evolved by hydrophobic substitution of the active site D541 to L and K593 to M.[36] These data further indicated that the complementary shape as well as the polar interactions have important influence on the selectivity of a DNA pol.

Increased selectivity for mismatch extension[35,37]: Further work again by the Marx group and based on Ref. 35 created a Taq DNA pol I that was randomly mutagenized in the motif C at the three amino acids Q782/V783/H784 and these multiple mutants had an increase in selectivity of mismatch extension.[37]

Increased elongation[38]: The Yoshizumi Ishino's laboratory made a nice approach to produce PCR enzyme with increased elongation properties. For this they amplified by PCR gene fragments of family A DNA pols from microorganism DNA's of environmental soil samples. A primer set for the two conserved regions was used. The corresponding region of the pol gene for the Taq DNA pol was substituted with the amplified gene fragments, and different chimeric DNA pols were thus prepared. The two Taq DNA pol residues E742 and A743 were critical for its elongation ability. In addition these mutants showed also a higher DNA affinity and a faster primer extension ability. In summary an improved PCR resulted with this chimeric mutant enzyme.[38]

Increased strand displacement activity[39]: Konstantin Ignatov *et al.* constructed a novel Taq DNA pol mutant, which has a strong strand displacement (SD) activity and called it SD DNA pol. This enzyme appears to be superior in PCR (also PCR of GC-rich and complex secondary structures), long-range PCR (LR PCR), loop-mediated amplification (LAMP, see below), and polymerase chain displacement reaction (PCDR). This mutant has promising properties for many applications in biotechnology[39] (see also Section 8.6 DNA polymerases and recent applications).

Increased tolerance against uracil[40]: Archaeal family B DNA pols are uniquely able to recognize unrepaired uracil in a template strand and stall polymerization upstream of the lesion, thereby preventing the irreversible fixation of an A-T mutation. A "pocket" in the N-terminal domains of archaeal DNA pols was identified for this property. The determination of its structure showed a conserved region in the enzyme that can discriminate uracil from the four normal DNA bases. By site-directed mutagenesis the authors confirmed the biological role of this pocket and they have engineered specific mutations in the Pfu DNA pol that confer the ability to read through uracils. Finally, PCR with dUTP in place of dTTP could be performed.[40]

Cold sensitive mutants[41,42]: The thermophilic bacterium *Thermus aquaticus* cannot grow at moderate temperatures. However, the DNA pol I has some activity at 20–37°C. This activity is bad for PCR, since it can produce non-specific priming. The Wayne Barnes laboratory constructed mutations of Klentaq DNA pol I that have markedly reduced activity at 37°C but indicated normal activity at 68°C and resistance at 95°C. This now "cold-sensitive" mutant enzymes can be used in hot-start PCR and moreover have an improved fidelity.[41] The same laboratory recently solved the X-ray crystal structure of the Taq DNA pol I containing the cold-sensitive mutation in the ternary (E-DNA-ddNTP) and binary (E-DNA) complexes. Their structural results might explain the cold-sensitive phenotype of the I707L mutation in Taq DNA pol I.[42]

Increased DNA binding[50]: Compartmentalized self replication (CSR) has been applied by Bahram Arezi *et al.* to evolve faster-cycling mutants of Taq DNA pol I. By testing shorter PCR extension times they identified, after five rounds of CSR selection, individual mutations in the fastest-cycling clones. Next they were then randomly combined by the so-called ligation-based multi-site mutagenesis. The best of these combinatorial mutants exhibit a 35–90 fold higher affinity (lower Kd) for primed template and a two-fold increase in extension rate compared to wild-type Taq DNA pol I.[50]

8.5.2 *DNA polymerases resistant to inhibitors*

Heparin[24]: An early Taq DNA pol I mutant was developed by compartmentalized self replication (CSR) in which the Holliger lab isolated a mutant

DNA pol with a more than 130-fold increased resistance to heparin. This is relevant to perform PCR in blood from patients that have been anti-coagulated.

Blood and components therein[49,50]: The Klentaq DNA pol mutant E708K showed enhanced resistance to whole blood, plasma, hemoglobin, lactoferrin and serum IgG. A more recent CSR study revealed a Tag DNA pol I multiple mutant (Taq2C2: G59W, V155I, L245M, L375V, E507K, K508R, E734G and F749I) that was effective in PCR from EDTA blood, heparinized blood and at physiological NaCl concentrations.

Crude clinical samples[51]: Another strategy to improve PCR from crude clinical samples was the identification of better PCR-enhancing cocktails with nonionic detergents, L-carnitine, D-(+)-trehalose, and heparin. Such cocktails in combination with the two inhibitor-resistant Taq DNA pol mutants (OmniTaq and Omni Klentaq) enabled efficient amplification of any type of DNA targets directly from crude samples containing human plasma, serum, and whole blood without the need of prior DNA purification.

Crude soil[49]: From fresh soils of different sources crude extracts were prepared and successfully tested with the Klentaq DNA pol mutant E708K in PCR. A similar result was detected with humic acid.

Environmental inhibitors[52]: Claudia Baar *et al.* described the engineering by CSR of DNA pols with a broad resistance to complex environmental inhibitors. Initially they selected for resistance to Neomylodon bone powder, thus isolating a chimeric DNA pol that contained sequences from *Thermus aquaticus*, *Thermus oshimai*, *Thermus thermophilus* and *Thermus brockianus*. This DNA pol, called 2D9, was resistant to humic acid, bone dust, coprolite, peat extract, clay-rich soil, cave sediment and tar. This DNA pol chimera can be applied in PCR for topics such as palaeobiology, archaeology, conservation biology, forensic and historic medicine.

8.5.3 *DNA polymerases handling altered DNA*

Amplification of damaged and ancient DNA[53–55]: A nice example where DNA pols with either higher or lower substrate specificity are wanted is the amplification from damaged DNA. Evolving the M747 to K in Taq DNA pol I resulted in the property to amplify from damaged DNA such as abasic sites and from oxidative damages such as 8-oxo-G and 8-oxo-A.[54] Such enzymes are of great value in forensic medicine to identify individuals that were dead since a long time. Another work resulted in evolving a Taq DNA pol I that retains the high turnover, processivity and fidelity, but was able to replicate from various lesions and non-natural substrates.[77] These included abasic sites, CDP dimers, 7-deaza dGTP, aS-dNTPs, FITC labeled dATP, Biotin16-dUTP and the base analog 5-nitroindol, suggesting that this evolved Taq DNA pol I, that contains seven mutations, is of great help for many applications in biotechnology and diagnostic medicine.

Another application was the evolving of a DNA pol that amplifies from ancient DNA.[78] In particular to amplify highly degraded DNA is of great importance in science to study ancient DNA. Holliger's lab performed molecular evolution to create a mixture of DNA pols from different species (Taq, Tth, Tfl) each containing multiple mutations that could amplify from damaged DNA.[55] Damaged DNA included abasic sites, 5-methyl-5-hydroxyhydantoin and 5-hydroxyhydantoin. Furthermore, extensions from multiple mismatches (up to four) and from misaligned DNA structures (e.g. hairpins after slippage) were possible with such multiple mutants. Finally, a cocktail of the seven most promising Taq DNA pol I mutants was chosen to amplify DNA from a 47,000-year-old cave bear called *Ursus spelaus*. When this blend was compared to wild type Taq DNA pol I the mutated cocktail amplified from 2 to 5 fold lower DNA starting material than the wild type enzyme. This suggested that the evolved DNA pol cocktail had a higher specificity with highly damaged ancient DNA.[55]

Increased translesion synthesis (TLS)[56,57]: The Marx laboratory, by using the directed evolution technology (see above), substituted in the Taq DNA pol I the neutral residues I614 and M747 into Ks. This resulted in a more competent bypass at abasic sites. However, the increased TLS resulted in a decrease of selectivity. So an increase in the positively charged surface potential in the neighborhood of the negative substrate DNA can promote TLS. Nevertheless this mutant enzyme is useful for forensic applications (e.g. dead bodies that were found after many days or even weeks) and for PCR of ancient DNA (see also above). A similar observation was made with the human DNA pol β. By screening a library of 11,000 DNA pol β variants the two mutations E232K and T233I enabled the enzyme to incorporate at an abasic site. Even though the enzyme had an increased activity a lower fidelity and processivity were measured when synthesizing undamaged DNA.

PCR from 5-methylcytosine and allele specific amplification[19,58,59]: The Marx laboratory developed DNA pols that can directly sense 5-methylcytosine by PCR. They evolved DNA pols from Family A (Taq Klentaq) and Family B (*Thermococcus kodakaraesnsis,* KOD) that are able to preferentially extend from a 5-methylcytosine (5mC). DNA pols with such properties are potentially very useful to detect epigenetic methylation marks in development and diseases such as cancer and neurological diseases. The development of such DNA pols is very promising, since so far detection of 5mC with single base resolution was restricted to the time consuming conversion of DNA by bisulfite or any other manipulation prior to analysis. Improved discrimination was found with the KOD DNA pol when G498 was mutated to M. The proof of principle was documented by amplifying from human genomic DNA.

Moreover, when the terminator DNA pol, a family B DNA pol, was mutated at L408 to Q, which is the conserved A motif in the DNA pol domain, this enzyme was able to synthesize from a C5-modified deoxyribonucleoside.

A very nice mutant of Taq DNA pol I was again evolved in the Marx laboratory with increased mismatch extension selectivity. It was shown that one mutant containing the single amino acid exchange R660V was suitable for application in allele-specific manner directly from whole blood without prior sample purification. Moreover, this mutant was also useful for methylation-specific amplification in the diagnostics of 5-methylcytosines.

SNP detection[30,62]: The double mutant D541L and K593M of *Pyrococcus furiosus* DNA pol was constructed by substituting two polar amino acid residues by hydrophobic amino acids with similar steric demands. This mutant DNA pol was successfully tested for improved detection of single-nucleotide variations in a microarrayed primer extension protocol.[30] The Benner laboratory, by reconstructing evolutionary adaptive paths, constructed Taq DNA pols that could be used for parallel SNP detection. It was found that the mutation L616A was the main amino acid for this property.[62]

Detection of DNA adducts[71]: Determination of DNA in specimens from nature and certain environments are important for toxicology and disease prevention. Consequently the detection of DNA modifications at single base resolution is an important development in DNA pol evolution. The Shana Sturla's laboratory in collaboration with the Marx laboratory studied the action of newly synthesized base-modified benzimidazole-derived 2′-deoxynucleoside-5′-O-triphosphates on DNA pols when performing translesion DNA synthesis past the pro-mutagenic DNA adduct O(6)-benzylguanine (O(6)-BnG). The mutant M747K of Klentaq DNA pol I, was able to catalyze O(6)-BnG adduct-specific processing of the artificial BenziTP in favor of the natural dNTPs. By combining the M747K mutant and BenziTP, a first round of DNA synthesis enabled subsequent amplification of Benzi-containing DNA, thus advancing the detection of DNA adducts in specimens from nature and the environment.

Synthesis of branched DNA[72]: A novel approach for the generation of DNA networks *in vitro* by PCR amplification of branched primer constructs suggests a flexible approach allowing tuning the meshes of the network by variation of the template size. Additionally, further diversification can be introduced by the employment of chemically modified nucleotides in PCR, allowing the introduction of functionalities and reporter moieties. Chemically modified nucleotides might be amplified by the Taq DNA pol I M747K mutant mentioned above.

8.5.4 *DNA polymerase hybrids and dimers*

DNA pol hybrids[52,65]: As indicated in Section 8.5.2. "DNA polymerases resistant to inhibitors", a chimeric DNA pol that contained sequences from *Thermus aquaticus*, *Thermus oshimai*, *Thermus thermophilus* and *Thermus brockianus* was engineered so that is was resistant to a variety of environmental inhibitors. Another strategy constructed family B DNA pol hybrids and mutants thereof from the *Pyrococcus furiosus*[79] and *Thermococcus kodakaraesnsis* (Tkod). It was found that some of the mutants in either the Pfu DNA pol or in the Tkod DNA pol or in both were more efficient in PCR, had a higher fidelity and a higher thermostability.

Heterodimeric DNA pols[66,67]: Fei Wang *et al.* linked DNA pol domains from both family A and family B DNA pols to a non-specific double-strand DNA binding protein from *Sulfolobus solfataricus*, called Sso7d. The so constructed "heterodimers" had an increased processivity, but retained the catalytic activity and the stability, thus establishing a superior PCR enzyme.

Ghislaine Henneke's laboratory studied the PCR performance of the proof-reading D-type DNA pol from *Pyrococcus abyssi* (Pab-polD). This DNA pol exists as a heterodimer of a small subunit possessing a 3′ to 5′ exonuclease proofreading activity and a large subunit being the DNA pol. It was for the first time demonstrated that an archaeal family D heterodimeric DNA pol is efficiently perfoming PCR.

8.5.5 *DNA polymerase with RNA polymerase properties*

A DNA pol becomes an RNA pol[19,47,63]: We will see in the next Section 8.5.6 that *Thermus* species DNA pols can be made into reverse transcriptase useful for RT-PCR.[48] Many publications exist that made a DNA pol into a RNA pol and four selected examples are discussed in more details here.

(1) By using genetic complementation, a library of approximately 8,000 active mutants Taq DNA pol I, was obtained and 350 were sequenced and analyzed resulting in a large collection of active polymerase mutants.[80] The high mutability of the polymerase active site *in vivo* and the ability to evolve altered enzymes may be required in nature for survival in environments that have increased mutagenesis. In a follow-up work a large library of 200,000 mutants Taq DNA pol I was created containing random substitutions within a part of the dNTP binding site (amino acids 605–617 in motif A), 291 selected active Taq DNA pol I enzymes containing multiple mutations were screened for the ability to incorporate successive ribonucleotides.[17] This led to the identification of 23 mutants that added rNMPs into a growing polynucleotide chain and suggested that DNA and RNA pols could have evolved by divergent evolution from an ancestor that shared a common mechanism for polynucleotide synthesis.

(2) The second approach utilized the activity-based selection method to evolve DNA pol with RNA pol activity.[28] For this the Taq DNA pol I Stoffel fragment was displayed on a filamentous phage by fusing it to a p3 coat protein, and the substrate DNA template/primer duplexes were attached to other adjacent p3 coat proteins (for details see methodology above). The phage particles displaying Taq DNA pol I Stoffel fragments, were then selected to incorporate rNMPs and biotinylated UMP. When the Taq DNA pol I Stoffel fragment library was screened four times three mutants could be harvested that incorporate rNTPs with an efficiency as the wild-type enzyme can incorporate dNTPs. In summary, the phage method created an RNA pol.

(3) The Holliger laboratory described the discovery of a distinct specificity checkpoint located over 25 Å from the active site in the polymerase thumb subdomain.[63] Together with the steric-gate mutation Y409G in the active site, E664K transforms the DNA pol from *Thermococcus gorgonarius*[81] into an efficient and processive RNA pol. Futhermore, they found that E664K enables RNA synthesis by selectively increasing polymerase affinity for the non-cognate RNA/DNA duplex as well as lowering the K_m for rNTPs.[63]

(4) By using the short-patch compartmentalized self-replication (spCSR, see methodology section), the Holliger laboratory targeted two specific motifs involved in substrate recognition in the active site of Taq DNA pol I and selected for incorporation of both NTPs and dNTPs. One mutant (E602V, A608V, I614M, E615G) was able to incorporate both NTPs and dNTPs with the same efficiency as the wild type Taq DNA pol I incorporated dNTPs. The spCSR thus was suggested to be a powerful tool for the generation of polymerases with altered substrate specificity for applications in nano- and biotechnology as well as in the enzymatic synthesis of antisense and RNAi molecules.[44]

8.5.6 *Improved reverse transcriptase PCR enzymes*

Improved RT-PCR enzymes[22,34,43–48]: About ten years ago several laboratories developed DNA pols that can perform reverse transcriptase (RT) PCR. The ideal enzyme should be used as a single enzyme in a one tube reaction (see e.g. refs above). Here three selected examples are chosen to be discussed in some more detail.

(1) Kranaster *et al.* described cloning and characterization of a mutated Taq DNA pol I that exhibited an increased RT activity and is therefore designated for one-step PCR pathogen detection. This multiple mutant (called Taq M1: L322M, L459M, S515R, I638F, S739G, L789F) had a similar PCR sensitivity and nuclease activity as the wild-type DNA pol. In contrast to the wild-type, Taq M1 exhibited a significantly increased RT activity at high temperatures. RT at high temperatures is desired to weaken or melt secondary structure motifs often occurring in many RNAs. As a proof of principle the TaqMan probe-based diagnostics of Dobrava viruses by one-step RT-PCR with the M1 mutant was demonstrated.[45]

(2) Nina Blatter *et al.* improved the Taq DNA pol multiple mutant M1 (see above) and recombined it with DNA shuffling with the M747K mutant to become KTqM1/M747K. A second multiple mutant, called KTq2, contained the four mutants S515R, I638F, M747K and L459M. Both enzymes were very efficient in RT-PCR in a single enzyme reaction. Next they determined the structure of these multiple mutants. Their conclusion was that this first determination of a DNA dependent DNA pol, the wild type of which did not contain RT activity, highlighted an altered structure of the DNA/RNA duplex compared to a DNA/DNA duplex. The M747K increases the binding affinity to negatively charged nucleic acids, the S515R stabilizes the nucleic acid binding motif in the thumb domain and the L459 results in an increased flexibility of the thumb domain, thus contributing to preventing a clash with a DNA/RNA hybrid complex. In summary this work illustrated in a nice way why these mutants have RT-PRC properties. Moreover the fast PCR production makes especially the RT-KTq2 enzyme a promising candidate for minimizing labor and time-consuming optimizations for reliable PCR.[47]

(3) Michael Moser *et al.* identified from a viral metagenomic DNA a thermostable DNA pol, called 3173 DNA pol. The 3173 DNA pol contained both a high thermostability and an RT activity. The 3173 DNA pol was efficient in single-enzyme RT-PCR. In addition it contains a proofreading 3′ to 5′ exonuclease activity thus conferring a high fidelity in PCR. The authors finally claim that the 3173 DNA pol is superior to the one enzyme Tth DNA pol and the two enzyme MMLV RT/Taq DNA pols systems.[46]

8.5.7 *DNA polymerases and expansion of the genetic code*

Expansion of the genetic code — general[82]: More than 20 years ago the Steven Benner and Hübscher labs tested the ability of DNA pols from the B, X and RT families to catalyze the template-directed synthesis of duplex oligonucleotides containing non-standard Watson-Crick base pairs.[83] They included a nucleotide bearing a 5-(2,4-diaminopyrimidine) heterocycle (dκ) and a nucleotide bearing either deoxyxanthosine (dX) or a N1-methyloxoformycin B (π). The dκ-X and dκ- π base pairs are joined by a hydrogen bonding pattern different from and exclusive of those joining the A-T and G-C base pairs (Figure 8.8.). HIV-1 RT incorporated dXMP into an oligonucleotide opposite dκ in a template with good fidelity. With lower efficiency and fidelity, HIV-1 reverse transcriptase also incorporated dκTP opposite dX in the template. With dπ in the template, no

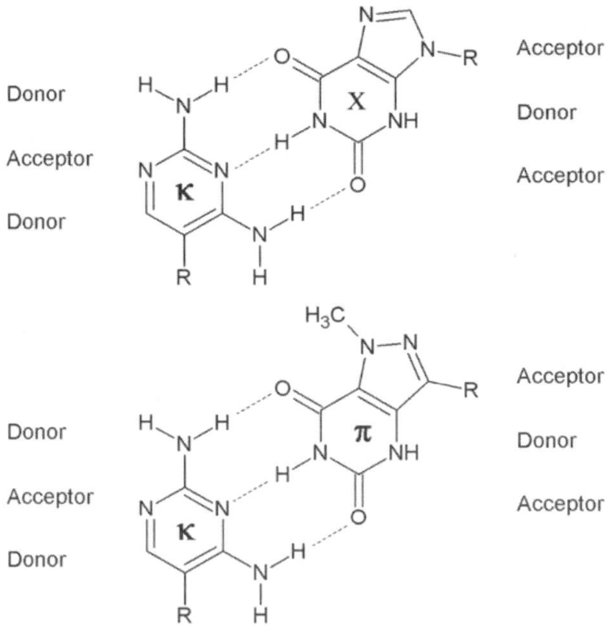

Figure 8.8 **Non-standard Watson-Crick base pairs**. Shown are κ, X and π including their donor hydrogens and their acceptor oxygens or nitrogens. For details see text and Ref. 83.

incorporation of dκTP was observed with HIV reverse transcriptase. The Klenow fragment of DNA pol I from *E. coli* did not incorporate dκMP opposite dX in a template but did incorporate dXMP opposite dκ. Mammalian DNA pols α, β and ε incorporated neither dXMP opposite dκ nor dκMP opposite dπ. The replicative DNA pols α and ε (but not the repair DNA pol β) incorporated dκMP opposite dX in a template but discontinued elongation after incorporating a single additional base. These results were discussed in the view of the ternary crystal structure of DNA pol β.[84]

Expansion of the genetic code — synthesis of DNA containing PICS self-pair[29,61]: The laboratory of Romesberg also made great efforts toward the expansion of the genetic alphabet. DNA pols were able to recognize a highly stable self-pairing hydrophobic base. For this a 7-propynyl isocarbostyril nucleoside was synthesized and converted into the nucleoside triphosphate.[85] This self-pairing base pair, called PICS, could be synthesized by *E. coli* DNA pol I and by Taq DNA pol I. However, the proteolytic fragment of Taq DNA pol I called the Stoffel fragment was unable to do this. The Stoffel fragment was subsequently evolved by using the phage display technology.[61] The same laboratory also synthesized 2′-O-methyl modified NTP and evolved with the same technology Stoffel fragment mutants that could efficiently synthesize an unnatural polymer from 2′-O-methyl-rNTPs[29] and 7-propynyl iso-carbostyril nucleoside triphosphate[85] (Figure 8.9.). One evolved Stoffel

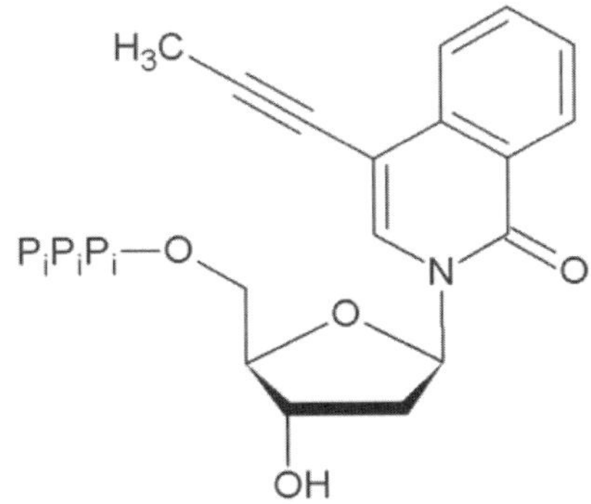

Figure 8.9 **7-propynyl iso-carbostyril nucleoside triphosphate**. For details see text and Ref. 29.

Figure 8.10 **Z:P pair from an artificially expanded genetic information system**. The two Taq DNA pol I mutants mentioned in the text could handle this nonstandard hydrogen pattern base pair. For details see Ref. 64. (reproduced with permission of S. Benner).

fragment could incorporate the 2′-O-methyl-rNTPs with an efficiency and fidelity equal to that of the wild-type Stoffel fragment with natural substrates. Such an enzyme is a great use for applications in biotechnology and therapeutic medicine.

Evolution from non standard hydrogen bond patterns[29,64]: Faithful to the concept "you get what you select for", the Benner laboratory selected a Taq DNA pol I able to synthesize nucleotides with nonstandard hydrogen bond patterns. By using the CSR technology (see above for methodology) the two Taq DNA pol mutants N580S, L628V, E832V and M444V, P527A, D551E, E832V were identified that could incorporate a dZMP into a dP-containing template (Figure 8.10). However the vice versa namely dPTP could not be incorporated into a Z-containing DNA, but as the authors claim this has not been selected for.[64]

8.5.8 *DNA polymerases and next-generation sequencing (NGS)*

Sequencing evolution[69] *(Next-generation sequencing, NGS)*: The Romesberg laboratory optimized by directed evolution the first DNA pol for next-generation sequencing (NGS). For this they employed Taq DNA pol I and their activity-based phage display method (see methodology above) and identified a mutant DNA pol, called Taq197 containing 14 different mutations, that can incorporate fluorophore-labeled dA, dT, dC and dG at least two orders of magnitude more efficiently than the wild-type enzyme in a DNA synthesis reaction. Moreover, it also had an improved performance in a sequencing reaction. It is conceivable that this phage display method was extremely useful to further evolve DNA pols for the NGS platforms.[69]

Reversible terminators[62]: As indicated in Section 8.5.3 Taq DNA pols that could be used for parallel sequencing were constructed. It was found that the mutation L616A was the main amino acid for this property.[62] For more information and reviews concerning NGS see Chapter 9.

8.5.9 *DNA polymerases and processing of modified dNTPs and rNTPs*

Synthesis of unnatural polymers from 2′O-methyl rNTP[29]: By activity base phage selection a DNA pol was evolved that could synthesize efficiently an unnatural polymer from the 2′O-methyl rNTP. The thus synthesized polymer appears to have better hybridization properties and is resistant to nucleases. Therefore such molecules are useful for therapeutic and biosensor applications.[29]

DNA synthesis with Cy-dyes, incorporation of non-standard dNTPs[68]: Incorporation of Cy-modified dNMP into DNA is of paramount interest to perform microarray assays and labeling of DNA for various purposes in biotechnology and diagnostics and as mentioned before for NGS. By using the short-patch compartmentalized self-replication system (spCSR, see above methods) the Holliger group evolved from *Pyrococcus furiosus*[79] a DNA pol variant that contained the five mutations V337I, E399D, N400D, R407I, Y546H. This mutant DNA pol synthesized processively brightly colored highly fluorescent DNA in PCR in which all dC were replaced either by C3-dC or C5-dC.[68]

Incorporation of two C5-modified nucleotides[70]: Samra Obeid *et al.* presented a study that gives first insight into the mechanism of how multiple modifications are processed by a DNA pol and this is important for understanding how this enzyme works with artificial substrates and leads to our understanding of DNA pol evolution.[70]

Incorporation of C5-modified pyrimidine and C7-modified 7-deazapurine dNMP[73]: Another approach by the Marx group presented the ternary structure of the Klentaq DNA pol I after incorporating in the presence of a template primer modified dNMPs carrying modification at the C5 position of pyrimidines and at the C7 position of 7-deazapurines.

This structural study is important for our understanding of the mechanism of how modified dNTPs can be accepted and incorporated by DNA pol I and evolved mutations thereof.[73]

Evolving DNA pols to handle hydrophobic analogs[75]: Hydrophobic base analogs (HBAs) are generally poor substrates for DNA pols. By using the compartmentalized self-replication system CSR (see above methods) and by selecting with archetypal HBA 5′nitroindole (d5N1) and a derivative thereof 5′nitroindole-3-carboxamide (d5NIC), dedicated pols were evolved to improve HBA substrate properties. The most promising was a Tth/Taq chimera with 14 additional mutations from the Tth/Taq consensus (V62I, Y78H, T88S, P114Q, P264S, E303V, G389V, E424G, E432G, E602G, A608V, I614M, M761T, M775T). This DNA pol chimera mutant might be useful for applications in various topics such as chemistry, biotechnology and nanotechnology.[75]

8.5.10 *DNA polymerases active on surfaces*

DNA pols active on surfaces[60]: The single-molecule DNA sequencing technology requires DNA pols that are attached to a surface. This in combination with the template/primer should result in a stable complex being able to efficiently and processively synthesize DNA. The authors inserted two biotinylated peptide "legs" at positions in the vicinity of the DNA-binding cleft of the archaeal DNA pol 9 degrees N. This allowed the Streptavidin binding on either side of the cleft, thus trapping the DNA template in the DNA pol and oriented the complex on a biotinylated surface.

8.6 DNA polymerases and recent applications

8.6.1 *DNA polymerases and loop-mediated isothermal amplification of DNA*

A technique that has been initially described in 2000 by Notomi *et al.* has been named loop-mediated isothermal amplification, (LAMP) of DNA.[86] LAMP functions as follows (for details see Ref. 86 and Figure 1 therein): four primers are designed so that they can recognize six distinct sequences on the DNA. An "inner" primer must contain sequences from both the sense and antisense target DNA. From these inner primers LAMP can start. Next a strand-displacement DNA pol from *Bacillus stearothermophilus* synthesizes DNA by the outer primers and thus releases single-stranded DNA. This single-stranded DNA serves as a template that will be primed by the outer primers, which can hybridize to the other end of the target DNA. This is important since a stem-loop is thus formed. In the subsequent reaction one inner primer hybridizes to the one on the loop of the product, thus starting stand-displacement synthesis by the DNA pol. This yields an original stem-loop and a new-stem loop DNA containing a stem twice as long. Within 60 min the authors claim that 10^9 target sequences were amplified.[86]

The fact that LAMP is not used so much so far is due to the complex design of primers, some limitation of assay conditions and most importantly the DNA pol employed. Yogesh Chander *et al.* have recently introduced their so called OmniAmp DNA pol, which is a thermostable viral enzyme that enables its use in the field by overcoming important barriers of LAMP.[74] Their data were optimized compared to earlier used LAMP DNA pols. In addition to the DNA pol, sample preparation speed, stability and reliability on the structured template (palindromes) were optimized. It appears that due to its better simplicity the OmniAmp DNA pol might be suited for low resource molecular detection of pathogens and thus making it conceivable that LAMP will have a renaissance in the near future.

8.6.2 *DNA polymerases and recombinase polymerase amplification*

Olaf Piepenburg *et al.* published in 2006 a novel approach, called recombinase polymerase amplification (RPA).[87] It can couple isothermal recombinase-driven primer targeting of DNA by strand-displacement DNA synthesis. Exponential amplification with no need for pretreatment of sample DNA can be obtained. What is important are the facts that the reactions are sensitive, specific, rapid and operate at constant low temperature. The authors showed as proof of principle a test for the medically very important pathogen methicillin-resistant *Staphylococcus aureus*. The sensitivity was down to fewer than ten copies of genomic DNA.

The technique works as follows[87]: Three proteins are necessary, namely a recombinase, a single-stranded DNA-binding protein (SSB) and a DNA pol that is able to strand displace DNA. The recombinase can pair primers within homologous sequences in the DNA. Next the SSB binds to single standed DNA thus stabilizing the primers. Finally, the strand-diplacing DNA pol can start DNA synthesis using two opposite primers. The advantage of this technique in amplifying DNA is the fact that no DNA melting (heating to 90°C) is required so that the reaction can be perfomed at temperatures of 37–43°C. In addition, the reaction is extremely fast so that in about 10 min DNA or even RNA can be specifically detected.[87] The link *www.twistdx.co.uk* explains in a nice and illustrated overview by Mattheu Forrest and Oliver Nentwich the RPA technology in more details.

8.6.3 *DNA polymerases and ultrafast PCR*

Bioengineers at the University of California Berkeley introduced a novel technology. It is based on accelerating heating and cooling of genetic samples with the switch of a light. This thermal cycling greatly expands the clinical and research applications of PCR tests, with results ready in minutes instead of an hour or more (Ref. 88 and Figure 1 therein). By using light-emitting diodes (LED) electrons were heated at the interface of thin films of gold and a DNA solution. The speed of heating the solution was around 55°F (12.8°C) s^{-1} and the rate of cooling in was about 43.9°F (6.6°C) s^{-1}. In summary this robust and low-cost approach to ultrafast PCR using an efficient photonic-based heating procedure could be generally integrated into many types of application of PCR including LAMP. Optimized DNA pols for this technology are certainly in the pipeline of many laboratories. More information can be obtained in http://news.berkeley.edu/2015/07/31/light-speeds-up-pcr-test.

8.7 Chapter summary

In this Chapter 8 we summarized the novel and exciting field of synthetic evolution of DNA pols with novel properties. Many biotechnological, forensic and medical applications call for DNA pols with certain improved catalytic properties.

First, the tight active site of DNA pols to which the substrates fit is discussed. These data can give prediction in which direction to evolve a DNA pol. Next five different methods are discussed: (i) detection and characterization of DNA pols and mutants thereof by functional complementation in *Escherichia coli*; (ii) DNA pol evolution by random point mutagenesis and screening; (iii) DNA pol evolution by compartmentalized self-replication (CSR); (iv) DNA pol evolution by phage display and (v) DNA pol evolution by oligonucleotide addressed enzyme assay (OAEA). Applications of DNA pols with novel properties include the construction of DNA pols for better PCR and sequencing properties, to DNA pols that are also RT's and RNA pols. In particular DNA pols are discussed that have increased fidelity, can amplify from ancient and damaged DNA, create RNA pols from DNA pols and evolving the dNTP substrates of the genetic code.

In summary the synthetic evolution already has a great impact in biotechnology but possibly also in biology and medicine. Moreover, the three promising techniques loop-mediated isothermal amplification (LAMP), recombinase polymerase amplification (RPA) and ultrafast PCR are briefly described. These techniques do already or will ask for novel DNA pol properties in the near future. In Chapter 9 we will go into more details to the expanding market and routine applications for evolved DNA polymerases.

References

1. Robertson, M. P. & Scott, W. G. Biochemistry: designer enzymes. *Nature* **448**, 757–758. (2007).
2. Cherry, J. R. & Fidantsef, A. L. Directed evolution of industrial enzymes: an update. *Curr Opin Biotechnol* **14**, 438–443, doi:S0958166903000995 [pii] (2003).
3. Brautigam, C. A. & Steitz, T. A. Structural and functional insights provided by crystal structures of DNA polymerases and their substrate complexes. *Curr Opin Struct Biol* **8**, 54–63. (1998).
4. Steitz, T. A. DNA polymerases: structural diversity and common mechanisms. *J Biol Chem* **274**, 17395–17398 (1999).
5. Steitz, T. A. Visualizing polynucleotide polymerase machines at work. *Embo J* **25**, 3458–3468 (2006).

6. Kool, E. T. Active site tightness and substrate fit in DNA replication. *Annu Rev Biochem* **71**, 191–219, doi:10.1146/annurev.biochem.71.110601.135453 110601.135453 [pii] (2002).
7. Di Pasquale, F. *et al.* Opposed steric constraints in human DNA polymerase beta and E. coli DNA polymerase I. *J Am Chem Soc* **130**, 10748–10757, doi:10.1021/ja8028284 (2008).
8. Kiefer, J. R., Mao, C., Braman, J. C. & Beese, L. S. Visualizing DNA replication in a catalytically active Bacillus DNA polymerase crystal. *Nature* **391**, 304–307, doi:10.1038/34693 (1998).
9. Johnson, S. J., Taylor, J. S. & Beese, L. S. Processive DNA synthesis observed in a polymerase crystal suggests a mechanism for the prevention of frameshift mutations. *Proc Natl Acad Sci USA* **100**, 3895–3900, doi:10.1073/pnas.0630532100 0630532100 [pii] (2003).
10. Johnson, S. J. & Beese, L. S. Structures of mismatch replication errors observed in a DNA polymerase. *Cell* **116**, 803–816, doi:S0092867404002521 [pii] (2004).
11. Henry, A. A. & Romesberg, F. E. The evolution of DNA polymerases with novel activities. *Curr Opin Biotechnol* **16**, 370–377, doi:S0958-1669(05)00099-6 [pii] 10.1016/j.copbio.2005.06.008 (2005).
12. Chen, T. & Romesberg, F. E. Directed polymerase evolution. *FEBS letters* **588**, 219–229, doi:10.1016/j.febslet.2013.10.040 (2014).
13. Sweasy, J. B. & Loeb, L. A. Detection and characterization of mammalian DNA polymerase beta mutants by functional complementation in Escherichia coli. *Proc Natl Acad Sci USA* **90**, 4626–4630 (1993).
14. Suzuki, M., Baskin, D., Hood, L. & Loeb, L. A. Random mutagenesis of Thermus aquaticus DNA polymerase I: concordance of immutable sites in vivo with the crystal structure. *Proc Natl Acad Sci USA* **93**, 9670–9675 (1996).
15. Kim, B., Hathaway, T. R. & Loeb, L. A. Human immunodeficiency virus reverse transcriptase. Functional mutants obtained by random mutagenesis coupled with genetic selection in Escherichia coli. *J Biol Chem* **271**, 4872–4878 (1996).
16. Loh, E. & Loeb, L. A. Mutability of DNA polymerase I: implications for the creation of mutant DNA polymerases. *DNA Repair (Amst)* **4**, 1390–1398, doi:S1568-7864(05)00257-0 [pii] 10.1016/j.dnarep.2005.09.006 (2005).
17. Patel, P. H. & Loeb, L. A. Multiple amino acid substitutions allow DNA polymerases to synthesize RNA. *J Biol Chem* **275**, 40266–40272, doi:10.1074/jbc.M005757200 M005757200 [pii] (2000).

18. Loh, E., Choe, J. & Loeb, L. A. Highly tolerated amino acid substitutions increase the fidelity of Escherichia coli DNA polymerase I. *J Biol Chem* **282**, 12201–12209, doi:M611294200 [pii] 10.1074/jbc.M611294200 (2007).
19. Staiger, N. & Marx, A. A DNA polymerase with increased reactivity for ribonucleotides and C5-modified deoxyribonucleotides. *Chembiochem : a European Journal of Chemical Biology* **11**, 1963–1966, doi:10.1002/cbic.201000384 (2010).
20. Kranaster, R. & Marx, A. Taking fingerprints of DNA polymerases: multiplex enzyme profiling on DNA arrays. *Angew Chem Int Ed Engl* **48**, 4625–4628, doi:10.1002/anie.200900953 (2009).
21. Arnold, F. & Georgiou, G. *Directed Evolution Library Creation.* (Totowa: Humana Press, 2003).
22. Sauter, K. B. & Marx, A. Evolving thermostable reverse transcriptase activity in a DNA polymerase scaffold. *Angew Chem Int Ed Engl* **45**, 7633–7635, doi:10.1002/anie.200602772 (2006).
23. Tawfik, D. S. & Griffiths, A. D. Man-made cell-like compartments for molecular evolution. *Nat Biotechnol* **16**, 652–656, doi:10.1038/nbt0798-652 (1998).
24. Ghadessy, F. J., Ong, J. L. & Holliger, P. Directed evolution of polymerase function by compartmentalized self-replication. *Proc Natl Acad Sci USA* **98**, 4552–4557, doi:10.1073/pnas.071052198 071052198 [pii] (2001).
25. Pinheiro, V. B. *et al.* Synthetic genetic polymers capable of heredity and evolution. *Science* **336**, 341–344, doi:10.1126/science.1217622 (2012).
26. Russel, M., Lowman, H. & Clackson, T. in *A practical apporach* Vol. 266 (eds T. Clackson & H. Lowman) 1–26 (Oxford University Press, 2004).
27. Brunet, E., Chauvin, C., Choumet, V. & Jestin, J. L. A novel strategy for the functional cloning of enzymes using filamentous phage display: the case of nucleotidyl transferases. *Nucleic Acids Res* **30**, e40 (2002).
28. Xia, G. *et al.* Directed evolution of novel polymerase activities: mutation of a DNA polymerase into an efficient RNA polymerase. *Proc Natl Acad Sci USA* **99**, 6597–6602, doi:10.1073/pnas.102577799 99/10/6597 [pii] (2002).
29. Fa, M., Radeghieri, A., Henry, A. A. & Romesberg, F. E. Expanding the substrate repertoire of a DNA polymerase by directed evolution. *J Am Chem Soc* **126**, 1748–1754, doi:10.1021/ja038525p (2004).
30. Kranaster, R., Ketzer, P. & Marx, A. Mutant DNA polymerase for improved detection of single-nucleotide variations in microarrayed primer extension. *Chembiochem* **9**, 694–697, doi:10.1002/cbic.200700609 (2008).

31. Bedford, E., Tabor, S. & Richardson, C. C. The thioredoxin binding domain of bacteriophage T7 DNA polymerase confers processivity on Escherichia coli DNA polymerase I. *Proc Natl Acad Sci USA* **94**, 479–484 (1997).
32. Davidson, J. F., Fox, R., Harris, D. D., Lyons-Abbott, S. & Loeb, L. A. Insertion of the T3 DNA polymerase thioredoxin binding domain enhances the processivity and fidelity of Taq DNA polymerase. *Nucleic Acids Res* **31**, 4702–4709 (2003).
33. Wang, Y. *et al.* A novel strategy to engineer DNA polymerases for enhanced processivity and improved performance in vitro. *Nucleic Acids Research* **32**, 1197–1207, doi:10.1093/nar/gkh271 (2004).
34. Barrioluengo, V., Alvarez, M., Barbieri, D. & Menendez-Arias, L. Thermostable HIV-1 group O reverse transcriptase variants with the same fidelity as murine leukaemia virus reverse transcriptase. *The Biochemical Journal* **436**, 599–607, doi:10.1042/BJ20101852 (2011).
35. Summerer, D., Rudinger, N. Z., Detmer, I. & Marx, A. Enhanced fidelity in mismatch extension by DNA polymerase through directed combinatorial enzyme design. *Angew Chem Int Ed Engl* **44**, 4712–4715, doi:10.1002/anie.200500047 (2005).
36. Rudinger, N. Z., Kranaster, R. & Marx, A. Hydrophobic amino acid and single-atom substitutions increase DNA polymerase selectivity. *Chem Biol* **14**, 185–194, doi:S1074-5521(07)00002-6 [pii] 10.1016/j.chembiol.2006.11.016 (2007).
37. Strerath, M., Gloeckner, C., Liu, D., Schnur, A. & Marx, A. Directed DNA polymerase evolution: effects of mutations in motif C on the mismatch-extension selectivity of thermus aquaticus DNA polymerase. *Chembiochem* **8**, 395–401, doi:10.1002/cbic.200600337 (2007).
38. Yamagami, T., Ishino, S., Kawarabayasi, Y. & Ishino, Y. Mutant Taq DNA polymerases with improved elongation ability as a useful reagent for genetic engineering. *Frontiers in Microbiology* **5**, 461, doi:10.3389/fmicb.2014.00461 (2014).
39. Ignatov, K. B. *et al.* A strong strand displacement activity of thermostable DNA polymerase markedly improves the results of DNA amplification. *BioTechniques* **57**, 81–87, doi:10.2144/000114198 (2014).
40. Fogg, M. J., Pearl, L. H. & Connolly, B. A. Structural basis for uracil recognition by archaeal family B DNA polymerases. *Nature Structural Biology* **9**, 922–927, doi:10.1038/nsb867 (2002).
41. Kermekchiev, M. B., Tzekov, A. & Barnes, W. M. Cold-sensitive mutants of Taq DNA polymerase provide a hot start for PCR. *Nucleic Acids Research* **31**, 6139–6147 (2003).

42. Wu, E. Y. *et al.* A conservative isoleucine to leucine mutation causes major rearrangements and cold sensitivity in KlenTaq1 DNA polymerase. *Biochemistry* **54**, 881–889, doi:10.1021/bi501198f (2015).
43. Vichier-Guerre, S., Ferris, S., Auberger, N., Mahiddine, K. & Jestin, J. L. A population of thermostable reverse transcriptases evolved from Thermus aquaticus DNA polymerase I by phage display. *Angew Chem Int Ed Engl* **45**, 6133–6137, doi:10.1002/anie.200601217 (2006).
44. Ong, J. L., Loakes, D., Jaroslawski, S., Too, K. & Holliger, P. Directed evolution of DNA polymerase, RNA polymerase and reverse transcriptase activity in a single polypeptide. *J Mol Biol* **361**, 537–550, doi:S0022-2836(06)00791-1 [pii] 10.1016/j.jmb.2006.06.050 (2006).
45. Kranaster, R. *et al.* One-step RNA pathogen detection with reverse transcriptase activity of a mutated thermostable Thermus aquaticus DNA polymerase. *Biotechnology Journal* **5**, 224–231, doi:10.1002/biot.200900200 (2010).
46. Moser, M. J. *et al.* Thermostable DNA polymerase from a viral metagenome is a potent RT-PCR enzyme. *PloS One* **7**, e38371, doi:10.1371/journal.pone.0038371 (2012).
47. Blatter, N. *et al.* Structure and function of an RNA-reading thermostable DNA polymerase. *Angewandte Chemie* **52**, 11935–11939, doi:10.1002/anie.201306655 (2013).
48. Schonbrunner, N. J. *et al.* Chimeric thermostable DNA polymerases with reverse transcriptase and attenuated 3′-5′ exonuclease activity. *Biochemistry* **45**, 12786–12795 (2006).
49. Kermekchiev, M. B., Kirilova, L. I., Vail, E. E. & Barnes, W. M. Mutants of Taq DNA polymerase resistant to PCR inhibitors allow DNA amplification from whole blood and crude soil samples. *Nucleic Acids Research* **37**, e40, doi:10.1093/nar/gkn1055 (2009).
50. Arezi, B. *et al.* Compartmentalized self-replication under fast PCR cycling conditions yields Taq DNA polymerase mutants with increased DNA-binding affinity and blood resistance. *Frontiers in Microbiology* **5**, 408, doi:10.3389/fmicb.2014.00408 (2014).
51. Zhang, Z., Kermekchiev, M. B. & Barnes, W. M. Direct DNA amplification from crude clinical samples using a PCR enhancer cocktail and novel mutants of Taq. *The Journal of Molecular Diagnostics: JMD* **12**, 152–161, doi:10.2353/jmoldx.2010.090070 (2010).
52. Baar, C. *et al.* Molecular breeding of polymerases for resistance to environmental inhibitors. *Nucleic Acids Research* **39**, e51, doi:10.1093/nar/gkq1360 (2011).

53. McDonald, J. P. *et al.* Novel thermostable Y-family polymerases: applications for the PCR amplification of damaged or ancient DNAs. *Nucleic Acids Research* **34**, 1102–1111, doi:10.1093/nar/gkj512 (2006).
54. Gloeckner, C., Sauter, K. B. & Marx, A. Evolving a thermostable DNA polymerase that amplifies from highly damaged templates. *Angew Chem Int Ed Engl* **46**, 3115–3117, doi:10.1002/anie.200603987 (2007).
55. d'Abbadie, M. *et al.* Molecular breeding of polymerases for amplification of ancient DNA. *Nat Biotechnol* **25**, 939–943 (2007).
56. Obeid, S. *et al.* Learning from directed evolution: Thermus aquaticus DNA polymerase mutants with translesion synthesis activity. *Chembiochem: A European Journal of Chemical Biology* **12**, 1574–1580, doi:10.1002/cbic.201000783 (2011).
57. Gieseking, S. *et al.* Human DNA polymerase beta mutations allowing efficient abasic site bypass. *The Journal of Biological Chemistry* **286**, 4011-4020, doi:10.1074/jbc.M110.176826 (2011).
58. Aschenbrenner, J., Drum, M., Topal, H., Wieland, M. & Marx, A. Direct sensing of 5-methylcytosine by polymerase chain reaction. *Angewandte Chemie* **53**, 8154–8158, doi:10.1002/anie.201403745 (2014).
59. Drum, M., Kranaster, R., Ewald, C., Blasczyk, R. & Marx, A. Variants of a Thermus aquaticus DNA polymerase with increased selectivity for applications in allele- and methylation-specific amplification. *PloS One* **9**, e96640, doi:10.1371/journal.pone.0096640 (2014).
60. Williams, J. G. *et al.* An artificial processivity clamp made with streptavidin facilitates oriented attachment of polymerase-DNA complexes to surfaces. *Nucleic Acids Res* **36**, e121, doi:gkn531 [pii] 10.1093/nar/gkn531 (2008).
61. Leconte, A. M., Chen, L. & Romesberg, F. E. Polymerase evolution: efforts toward expansion of the genetic code. *J Am Chem Soc* **127**, 12470–12471, doi:10.1021/ja053322h (2005).
62. Chen, F. *et al.* Reconstructed evolutionary adaptive paths give polymerases accepting reversible terminators for sequencing and SNP detection. *Proceedings of the National Academy of Sciences of the United States of America* **107**, 1948–1953, doi:10.1073/pnas.0908463107 (2010).
63. Cozens, C., Pinheiro, V. B., Vaisman, A., Woodgate, R. & Holliger, P. A short adaptive path from DNA to RNA polymerases. *Proceedings of the National Academy of Sciences of the United States of America* **109**, 8067–8072, doi:10.1073/pnas.1120964109 (2012).
64. Laos, R., Shaw, R., Leal, N. A., Gaucher, E. & Benner, S. Directed Evolution of Polymerases To Accept Nucleotides with Nonstandard Hydrogen Bond Patterns. *Biochemistry*, doi:10.1021/bi400558c (2013).

65. Elshawadfy, A. M. *et al.* DNA polymerase hybrids derived from the family-B enzymes of Pyrococcus furiosus and Thermococcus kodakarensis: improving performance in the polymerase chain reaction. *Frontiers in Microbiology* **5**, 224, doi:10.3389/fmicb.2014.00224 (2014).
66. Killelea, T., Ralec, C., Bosse, A. & Henneke, G. PCR performance of a thermostable heterodimeric archaeal DNA polymerase. *Frontiers in Microbiology* **5**, 195, doi:10.3389/fmicb.2014.00195 (2014).
67. Wang, F. *et al.* Expression and Characterization of the RKOD DNA Polymerase in Pichia pastoris. *PloS One* **10**, e0131757, doi:10.1371/journal.pone.0131757 (2015).
68. Ramsay, N. *et al.* CyDNA: synthesis and replication of highly Cy-dye substituted DNA by an evolved polymerase. *Journal of the American Chemical Society* **132**, 5096–5104, doi:10.1021/ja909180c (2010).
69. Leconte, A. M. *et al.* Directed evolution of DNA polymerases for next-generation sequencing. *Angewandte Chemie* **49**, 5921–5924, doi:10.1002/anie.201001607 (2010).
70. Obeid, S., Busskamp, H., Welte, W., Diederichs, K. & Marx, A. Snapshot of a DNA polymerase while incorporating two consecutive C5-modified nucleotides. *Journal of the American Chemical Society* **135**, 15667–15669, doi:10.1021/ja405346s (2013).
71. Wyss, L. A. *et al.* Specific incorporation of an artificial nucleotide opposite a mutagenic DNA adduct by a DNA polymerase. *Journal of the American Chemical Society* **137**, 30–33, doi:10.1021/ja5100542 (2015).
72. Keller, S., Wang, J., Chandra, M., Berger, R. & Marx, A. DNA polymerase-catalyzed DNA network growth. *J Am Chem Soc* **130**, 13188–13189, doi:10.1021/ja8045348 (2008).
73. Bergen, K. *et al.* Structures of KlenTaq DNA polymerase caught while incorporating C5-modified pyrimidine and C7-modified 7-deazapurine nucleoside triphosphates. *Journal of the American Chemical Society* **134**, 11840–11843, doi:10.1021/ja3017889 (2012).
74. Chander, Y. *et al.* A novel thermostable polymerase for RNA and DNA loop-mediated isothermal amplification (LAMP). *Frontiers in microbiology* **5**, 395, doi:10.3389/fmicb.2014.00395 (2014).
75. Loakes, D., Gallego, J., Pinheiro, V. B., Kool, E. T. & Holliger, P. Evolving a polymerase for hydrophobic base analogues. *Journal of the American Chemical Society* **131**, 14827–14837, doi:10.1021/ja9039696 (2009).
76. Fuso, A., Scarpa, S., Grandoni, F., Strom, R. & Lucarelli, M. A reassessment of semiquantitative analytical procedures for DNA methylation: comparison of bisulfite- and HpaII polymerase-chain-reaction-based methods.

Anal Biochem **350**, 24–31, doi:S0003-2697(05)00883-3 [pii] 10.1016/j.ab.2005.12.008 (2006).
77. Ghadessy, F. J. *et al.* Generic expansion of the substrate spectrum of a DNA polymerase by directed evolution. *Nat Biotechnol* **22**, 755–759, doi:10.1038/nbt974 nbt974 [pii] (2004).
78. Gilbert, M. T. & Willerslev, E. Rescuing ancient DNA. *Nat Biotechnol* **25**, 872-874, doi:nbt0807-872 [pii] 10.1038/nbt0807-872 (2007).
79. Barth, S. *et al.* Epstein-Barr virus-encoded microRNA miR-BART2 down-regulates the viral DNA polymerase BALF5. *Nucleic Acids Res* **36**, 666-675, doi:gkm1080 [pii] 10.1093/nar/gkm1080 (2008).
80. Patel, P. H. & Loeb, L. A. DNA polymerase active site is highly mutable: evolutionary consequences. *Proc Natl Acad Sci USA* **97**, 5095–5100, doi:97/10/5095 [pii] (2000).
81. Hanson, L. A., Rudis, M. R., Vasquez-Lee, M. & Montgomery, R. D. A broadly applicable method to characterize large DNA viruses and adenoviruses based on the DNA polymerase gene. *Virol J* **3**, 28, doi: 10.1186/1743-422X-3-28 (2006).
82. Laos, R., Thomson, J. M. & Benner, S. A. DNA polymerases engineered by directed evolution to incorporate non-standard nucleotides. *Frontiers in Microbiology* **5**, 565, doi:10.3389/fmicb.2014.00565 (2014).
83. Horlacher, J., Hottiger, M., Podust, V. N., Hubscher, U. & Benner, S. A. Recognition by viral and cellular DNA polymerases of nucleosides bearing bases with nonstandard hydrogen bonding patterns. *Proc Natl Acad Sci USA* **92**, 6329–6333 (1995).
84. Pelletier, H., Sawaya, M. R., Kumar, A., Wilson, S. H. & Kraut, J. Structures of ternary complexes of rat DNA polymerase beta, a DNA template-primer, and ddCTP. *Science* **264**, 1891–1903 (1994).
85. McMinn, D. l. *et al.* Efforts toward Expansion of the Genetic Alphabet: DNA Polymerase Recognition of a Highly Stable, Self-Pairing Hydrophobic Base. *J Am Chem Soc.* **121**, 11585–11586 (1999).
86. Notomi, T. *et al.* Loop-mediated isothermal amplification of DNA. *Nucleic Acids Research* **28**, E63 (2000).
87. Piepenburg, O., Williams, C. H., Stemple, D. L. & Armes, N. A. DNA detection using rec0204 (2006).ombination proteins. *PLOS Biology* **4**, e204, doi:10.1371/journal.pbio.0040204 (2006).
88. Son *et al.* Ultrafast photonic PCR, *Light: Science & Application*, 4, e280; doi:10.1038/lsa.2015.53 (2015).

Chapter 9

Market for Evolved DNA Polymerases in Routine and Medical Applications

9.1 Introduction: The market

The German journal *Laborjournal* published in the last five years lists of companies that sell products for PCR and Sequencing (http://www.laborjournal.de/rubric/produkte/index.lasso). In the March 2011 issue (*Laborjournal* 3, 63–69, 2011) there are 30 companies listed that are selling 220 different products for PCR either as DNA pols or as special PCR kits. In the January/February 2013 issue of this journal (*Laborjournal* 1–2, 54–62, 2013) quantitative real-time PCR (qPCR) kits and master mixes are listed from 24 companies selling 166 different kits and mixes. Finally, in the 2015 December issue of the same journal (*Laborjournal* 12, 52–54, 2015) the names of 16 companies that sell real-time-PCR-thermocycler are listed. They offer 38 analysing systems for various applications such as in biology, biotechnology, diagnostics, forensic and many others. This is self explanatory for the tremendous application potential for DNA pols in many PCR and related technologies. In addition, we will discuss below technologies such as: next-generation sequencing, single-cell sequencing, synthetic biology, loop-mediated isothermal amplification, rolling circle amplification, recombinase polymerase amplification, digital and direct PCR and -omic analysis (e.g. microbiomics, metagenomics, personal genomics). In Chapter 8 we have already described the novel properties of DNA pols and briefly mentioned their application potential.

According to the *Polymerase Chain Reaction (PCR) in Medical Application — An Analytical Report*, 2009–2015 (Research & Markets, Dublin): "the Global Polymerase Chain Reaction (PCR) market is expected to reach US$27.4 billion by 2015 on positive growth prospects of q-PCR segment. The Compounded Annual Growth Rate (CAGR) will be 13.9% for the analysis period, 2009–2015. North America is estimated as the largest market with a market share of 40% of global market and projects US$10.7 billion by end-2010. Europe and Asia-Pacific together account for more than 50% of the global market share. Interestingly, Latin America is seen as the fastest growing market with a CAGR of 14.8%, to worth around US$1.3 billion in 2015. Among the techniques used, q-PCR appears to be the fastest growing market with a CAGR of 14.9%. Reverse-Transcription PCR (RT-PCR) and Assembly PCR are expected to grow with a CAGR of 14.3% and 14%, respectively. Based on the classification by products/tools, PCR Reagents are expected to be fastest growing market with a CAGR of 14.6%. PCR Consumables and PCR Machines are expected to grow with a CAGR of 14.2% and 13.1%, respectively. Global PCR Research market forecasts a CAGR of 13.3% during the analysis period 2009–2015. PCR Diagnosis market is expected to represent the market share of 29.9% by 2015. PCR Infectious Diseases stand as the fastest growing market."

This statement, as well as the many companies that sell apparatus for PCR, is in itself a clear demonstration that there is an enormous market for DNA pols and their selected mutants.

9.2 Next-generation sequencing (NGS): Prices drop while speed increases

First sequencing strategies: The development of the dideoxy sequencing by Sanger and collaborators used DNA pols to synthesize a complementary copy of the DNA to be sequenced. From a 3′primed primer end the

DNA is synthesized. The key issue here was that a DNA pol can incorporate 2′3′-dideoxynucleoside monophosphates[1,2] (for more details see the Sanger review of 1988).[3] Since this first DNA sequencing by a DNA pol was developed by Sanger many different approaches were developed in the subsequent two decades (reviewed in Ref. 4). They include sequencing on solid phase, sequencing by hybridization, mass spectrometry, cyclic array sequencing, micro-electrophoresis and nanopore sequencing.

A first engineered DNA pol was achieved by the Richardson lab in 1989 when the *E. coli* bacteriophage T7 DNA pol was mutagenized so that the 3′→5′ exonuclease was eliminated and the DNA pol activity increased. Together with the *E. coli* protein thioredoxin, which made the T7 DNA pol very processive (e.g able to synthesize very long DNA fragments with a single binding event), this mutant DNA pol was the basis to develop the successful "Sequenase". It can be stated that the "Sequenase" somehow pioneered the development of heat-stable DNA pol enzymes and might thus have facilitated automation of high-throughput sequencing.[5]

Next generation sequencing (NGS): As indicated in a review by Michael Metzker in 2011 the Human Genome Research Institute has given about 100 million US$ for the development of NGS technologies.[6] It was demanding in the last 15 years to develop technologies that can provide faster, cheaper and more accurate information about genomes of any sources. Here, besides high-tech equipment, again DNA pols were important to achieve this ambitious goal. In the above mentioned review the technology developments, the broad range of applications as well as the platform selection guidelines were discussed.[6] In brief the technologies are mostly on surfaces (for details see Nature Perspective by Elaine Mardis[7]) as e.g. documented by the emulsion PCR (Roche 454, Life Technologies), by solid phase amplification (Illumina/Sloexa), and single-molecule sequencing by using immobilized primers and immobilized templates (Helicos Bioscience) or immobilized DNA pols (Pacific Bioscience, Life Visigen and Li-COR Bioscience). It is important to amplify the DNA fragments by efficient and accurate DNA pols so that there are enough products to be analysed by the optical system of the respective apparatus.

As indicated above DNA pols were essential for DNA sequencing from the very beginning as exemplified by the Sanger methods (see above). In a 2014 review Cheng-Yao Chen summarized how family A/B DNA pols (for the different DNA pol families see Chapter 2) from mesophilic/thermophilic bacteria/archaea were modified and tested in today's standard capillary electrophoresis (CE) and NGS sequencing platforms.[8] Subsequently DNA pols were selected for very efficient incorporation of bulky dye-terminator and reversible dye-terminator nucleotides, respectively. On the other hand single-molecule sequencing platforms require highly processive DNA pols that can perform strand displacement synthesis and can incorporate terminally phosphate-labeled nucleoside polyphosphate (for details see Figure 2 in Ref. 8). The DNA pol from *E. coli* bacteriophage ϕ29 has these properties as exemplified by structural studies of this enzyme.[9] In summary it is needless to state that DNA pols will play, besides technology development, the key issue for future progress in NGS.

9.3 Single-cell sequencing

In a Technology Feature published in *Nature* Charlotte Schubert argues that many technologies that are established can now be applied to individual cells.[10] This also includes PCR. Two examples are mentioned: first, transcriptome analysis on single cells (e.g. during development of the brain and other organs) and second, fast sequencing of individual cells in cancer tissues (e.g. cancer stem cells).

In a review Nicholas Navin and James Hicks discussed the challenges and technical aspects of single-cell sequencing, with a strong focus on genomic copy number, and they discuss how this information can be used to diagnose and treat cancer patients.[11] Single cancer cells from a tumor can be isolated by micromanipulation and the microscope, by laser capture microdisection (LCM) and by fluorescence activated cell sorting (FACS). Analysis is then performed (i) by cytological methods of Giemsa staining and spectral caryotyping on chromosomes, (ii) by whole-genome amplification and microarray comparative genome hybridization (CGH) and by whole-genome amplification and next-generation sequencing (NGS). Applications of these technologies include (i) the detection of rare tumor cells in clinical samples, (ii) early detection of circulating tumor cells to prevent metastasis, (iii) investigating the heterogeneity within a tumor and (iv) response to chemotherapy. In conclusion these applications are only useful since the costs of sequencing dramatically dropped two to three orders of magnitude in the last few years.

The evolution of a tumor by single-cell sequencing was analysed by Navin *et al.*[12] Since tumors are in most cases heterogeneous it might be difficult to trace the history of how a tumor evolved during a long time. In human it can last several decades until a tumor develops into clinical signs of the disease. These authors showed that with flow-sorted nuclei, whole-genome amplification and next-generation sequencing (NGS, see above) could accurately quantify the genomic copy number within an individual nucleus.[12] As proof of principle they tested the tumor population structure and the evolution of two human breast cancer cases. Their data suggested that in contrast to the gradual models of tumor progression, tumors grow by punctuated clonal expansions with few persistent intermediates.[12]

9.4 Synthetic biology

Synthetic biology has a great potential for the benefit of humankind. The beginning of synthetic biology was an 1995 article published by Willem Stemmer *et al.* in which they single-step assembled a gene and an entire plasmid from large numbers of oligodeoxyribonucleotides.[13] The method they used was assembly PCR for the synthesis of long DNA sequences from large numbers of oligodeoxyribonucleotides and is derived from DNA shuffling.[14] It does not rely on DNA ligase but instead on DNA pol to build increasingly longer DNA fragments during the assembly process.

During the last 20 years a lot of progress in synthetic biology and synthetic genomics has been reported. Here we only summarize three reviews as proof of principle. What is clear, is that in this research area the investigators all rely on efficient DNA pols with various properties.

In a *Nature Reviews Genetics* article Steven Benner and Michael Sismour[15] argue that synthetic biology can be divided into two classes. The first is to create artificial life by using unnatural molecules that can be replicated by DNA pols (see also Section 8.5.7 *DNA polymerases and expansion of the genetic code*). The second contains parts from natural organisms and these parts are assembled into systems that can function in a manner not found in nature. The authors state: "Synthetic biology has generated diagnostic tools that improve the care of patients with infectious diseases, as well as devices that oscillate, creep and play tic-tac-toe". This means that this discipline is not free from controversy. In an Expert Opinion article the same authors claim that chemistry contributed a lot to the development of synthetic biology.[16] Synthetic biology has the following ambitious goal: it should recreate in artificial systems the emergent properties found in nature. In the last 10 years since these reviews appeared many institutions improved the technologies of synthetic biology. One statement in the view of this book is clear: they all rely on efficient DNA pols.

Michael Montague *et al.* describe the potential and the limitation of synthetic genomics, another term for synthetic biology.[17] It is possible to assemble chromosome-sized DNA fragments and one can make thousands of simultaneous mutations to existing genomes. The possibility of synthetic genomics are not limited to engineering of chromosomes, but can be extended to the creation of artificial organisms such viruses, bacteria and even single-cell nuclear organisms such as the baker's yeast *Saccharomyces cerevisiae*. Other possibilities are entire metabolisms, regulatory networks and ecosystems (see also Section 9.6 metagenomics). The authors caution, however, that existing barriers are: (i) the limited number of organisms suitable for hosting functional synthetic genomics and (ii) the still limited software presently available for designing entire synthetic genomic sequences.

Two examples of successful synthetic biology are the following: first a 582,970-bp *Mycoplasma genitalium* genome[18] and second the creation of a bacterial cell controlled by a chemically synthesized genome.[19]

9.5 Direct and digital PCR

Direct PCR: As described in detail in Section 8.5.2. many mutant DNA pols, mainly from *Thermus aquaticus* are available that allow efficient direct PCR amplification to common inhibitors (for details see the references in Section 8.5.2). Such inhibitors include e.g. blood, heparin, buccal swabs, soil, feces, urine animal tissues, plant and any type of crude extract from the environment. Needless to say that these DNA pol mutants have a great commercial potential for the wide application in medicine, environment, criminal applications, evolutionary studies, just to mention a few.

Digital PCR[20]: In contrast to conventional PCR, digital PCR (also called dPCR) is achieved by microfluidic dilution in many droplets with femtoliter volumes. Then, amplification is achieved by DNA pols and the quantification of low abundant targets can be achieved. The result is either positive or negative leading to a digital output. With high-throughput screening final results can be evaluated by Poisson distribution and the abundance of the target to be analyzed can be estimated in a material of choice. The dPCR technique is suited to detect rare pathogens, DNA sequences, gene copy variants, seldom mutation e.g. in cancer and quantitative gene expression in diseases of any kind.

In summary while direct PCR has a great potential to detect specific DNA from everywhere and has thus a great commercial potential, the digital PCR is more restricted to basic research in combination with clinical testing or screening procedures.

9.6 Microbiomics, metagenomics and personal genomics

Microbiomics, metagenomics and personal genomics are upcoming technologies that will require efficient DNA pols, since in both areas large next-generation sequencing (NGS) technologies will be required.[6,7]

Microbiomics searches for the complexity of microorganisms in a given body environment such as stomach, gut, skin, vagina and mouth. Special conditions might be important such as treatment with antibiotics.

Metagenomics describes the sequencing of DNA and RNA from a mixed organism population from a natural habitat (e.g. soil).

In *personal genomics* massive parallel sequencing is required. A new initiative on precision medicine was announced in the United States in 2015 with an immediate focus on cancers and with a longer-term aim to generate knowledge applicable to the whole range of health and disease.[21] Eric Green *et al.* described the future of genomics research and the path towards an era of genomic medicine.[22] Areas that might be important include the following[7]: (i) determining genomic differences between chromosomal and mitochondrial DNA from different tissue in a given human or animal body; (ii) determining the gene expression profiles of developmental and differentiated (adult) tissues;[23] (iii) epigenetic approaches that determine the kinetics of the methylation patterns of histones in a given tissue and (iv) identification and variation of non-coding RNA expression.

9.7 Additional amplification techniques

Further amplification procedures depending on DNA pols have been described. It is difficult to estimate which of those techniques will be economically important in the future.

9.7.1 *Loop-mediated isothermal amplification*

The technique for loop-mediated isothermal amplification (LAMP) of DNA has been introduced by Notomi *et al.*[24] and described in Section 8.6.1. Moreover, as also described in Section 7.6.1., Chander *et al.* have recently introduced their so-called OmniAmp DNA pol, which is a thermostable viral enzyme that enables its use in the field by overcoming important barriers of LAMP.[25] Zablon Kithinji Njiru summarized the LAMP technology in his view of "point of care diagnostic".[26] The advantages of the LAMP technology are as follows[26]: (i) less instrumentation is required and thus LAMP is less expensive; (ii) LAMP is rapid (less than 60 min);[23] (iii) the sensitivity is equal to or higher than classical PCR; (iv) LAMP can also amplify from crude material; (v) LAMP has a high specificity since four to six primers are used to identify six to eight DNA target regions; (vi) LAMP is easy due to the detection of pyrophosphate by subsequent signal generating transformations[27] and (vii) LAMP is achieved at isothermal conditions, since low heat is required. More information about the three DNA pols from *Bacillus steratothermophilus* (Bst) and their use in today's LAMP technologies are offered by *New England Biotechnology* and details can be seen form a feature article published online under www.neb-online.de.

9.7.2 *Rolling circle amplification*

Rolling Circle Amplification (RCA) can amplify oligonucleotides containing tandemly repetitive sequence motifs.[28] Different repetitive sequences of which the motif unit is the circular template can be synthesized by extending the primer. In contrast to PCR, RCA requires no change in the temperature. RCA appears to have an amplification efficiency like PCR. The authors comparatively estimated five different DNA pols to improve RCA (for more details see Ref. 28). In a recent work the same laboratory described that RCR can be improved by controlled microwave heating. They showed that microwave-assisted RCA reactions catalyzed by different heat-stable DNA pols were accelerated more than four fold and this in comparison to conventional RCA. The conclusion is that microwave heating accelerated isothermal RCA of DNA, since the different heating mechanisms of microwaves on the temperatures of the reaction components occurred, although the overall reaction temperature was the same.[29]

9.7.3 *Recombinase polymerase amplification*

A new and easy amplification approach was proposed by Piepenburg *et al.* and is called recombinase polymerase amplification (RPA).[30] RPA requires the following three proteins: a *recombinase*, a *single-stranded DNA-binding protein* and *strand-displacing DNA pol*. The recombinase is capable of pairing oligonucleotide primers with homologous sequence in duplex DNA. The SSB can bind to displaced single-strand DNA thus avoiding the primers from being displaced. Finally, a strand-displacing DNA pol begins DNA synthesis. With the two opposing primers, much like in classical PCR, exponential DNA amplification is initiated. The advantage is that at optimal temperatures 37–42°C, the amplification progresses rapidly and results can be obtained within 10 minutes. The proof of principle was achieved by these authors for the medically extremely important pathogen methicillin-resistant *Staphylococcus aureus*.[30]

Later RPA was adapted also for RNA amplification by first developing a two-step and then a one-step-RT-RPA protocol and was exemplified by the Rift Valley fever virus (RVFV).[31] Two other important RNA viruses in veterinary medicine were detected by RT-RPA. First, the Bovine Coronavirus (BCoV),[32] where the authors used a portable fluorescence reading device thus extending its application potential to use in the field and point-of-care diagnosis. Second, the Foot-and-Mouth Disease virus (FMD)[33] in which the authors argue that their RT-PCR assay is quicker and much easier to handle in the field than real-time RT-PCR. Thus RT-RPA might be implemented to perform diagnostics at quarantine stations or farms for rapid spot-of-infection detection.

In summary RPA has a great potential for medical applications where money is limited and rapid field detection could be essential to safe lives of men and animals.

9.7.4 *Nicking enzyme amplification reaction*

Nicking Enzyme Amplification Reaction (NEAR) is an isothermal DNA amplification at a constant temperature between of 55°C to 59°C by using a DNA pol and a nicking enzyme (an endonuclease) to exponentially amplify the DNA of choice. Whether this technique has a commercial future remains to be seen (see Ref. 34 for a discussion of this technique).

9.7.5 *Helicase-dependent amplification*

Helicase-dependent amplification (HDA) utilizes a DNA helicase to generate single-stranded templates for primer hybridization and subsequent primer extension by a DNA pol.[35] The advantage is that it does not require thermocycling and can be performed at a certain temperature. HAD has a great potential as simple portable DNA diagnostic device to be used in the field and at the point-of-care.[35]

9.8 Chapter summary

In this chapter we highlighted the tremendous commercial value of DNA pols in biotechnology and medicine. The fact that the annual sale of DNA pols is over 27 billion dollars and the growth rate is more than 10% annually render these enzymes in the centre of commercial interests. The following topics were exemplified in this chapter. Next-generation sequencing, single-cell sequencing, synthetic biology, direct PCR, digital PRC, microbiomics, metagenomics and personal genomics. In addition some further amplification technologies are briefly touched. They include (i) loop-mediated isothermal amplification, (ii) rolling circle amplification, (iii) recombinase polymerase amplification, (iv) nicking enzyme amplification and (v) helicase dependent amplification.

It is conceivable that the development and the evolvement of further DNA pol mutants with novel properties will allow the technologies to become more efficient, more accurate and much cheaper, what is important for the developing world. Fast and economic screening of novel epidemics such as the zika virus can help to prevent the spreading of the disease.

References

1. Sanger, F. *et al.* Nucleotide sequence of bacteriophage phi X174 DNA. *Nature* **265**, 687–695 (1977).
2. Sanger, F., Nicklen, S. & Coulson, A. R. DNA sequencing with chain-terminating inhibitors. *Proceedings of the National Academy of Sciences of the United States of America* **74**, 5463–5467 (1977).
3. Sanger, F. Sequences, sequences, and sequences. *Annual Review of Biochemistry* **57**, 1–28, doi:10.1146/annurev.bi.57.070188.000245 (1988).
4. Shendure, J. A. *et al.* Overview of DNA sequencing strategies. *Current protocols in molecular biology / edited by Frederick M. Ausubel ... [et al.]* **Chapter 7**, Unit7 1, doi:10.1002/0471142727.mb0701s96 (2011).
5. Zhu, B. Bacteriophage T7 DNA polymerase — sequenase. *Frontiers in Microbiology* **5**, 181, doi:10.3389/fmicb.2014.00181 (2014).
6. Metzker, M. L. Sequencing technologies — the next generation. *Nature Reviews Genetics* **11**, 31–46, doi:10.1038/nrg2626 (2010).
7. Mardis, E. R. A decade's perspective on DNA sequencing technology. *Nature* **470**, 198–203 Mardis, doi:10.1038/nature09796 (2011).
8. Chen, C. Y. DNA polymerases drive DNA sequencing-by-synthesis technologies: both past and present. *Frontiers in Microbiology* **5**, 305, doi:10.3389/fmicb.2014.00305 (2014).
9. Berman, A. J. *et al.* Structures of phi29 DNA polymerase complexed with substrate: the mechanism of translocation in B-family polymerases. *EMBO J* **26**, 3494–3505, doi:7601780 [pii]10.1038/sj.emboj.7601780 (2007).

10. Schubert, C. Single-cell analysis: The deepest differences. *Nature* **480**, 133–137, doi:10.1038/480133a (2011).
11. Navin, N. & Hicks, J. Future medical applications of single-cell sequencing in cancer. *Genome Medicine* **3**, 31, doi:10.1186/gm247 (2011).
12. Navin, N. *et al.* Tumour evolution inferred by single-cell sequencing. *Nature* **472**, 90–94, doi:10.1038/nature09807 (2011).
13. Stemmer, W. P., Crameri, A., Ha, K. D., Brennan, T. M. & Heyneker, H. L. Single-step assembly of a gene and entire plasmid from large numbers of oligodeoxyribonucleotides. *Gene* **164**, 49–53 (1995).
14. Stemmer, W. P. Rapid evolution of a protein in vitro by DNA shuffling. *Nature* **370**, 389–391, doi:10.1038/370389a0 (1994).
15. Benner, S. A. & Sismour, A. M. Synthetic biology. *Nature Reviews Genetics* **6**, 533–543, doi:10.1038/nrg1637 (2005).
16. Sismour, A. M. & Benner, S. A. Synthetic biology. *Expert opinion on biological Therapy* **5**, 1409–1414, doi:10.1517/14712598.5.11.1409 (2005).
17. Montague, M. G., Lartigue, C. & Vashee, S. Synthetic genomics: potential and limitations. *Current Opinion in Biotechnology* **23**, 659–665, doi:10.1016/j.copbio.2012.01.014 (2012).
18. Gibson, D. G. *et al.* Complete chemical synthesis, assembly, and cloning of a Mycoplasma genitalium genome. *Science* **319**, 1215–1220, doi:10.1126/science.1151721 (2008).
19. Gibson, D. G. *et al.* Creation of a bacterial cell controlled by a chemically synthesized genome. *Science* **329**, 52–56, doi:10.1126/science.1190719 (2010).
20. Perkel, J. Guiding our PCR experiments. *BioTechniques* **58**, 217–221, doi:10.2144/000114283 (2015).
21. Collins, F. S. & Varmus, H. A new initiative on precision medicine. *The New England Journal of Medicine* **372**, 793–795, doi:10.1056/NEJMp1500523 (2015).
22. Green, E. D. & Guyer, M. S. Charting a course for genomic medicine from base pairs to bedside. *Nature* **470**, 204–213, doi:10.1038/nature09764 (2011).
23. Zan, H. *et al.* The translesion DNA polymerase theta plays a dominant role in immunoglobulin gene somatic hypermutation. *EMBO J* **24**, 3757–3769, doi:7600833 [pii]10.1038/sj.emboj.7600833 (2005).
24. Notomi, T. *et al.* Loop-mediated isothermal amplification of DNA. *Nucleic Acids Research* **28**, E63 (2000).
25. Chander, Y. *et al.* A novel thermostable polymerase for RNA and DNA loop-mediated isothermal amplification (LAMP). *Frontiers in Microbiology* **5**, 395, doi:10.3389/fmicb.2014.00395 (2014).

26. Njiru, Z. K. Loop-mediated isothermal amplification technology: towards point of care diagnostics. *PLoS Neglected Tropical Diseases* **6**, e1572, doi:10.1371/journal.pntd.0001572 (2012).
27. Ihalainen, P. *et al.* Application of paper-supported printed gold electrodes for impedimetric immunosensor development. *Biosensors* **3**, 1–17, doi:10.3390/bios3010001 (2013).
28. Yoshimura, T., Arikado, S. & Ohuchi, S. Estimation of DNA polymerase for improvement of rolling circle amplification. *Nucleic Acids Symp Ser (Oxf)*, 303–304, doi:50/1/303 [pii]10.1093/nass/nrl151 (2006).
29. Yoshimura, T., Suzuki, T., Mineki, S. & Ohuchi, S. Controlled Microwave Heating Accelerates Rolling Circle Amplification. *PloS One* **10**, e0136532, doi:10.1371/journal.pone.0136532 (2015).
30. Piepenburg, O., Williams, C. H., Stemple, D. L. & Armes, N. A. DNA detection using recombination proteins. *PLoS Biology* **4**, e204, doi:10.1371/journal.pbio.0040204 (2006).
31. Euler, M. *et al.* Recombinase polymerase amplification assay for rapid detection of Rift Valley fever virus. *Journal of Clinical Virology: The Official Publication of the Pan American Society for Clinical Virology* **54**, 308–312, doi:10.1016/j.jcv.2012.05.006 (2012).
32. Amer, H. M. *et al.* A new approach for diagnosis of bovine coronavirus using a reverse transcription recombinase polymerase amplification assay. *Journal of Virological Methods* **193**, 337–340, doi:10.1016/j.jviromet.2013.06.027 (2013).
33. Abd El Wahed, A. *et al.* A portable reverse transcription recombinase polymerase amplification assay for rapid detection of foot-and-mouth disease virus. *PloS One* **8**, e71642, doi:10.1371/journal.pone.0071642 (2013).
34. Zanoli, L. M. & Spoto, G. Isothermal amplification methods for the detection of nucleic acids in microfluidic devices. *Biosensors* **3**, 18–43, doi:10.3390/bios3010018 (2013).
35. Vincent, M., Xu, Y. & Kong, H. Helicase-dependent isothermal DNA amplification. *EMBO Reports* **5**, 795–800, doi:10.1038/sj.embor.7400200 (2004).

Subject Index

Author Index